Electronic, Magnetic, and Thermal Properties of Solid Materials

ELECTRICAL ENGINEERING AND ELECTRONICS

A Series of Reference Books and Textbooks

Editors

Marlin O. Thurston
Department of Electrical
Engineering
The Ohio State University
Columbus, Ohio

William Middendorf
Department of Electrical
and Computer Engineering
University of Cincinnati
Cincinnati, Ohio

Electronics Editor

Marvin H. White
Westinghouse Electric Corporation
Defense and Space Center
Baltimore, Maryland

1. Rational Fault Analysis, *edited by Richard Saeks and S. R. Liberty*

2. Nonparametric Methods in Communications,
   edited by P. Papantoni-Kazakos and Dimitri Kazakos

3. Interactive Pattern Recognition, *Yi-tzuu Chien*

4. Solid-State Electronics, *Lawrence E. Murr*

5. Electronic, Magnetic, and Thermal Properties of Solid Materials,
   Klaus Schröder

Other Volumes in Preparation

Electronic, Magnetic, and Thermal Properties of Solid Materials

KLAUS SCHRÖDER

Department of Chemical Engineering and Materials Science
L. C. Smith College of Engineering
Syracuse University
Syracuse, New York

MARCEL DEKKER, INC.   New York and Basel

Library of Congress Cataloging in Publication Data

Schroder, Klaus, [Date]
    Electronic, magnetic, and thermal properties of solid
materials.

    (Electrical engineering and electronics ; v. 5)
    Bibliography: p.
    Includes indexes.
    1. Solids. 2. Materials. 3. Solid state physics.
I. Title. II. Series.
QC176.S3   530.4'1   78-8759
ISBN 0-8247-6487-0

MARCEL DEKKER, INC.
270 Madison Avenue, New York, New York 10016

Current printing (last digit):
10 9 8 7 6 5 4 3 2 1

PRINTED IN THE UNITED STATES OF AMERICA

PREFACE

Elements are discussed in numerous textbooks. Alloys, compounds and intermediate phases which are technically much more important are discussed less frequently. Therefore, this book was written to introduce students to the structure and properties of these often ignored materials. I emphasize in the analysis of alloy properties their composition dependence and describe also the effect of atomic ordering on the electronic, magnetic and thermal properties of materials.

It is not possible to start a textbook with a discussion of virtual impurity states or with a study on the effect of alloying on the energy gaps in the nearly free electron model. It is necessary to review briefly models of perfect crystals. Therefore, the first part of Chapter 1 contains a short description of simple models used in the discussion of elements and compounds (the reader may consult books like Kittel: <u>Introduction to Solid State Physics</u>, Ziman: <u>Principles of the Theory of Solids</u> or Hutchinson and Baird: <u>The Physics of Engineering Solids</u> for more details). Following Chapter 1, energy band calculations and Fermi surfaces are given for representative elements, compounds and alloys.

Chapters 2 through 5 describe experimentally determined properties of materials and models used to analyze the results. The concepts developed in Chapter 1 are illustrated with experimental data. Typical examples of alloy systems of A- and B-group metals,

transition element alloys, and semiconductor and semimetal alloys
are discussed. The presentation of the data is uneven since some
properties, like the susceptibility of gold, can be analyzed with
simple concepts, whereas the same properties for bismuth can be
explained only with complex calculations.

Properties of materials change frequently in a systematic way
with the composition of alloys. These changes sometimes give an
insight into the electronic structure of the materials. The
residual resistivity in alloy systems allows the evaluation of the
electron distribution around impurity atoms. Small amounts of impurities may act as "probes" and can show electric charge and magnetic moment fluctuations in solid solutions. Sometimes properties
of alloys can be more easily described than the properties of the
pure constituents. The Seebeck coefficient of alloys follows the
Nordheim-Gorter rule over wide ranges. Variations of the Seebeck
coefficient with alloying can be easily calculated even if the
Seebeck coefficient of the metal does not follow predictions from
simple models. Optical measurements show that the correlation
between optical absorption and the energy band structure is frequently easier if one knows how the shape and position of absorption edges and peaks change with alloying.

The powerful tools available for the study of pure elements
cannot always be used for the investigation of alloys. Most measurements which are sensitive to the shape of the Fermi surface
usually require very pure elements as samples. It is therefore
frequently necessary to investigate several properties of alloy
systems to obtain a good model of their electron energy states.

Naturally it is impossible in an introductory book like this
to give an adequate survey of the properties of elements and alloys.
Therefore, this volume lists only about 300 references. Some interesting fields like ferroelectricity have been completely neglected,
and other sections of the book give only brief comments on areas
which have been extensively investigated.

The author would like to thank various colleagues for their

suggestions and criticism. Unfortunately, not all can be mentioned.
Dr. Neal P. Baum, University of New Mexico, gave detailed comments
to the manuscript. Professor John Orehotsky, Wilkes College,
Dr. Edward Yen, IBM, and Professors Douglas Keller and
Harvey Kaplan, Syracuse University, made valuable suggestions.
Dr. John C. McClure, NASA, Huntsville, Alabama, read carefully the
last draft of the manuscript. I appreciated very much discussions
I had with the faculty and students in our Materials Science
Department at Syracuse University. Finally, I would like to thank
Professor Paul A. Beck, University of Illinois, who guided me in my
first work in the field of the electronic structure of transition
element alloys. I am also very grateful to the National Science
Foundation, which supported me and my students in research programs
in alloy studies. These investigations broadened my experience in
the field of alloy physics.

Special thanks are due to Mrs. Helen Turner for her patience
in assembling this manuscript, to Mr. Rolf Ziemer for his help
in preparing some of the diagrams, and to Mrs. M. Neuhierl for
proofreading.

Klaus Schröder

CONTENTS

Electronic, Magnetic, and Thermal Properties of Solid Materials

Chapter 1

LATTICE AND ELECTRONIC STRUCTURE OF ORDERED AND DISORDERED CRYSTALS

I. INTRODUCTION

The external shape of minerals suggested centuries ago that crystals are built up from a regular arrangement of basic building blocks. In 1690, Huyghens showed such a model for calcite in his Traite de la Lumiere. Convincing proof that this concept was correct was given by von Laue in 1912. He explained the diffraction of X-rays by solids with the assumptions that a solid is a periodic arrangement of atoms, and that X-rays are electromagnetic waves of extremely short wavelength. The properties of most simple solids are therefore frequently analyzed with mathematical models, which are based on the symmetrical arrangement of atoms.

Therefore, we begin this book with a brief description of the crystalline structure of solids. Then follows a discussion of their elastic properties. The static elastic deformations were already studied by Hook in 1660. Sound waves in solids were investigated extensively in the 19th century. A correlation of the elasticity with atomic models was given by Born and von Karman who analyzed the oscillations of spheres tied together with springs in 1912. High speed computers make it now possible to obtain the elastic energies of solids.

In the 1920's, the electron energies in metals were calculated with the model of a constant potential energy for electrons. The

1

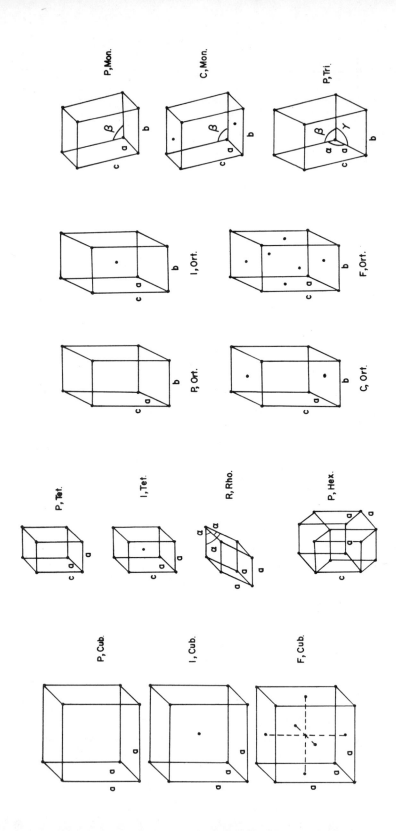

FIG. 1-1. Conventional unit cells of the 14 Bravais lattices. P: primitive cell; C: cell with a lattice point in the center of two parallel faces; F: lattice point in the center of each face; I: cell with a lattice point in the center of the interior; Cub: cubic; Tet: tetragonal; Rho: rhombohedral; Hex: hexagonal; Mon: monoclinic; Tri: triclinic. After Ref. 1, with permission.

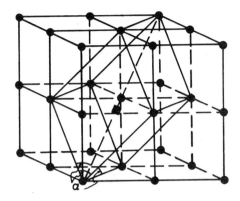

FIG. 1-2. Bi structure: Move center atom of group of eight
simple cubic lattices to the black square, and reduce the angles
α from their 60° value. After Ref. 2, with permission.

interaction between one electron and ions was first considered by
using a very weak perturbation potential with the periodicity of
the lattice. This rather unrealistic potential model explained the
basic properties of electrons in crystalline solids. Much theoretical effort has been spent in recent years to show how such weak
potentials are related to the real potential inside a solid.

The last part of this chapter gives a summary of results obtained both theoretically and experimentally on the electron states
in metals, alloys and compounds.

II. CRYSTAL STRUCTURE

A. Elements and Compounds

Figure 1 shows the 14 Bravais lattice spaces which are the
building blocks for solid materials in crystalline form. They
were obtained from symmetry and space filling considerations.
Typical unit cells for metals are the body centered cubic (bcc)
structure found in chromium and tungsten, the face centered cubic

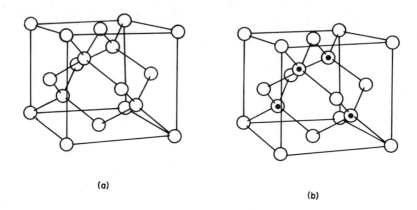

FIG. 1-3. Diamond lattice (a) and cubic zinc sulfide lattice (b). After Ref. 3.

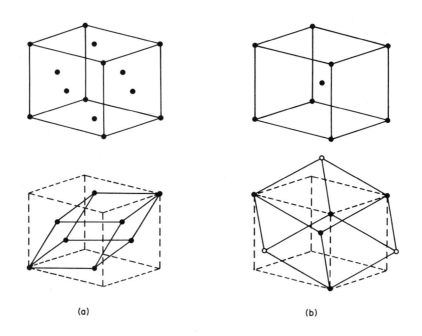

FIG. 1-4. Unit cells and primitive cells of a face centered cubic (a) and body centered cubic (b) lattice.

(fcc) structure, found in nickel and copper, and the hexagonal
closed packed (hcp) structure found in zinc and cadmium. The unit
cell of semimetals like bismuth or antimony can be quite complex.
One can describe this cell for the case of bismuth in the following
way. One takes a group of eight simple cubic unit cells which form
one cube (1-2). Then one distorts this cube by increasing the
length of one body diagonal, and by decreasing the length of the
other two body diagonals. This gives the rhombohedral lattice.
The atom in the center of this system is then pushed along the longest body diagonal toward one corner. The full lines in Fig. 1-2
show the resulting unit cell.

The unit cell used to describe the crystallographic structure
of the semiconductors germanium and silicon, or of the insulator
diamond, is given in Fig. 1-3a. It consists of two intersecting
face centered cubic lattices. The corner of one of the fcc lattices
is placed at the center between the corner atom, and adjacent atoms
in the center of faces. The unit cell of cubic ZnS is given in
Fig. 1-3b. It is sometimes convenient to use other types of unit
cells. Figure 1-4 gives us an example the "primitive cells" for
the fcc and bcc; they contain only one atom per unit cell. The
Bravais lattice has 4 or 2 atoms per unit cell respectively for
these atomic structures. These primitive cells have the disadvantage that their three crystal axes are not perpendicular to each
other. Another unit cell, again not cubic, is the "Wigner-Seitz
cell", which is given for the bcc system in Fig. 1-5. This cell is
obtained in the following way. One bisects the connecting lines
between one atom and all its neighboring atoms and constructs at
each of these points a plane which is perpendicular to the connecting line. These planes form the boundaries of the Wigner-Seitz
cell which contains only one atom and usually has a rather complex
shape. It is therefore not always a convenient description of the
arrangement of atoms in crystals. However, the Wigner-Seitz cell
may be a reasonable approximation of the shape of an atom. It
gives the "sphere of influence" in a way intuitively expected for

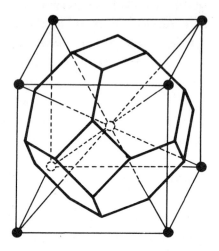

FIG. 1-5. Wigner-Seitz cell for a body centered cubic lattice. After Ref. 2, with permission.

each atom, since each point in the Wigner-Seitz cell is closer to the center of this cell than to the center of any other cell. Figure 1-6 gives the face centered cubic lattice showing together the cubic unit cell, the primitive cell, and the Wigner-Seitz cell. The radius of a sphere with the same volume as the Wigner-Seitz cell is the "atomic radius" R_a.

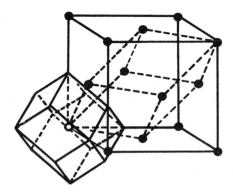

FIG. 1-6. Face centered cubic lattice showing cubic unit cell, primitive cell, and Wigner-Seitz cell. After Ref. 2, with permission.

An ideal crystal is built up of individual unit cells. If one knows the positions of atoms in one unit cell, one can determine the positions of all atoms in a crystal. The three edges of the unit cell can be given by the vectors $\vec{a}_1$, $\vec{a}_2$, and $\vec{a}_3$. Any point inside the unit cell is given by:

$$\vec{r} = \sum_{i=1}^{3} m_i \vec{a}_i, \qquad\qquad (1-1)$$

with $0 < m_i < 1$. A set of three numbers m_i characterizes the position of each lattice point uniquely. This point is given by $[m_1\ m_2\ m_3]$. More than one type of atom may be found in a unit cell. Different types of atoms may be distributed randomly over available lattice sites. The crystal is then not really a periodic structure. However, different types of atoms may occupy well defined subsets of lattice points. Then the crystal is a "lattice with base" in such a multi-atom system. It consists of several "sublattices".

The direction between two points $[m_1\ m_2\ m_3]$ and $[m_1'\ m_2'\ m_3']$ is given by $[m_1-m_1'\ m_2-m_2'\ m_3-m_3']$. The symbol for a point is also the symbol for the direction from the origin to this point. The position for any arbitrary lattice point is given by:

$$\vec{r} = \sum_{i=1}^{3} (n_i + m_i)\ \vec{a}_i, \qquad\qquad (1-2)$$

where each n_i is an integer. "Equivalent atoms" are connected by the vector:

$$\vec{r} = \sum_{i=1}^{3} n_i \vec{a}_i\ . \qquad\qquad (1-3)$$

All vectors or lattice points with the same $|m_i|$ values form a "family" of vectors or lattice points given by $< m_1\ m_2\ m_3 >$. Planes in a crystal can be also characterized by a set of three numbers. The coordinates of points in a plane are correlated by a linear function of the three coordinates x_i (i = 1, 2, 3), which can be written as:

$$x_1\, h/a_1 + x_2\, k/a_2 + x_3\, \ell/a_3 = 1. \tag{1-4}$$

h, k, and ℓ are the "Miller indices" of the plane. Miller indices
are given in parentheses: (h, k, ℓ). Equation (1-4) gives for
$x_1 = x_2 = 0$: $x_3/a_3 = 1/\ell$. This means that the reciprocal values of
the Miller indices give the positions of the intercepts of the plane
with the three basic directions $\vec{a}_i$ of the unit cell. The distances
are measured in units of a_i. Planes parallel to each other are
usually characterized by the same set of Miller indices. One takes
the lowest set of integers for h, k, and ℓ. A vector perpendicular
to a plane has the same indices as the plane in the cubic lattice.
A "family of planes" is defined in the same way as a family of
points. All planes which have the same absolute values of Miller
indices belong to one family and are given by $\{h\ k\ \ell\}$.

Since planes are uniquely described by a set of three numbers,
it is possible to characterize all planes in a crystal by vectors
with the three coordinates h, k, and ℓ. h, k, and ℓ would be the
three coordinates in a 3-dimensional space which is reciprocal to
the "real space." The vector in "reciprocal space", which characterizes a plane in real space, is sometimes given in a coordinate
system in which the basic vector, $\vec{b}_1'$, is defined with the equation
$\vec{b}_1' = \vec{a}_2 \times \vec{a}_3 / \vec{a}_1 \cdot \vec{a}_2 \times \vec{a}_3$. $\vec{b}_2'$ and $\vec{b}_3'$ are defined similarly. However, it is frequently more convenient to use vectors 2π times
larger for the description of the interaction of lattice planes
with electron waves or electromagnetic waves. Therefore, the
fundamental vectors $\vec{b}_1$, $\vec{b}_2$ and $\vec{b}_3$ of the reciprocal lattice are
defined by the equation:

$$\vec{b}_1 = 2\pi\, \vec{a}_2 \times \vec{a}_3 \,/\, \vec{a}_1 \cdot \vec{a}_2 \times \vec{a}_3, \quad \vec{b}_2 = 2\pi\, \vec{a}_3 \times \vec{a}_1 \,/\, \vec{a}_2 \cdot \vec{a}_3 \times \vec{a}_1$$

$$\vec{b}_3 = 2\pi\, \vec{a}_1 \times \vec{a}_2 \,/\, \vec{a}_3 \cdot \vec{a}_1 \times \vec{a}_2. \tag{1-5}$$

A periodic function in real space, $f(\vec{r})$, with the periodicity of
the lattice, can be given (pathological cases are excluded) by a

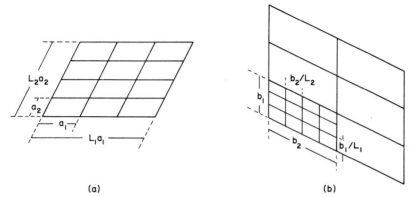

(a)   (b)

FIG. 1-7. Two dimensional lattice in "real space" (a) and its
reciprocal space (b).

series expansion of the form:

$$f(\vec{r}) = f(\vec{r} + \sum_{i=1}^{3} n_i \vec{a}_i) = \sum_{n} A_n \exp(i\vec{r} \cdot \vec{q}_n),$$ (1-6a)

where n is the summation over all vectors in reciprocal space. $\vec{q}_n$ is
given by the equation:

$$\vec{q}_n = \sum_{j=1}^{3} p_j \vec{b}_j.$$ (1-6b)

p_j is an integer. Figure 1-7 gives an example of a two dimensional

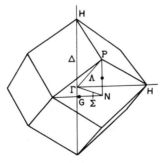

FIG. 1-8. Brillouin zone for body centered reciprocal lattice
with important points and lines of symmetry. See Ref. 5 for more
details.

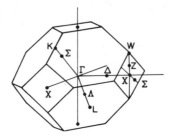

FIG. 1-9. Brillouin zone for face centered reciprocal lattice
with important points and lines of symmetry. See Ref. 5.

real space lattice and its reciprocal lattice. It is usually sufficient for the discussion of the interaction of lattice planes with
electromagnetic or electron waves to take only a small volume of the
reciprocal lattice into consideration. One takes one point on the
reciprocal lattice as origin and constructs around it unit cells in
the same way as the Wigner-Seitz cells in real space. These unit
cells are the "Brillouin zones." Symmetry points in reciprocal
space are usually characterized by letters of the Greek alphabet.
They are shown in Figs. 1-8 to 10 for bcc, fcc and hcp lattices.

B. Solid Solutions

The crystal structure of a material remains unchanged, if small

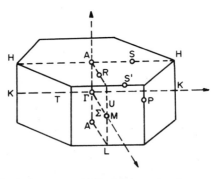

FIG. 1-10. Brillouin zone for hexagonal closed packed reciprocal lattice with important points and lines of symmetry. See Ref. 5.

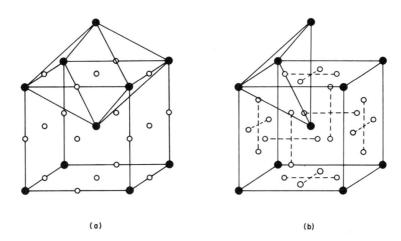

FIG. 1-11. Octahedral (a) and tetrahedral interstices (b) given as open circles in a body centered cubic crystal. After Ref. 6, with permission.

amounts of impurities are added to the pure element. The maximum impurity concentration under equilibrium conditions at room tempera- ture is less than a part per million for carbon in iron. It is also possible that two elements may be miscible over the total composition range. This is found in copper-nickel and copper-gold alloys.

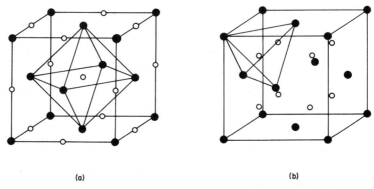

FIG. 1-12. Octahedral (a) and tetrahedral interstices (b) given as open circles in a face centered cubic crystal. After Ref. 6 with permission.

Complete solubility is only possible if both elements have the same
crystal structure, and their atomic radii differ by less than 15%.
This is not a sufficient condition for complete solubility. The
solubility of copper in silver and vice versa is less than 0.2 at.%
at room temperature, in spite of the fact that both metals have the
same crystal structure, are electronically similar, and that the
atomic radius of silver is about 13% larger than the atomic radius
of copper.

One has a "solid solution" if one can add a second type of atom
to a pure element without changing the crystal structure. One distinguishes between two types of solid solutions. One has a "substitutional solid solution" if an atom A of the original lattice is
replaced by an atom B; one has an "interstitial solid solution" if
a B-atom is squeezed into an extra place between original A-sites.
This second type of a solid solution is only possible for small B
atoms. The interstitial impurity would fit into the $[1/2 \ 0 \ 1/4]$
position (tetrahedral site, see Fig. 1-11b) of a bcc lattice, if one
assumes that A and B atoms are hard spheres, that the A-atoms touch
each other, and that the ratio of the radius of B-atoms to A-atoms
(radius ratio or RR) is less than 0.095. $[1/2 \ 1/2 \ 0]$ would be the
center of an octahedral site in the bcc lattice (1-11a). Octahedral
sites in the center of the fcc unit cell with coordinates $[1/2 \ 1/2 \ 1/2]$
and tetrahedral sites with coordinates $[1/4 \ 1/4 \ 1/4]$ are available
for interstitional atoms in fcc crystals. Different types of atoms
may be randomly distributed in the crystal, or they may be regularly
arranged. The last case represents a "super lattice." It is equivalent to a "lattice with base." Most solid solutions show neither
complete atomic order or disorder, but some state of partial order.

Several factors determine the range of solubility in alloys.
Important are the relative atomic radii of the different atoms in
the crystal, their valency, and their electrochemical difference.
Phase stability criteria, based on these concepts, are called the
"Hume-Rothery rules" for solid solutions. These concepts have been
expanded in the "Brewer-Engel" theory, in which one compares energy

states of free atoms with bonding and antibonding electron energies
in the crystalline state.

The first Hume-Rothery rule states that extended solid solubility can be expected only if the relative difference of the atomic
radii $\Delta R/R$(average) is less than 0.15. This rule is probably related
to the maximum elastic strain which a solid solution may sustain
and still be stable. The definition of the size of an atom is relatively simple in pure metals with one lattice structure. One assumes
only that the atoms are spheres which touch each other. The radius
of such a sphere would be equal to 1/2 of the lattice parameter in
the simple cubic lattice, and $\sqrt{1/8}$ of the lattice parameter in the
fcc crystal. This radius R is not a simple function of temperature
for elements which exist in different crystal structures or "allo-
tropic modifications". For instance, iron is fcc between 906°C and
1401°C; below and above this temperature range it is bcc. R changes
discontinuously during transformation. The change in R should be
attributed to the change of the number of nearest neighbors during
the transformation. In the bcc crystal, each iron atom has eight
nearest neighbors; in the fcc crystal it has twelve. The bonding
between iron atoms in the bcc structure should be partially directional, which is responsible for an atomic packing of lower density.
This shows that the radius of an atom depends on the interaction
with neighbors. One would therefore expect that this radius will
also depend on the types of atoms which are nearest neighbors.

A simple composition dependence of the atomic radius is rarely
found in metallic systems. Vegard's law, which states that the lattice parameter of ionic crystals is a linear function of the ionic
concentration, cannot be applied to metals. Zen's law, which predicts
that the atomic volume is a linear function of composition, does not
seem to fare much better. However, deviations from such "laws" have
usually some physical significance. Figure 1-13 gives a plot of the
atomic volume as a function of composition for the copper-zinc alloy
system. This figure shows that both the volume of atoms, and the
distance between atoms are not linear functions of composition.

It has been attempted to calculate the change of the lattice
parameter with composition from known physical properties of the
elements, such as the compressibility, the modulus of elasticity, the
shear modulus, etc.   These data were used to calculate the elastic
distortion in a crystal due to impurity atoms which are either smaller or larger than the atoms they replace. Good agreement between
theory and experiments is obtained only for systems in which the
impurity atom is larger than the host atom.

It is surprising that a large number of solid solutions of
noble metals with elements in other groups of the periodic system
have a stability range which depends frequently only on the number
of conduction or valency electrons per atom (both names are used
to describe outer electrons of atoms in metals). Figure 1-14 shows
experimentally observed solid solution ranges. They end for noble
metal base alloys at the conduction electron number per atom, el./at.,
between 1.3 and 1.4. Here the el./at. ratio is defined as the number of electrons in unfilled shells per atom.

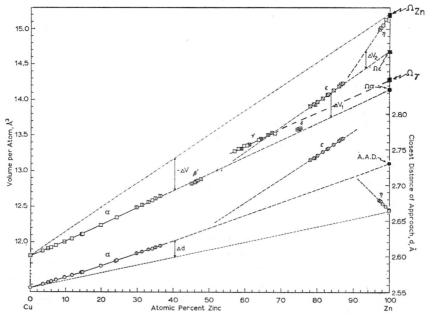

FIG. 1-13. Trends in lattice spacings and volume per atom in the
Cu-Zn system. Reproduced with permission from Ref. 7.

C. Intermediate Phases and Compounds

The attraction between different types of atoms in a solid
may be stronger than the attraction between identical atoms. This
attraction between atoms A and atoms B is due to forces between
electrons and nucleii. It is possible to evaluate this interaction
by semiquantitative parameters. The ability to attract an electron
is given as the "electronegativity" of the atom. In the NaCl crystal,
the "electronegative" chlorine atom absorbs an electron from an
"electropositive" sodium atom. The crystal is therefore composed of
negatively charged chlorine ions, and positively charged sodium ions.
The difference in the electronegativity of two elements gives the
stability of "compounds". Compounds are described with symbols of the
form $A_n B_m$ just as in molecules. However, the NaCl crystal cannot be
described as a system consisting of parts of sodium and chlorine

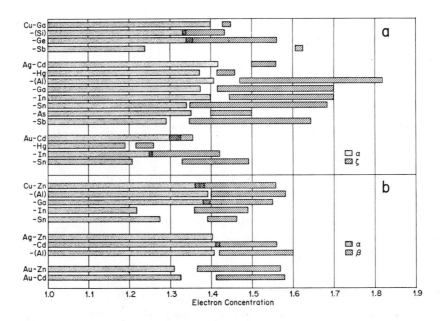

FIG. 1-14. Primary solid solution range and intermediate phases
for noble metal alloys. Reproduced with permission from Ref. 8.

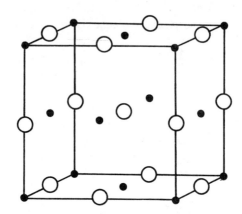

FIG. 1-15. Structure of NaCl. Cl: open circles. Na: filled
circles. After Ref. 9.

ions, which form individual molecules with strong bonding between one
pair of ions and weak bonding between different pairs.

NaCl is a crystal which has a lattice with base. Its unit cell,
given in Fig. 1-15, contains four sodium and four chlorine ions. Both
types of ions occupy a "sublattice" of the crystal. An example of an
ionic crystal of composition AB_2 is given in Fig. 1-16. This is CaF_2.
The sublattices given in these two examples are typical for ionic
crystals, where electrostatic Coulomb forces are very strong. These
forces are responsible for the cohesion in ionic lattices. They are
long range forces. The electrostatic energy between two ions "i" and
"j" is given by $U_{ij} = q_i q_j / 4\pi\varepsilon\varepsilon_o r_{ij}$. q_i is the charge of ion "i",
ε is the dielectric constant, ε_o the permittivity of vacuum, and r_{ij}
the distance between the two ions. The repulsive forces between
nearest neighbors are only of short range order. Quantum theory suggests for the repulsive energy an expression of the form U_{ij}(rep.) =
$A \exp[-r_{ij}/r]$. However, U_{ij}(rep.) = A'/r_{ij}^s, with $s \simeq 9$, describes
experimental data adequately. The electrostatic energy of an ionic
crystal U is obtained by summing overall ion pairs. This gives
$U = (1/2) \Sigma_{i \neq j} U_{ij}$. The factor 1/2 takes into consideration that each
ion pair is counted twice. $U = (1/2)(q^2/4\pi\varepsilon\varepsilon_o R) \Sigma_{i \neq j} \pm 1/(r_{ij}/R)$

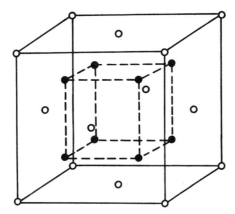

FIG. 1-16. Structure of CaF_2. F: full circles. Ca: open circles. After Ref. 9.

in crystals, in which the absolute values of all charges are the same. R is the equilibrium distance between nearest neighbors. $\pm$ indicates the charge of the ion. $\alpha = \Sigma_{i \neq j} \pm 1/(r_{ij}/R)$ is the Madelung constant. For a crystal like NaCl it is: $\alpha = 1.7475....$, for CsCl: $\alpha = 1.7626....$, and for cubic ZnS: $\alpha = 1.6381....$

The difference of the electronegativity of elements in typical metallic compounds is only a fraction of that found in ionic crystals. Metallic compounds with extended stability ranges, provided the stability range does not extend to one of the pure elements, are defined as "intermediate phases". The distinction between intermediate compound and intermediate phase is to a certain degree arbitrary. "Intermediate phases" and "intermediate compounds" are defined similarly. Elements in these systems have to be metals.

Electrostatic forces are orientation independent. Orientation dependent bonding is found if electrons are not exchanged between atoms but are shared by neighboring atoms. This represents "covalent bonding". Crystals with covalent bonding are characterized by a relatively open lattice. [volume of atoms]/[volume of crystal], and also the number of nearest neighbors, are low. A typical example for such a structure is the diamond lattice, as shown in Fig. 1-3a. In this system, each atom has four nearest neighbors, which form the corners of a tetrahedra.

All crystal positions are occupied by one type of atom in diamond, silicon and germanium. The dotted lattice positions in Fig.
1-3b are occupied by one type of atom, and the open circles by the
second type of atom in III-V or II-VI compounds (the Roman letters
indicate the group of the periodic system to which the atom belongs)
like GaAs or ZnS. Figure 1-3b is the "Zincblende" structure. This
arrangement gives an electron distribution in which two outer electrons are always shared between neighboring atoms. It is possible
in ternary alloy systems like ZnS-ZnSe that the sulphur and selenium
atoms occupy randomly the sublattice with open circles as shown in
Fig. 1-3b, and the zinc atoms occupy the other sublattice. This is
a "pseudo-binary" alloy system. One can mix arbitrary amounts of
ZnS and ZnSe to obtain this crystal. One writes it as $(ZnS)_c(ZnSe)_{1-c}$,
with c as concentration, or as $Zn_2S_cSe_{1-c}$. This is a "mixed crystal".
In some alloys with four components metallic ions can be randomly distributed in one sublattice, and non-metallic ions in the other.

Bonding in III-V and II-VI compounds is not purely covalent.
There exists also an ionic component since the group III elements
(or II elements) have a different electronegativity than the group
V elements (or group VI elements). Just as all pure elements can
absorb at least a few impurity atoms, compounds may deviate slightly
from the ideal (nominal) composition and may still be referred to as
intermediate compounds.

The stability of intermediate compounds and intermediate phases
is determined by several factors. Important are the ratio of valency
electrons per atom, the size of atoms, and the packing density. Phases
are called "Hume-Rothery phases" if the valency electron per atom
ratio, el./at., determines the stability, and if this ratio is 3/2,
21/13, or 7/4. The crystal structure for the 3/2 ratio is either
the complicated β-manganese structure or it is bcc with the composition AB. The center atom in this bcc cell is occupied by the A atom,
and the corners are occupied by B atoms. This is sometimes called
the CsCl-structure. It occurs, for instance, in the copper-zinc system
as CuZn. Tables 1-1 to 3 list examples of these phases. One can see

that the el./at. ratio in the stability range is spread around 3/2, 21/13 or 7/4. However, it is sometimes possible, as the HgMg example shows, that the stability range does not include one of these values.

The packing density in compounds and intermediate phases can also be an important stability criterion. The packing density is defined as the volume of all atoms, $\Sigma \frac{4\pi}{3} R_i^{\ 3}$, divided by the volume of the crystal. The sum of the radii of nearest neighbor atoms, $R(A) + R(B)$, is equal to or smaller than the distance between the centers of these atoms. "Laves phases" are examples in which both

TABLE 1-1
Hume-Rothery Phases with β-Manganese or β-Brass Structure

Composition	At.% of second component	Stability range outer electrons/atom
Ag_3Al	25	1.41 - 1.62
AgCd	50	1.41 - 1.56
AgMg	50	1.41 - 1.66
AuCd	51	1.44 - 1.58
AgZn	50	1.35 - 1.57
AlFe	50	1.50
AlNd	50	1.50
BeCo	50	1.00
BeCu	50	1.50 - 1.47
BeNi	50	1.00
Cu_5Sn	50	1.40 - 1.66
CuZn	50	1.36 - 1.56
FePt	50	1.00
HgMg	50	2.00
Mn_3Si	25	1.00

TABLE 1-2

Hume-Rothery Phases with γ-Brass Structure

Composition	At.% of second component	Stability range outer electrons/atom
Ag_5Cd_8	61.5	1.58 – 1.63
Ag_5Hg_8	61.5	1.60
$AgLi_3$	75	1.00
Al_4Cu_9	69.5	1.62 – 1.74
$Be_{21}Ni_5$	19.2	1.64 – 1.65
$Be_{21}Pt_5$	19.2	1.62
$Cd_{21}Co_5$	19.2	1.62
$Cd_{21}Pt_5$	19.2	1.62
Cu_9Ga	30.8	1.59 – 1.76
Fe_5Zn_{21}	81	1.40 – 1.55

TABLE 1-3

Hume-Rothery Phases with ε-Brass Structure

Composition	At.% of second component	Stability range outer electrons/atom
Ag_5Al_3	37.5	1.54 – 1.80
$AgCd_3$	75	1.65 – 1.81
Ag_3Sn	25	1.37 – 1.69
$AgZn_3$	75	1.67 – 1.90
$AuCd_3$	75	1.62 – 1.65
Au_3Sn	25	1.36 – 1.48
$AuZn_3$	75	1.89
Cu_3Ge	25	1.75
Cu_3Si	25	1.44
Cu_3Sn	25	1.73 – 1.75

the packing density and the el./at. ratio determine the phase stability. Figures 1-17 and 18 give examples of crystals belonging to
the group of Laves phases, which have the composition AB_2 like $MgZn_2$
(hcp) or $MgCu_2$ (cubic). Table 1-4 gives the radius ratio of several Laves phases. Densest packing of atoms in the $MgZn_2$ and $MgCu_2$
type phases is obtained with a radius ratio of 1.225 = R(A)/R(B).
This is close to the values found in most Laves phases, confirming
that a "geometrical factor", namely a high packing density, is important for the stability of these phases. The highest packing density
possible in Laves phases is 0.71. This is higher than the maximum
packing density of 0.68 possible in bcc crystals and only slightly
lower than the packing density of fcc and hcp crystals.

TABLE 1-4

Laves Phases

$MgCu_2$- Type	Radius ratio	$MgZn_2$ - Type	Radius ratio
$CaAl_2$	1.38	KNa_2	1.23
$MgCu_2$	1.25	$MgZn_2$	1.17
Mg(NiZn)	1.23	Mg(CuAl)	1.18
$CeAl_2$	1.27	$CaMg_2$	1.23
$LaAl_2$	1.30	Ca(AgAl)	1.37
$TiBe_2$	1.28	$CrBe_2$	1.13
$(FeBe)Be_4$	1.06	$MnBe_2$	1.16
$(PdBe)Be_4$	1.11	$FeBe_2$	1.12
$CuBe_{2.35}$	1.13	VBe_2	1.20
$AgBe_2$	1.27	$ReBe_2$	1.21
$(AuBe)Be_4$	1.14	$MoBe_2$	1.24
Cd(CuZn)	1.15	WBe_2	1.25
α-$TiCo_2$	1.15	WFe_2	1.11
$ZrFe_2$	1.26	$TiFe_2$	1.14
$ZrCo_2$	1.27	$TiMn_2$	1.11

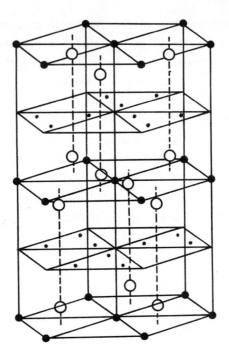

FIG. 1-17. Structure of $MgZn_2$. Zn: full circles; Mg: open circles. After Ref. 9.

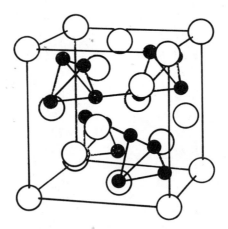

FIG. 1-18. Structure of $MgCu_2$. Cu: full circles; Mg: open circles. After Ref. 10.

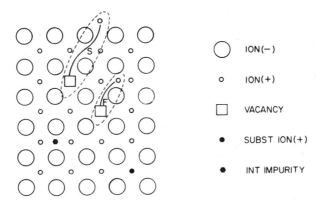

FIG. 1-19. Defects in an ionic lattice. S: Schottky defect.
F: Frenkel defect.

D. Lattice Defects

Even the most carefully prepared single crystal is not perfect.
Atoms vibrate around their equilibrium position, and some lattice
positions will not be occupied at all by atoms. Such an empty lattice position is a lattice defect called a "vacancy". It is an example of a "point defect" which is an incorrectly occupied or an
empty lattice site position. A vacancy in an ionic crystal, created
by the movement of an ion to the surface (this insures electrical
neutrality) is a Schottky defect. Metallic ions in an ionic or covalent crystal are usually much smaller than the cations. Therefore,
metallic ions may move into "interstitial positions". The vacancy
and interstitial form a "Frenkel pair" (1-19). A Frenkel pair should
not change the volume of the crystal, provided that the metal ion is
so small that the lattice deformation may be treated by the laws of
linear elasticity. On the other hand, a Schottky defect leads to a
decrease of density of the crystal.

One can influence the vacancy concentration in a crystal by
alloying. For instance, if an impurity atom is large, it is likely
that a vacancy will form next to this impurity, because the large
impurity will squeeze the neighboring atom out of its place. In ionic

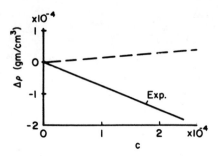

FIG. 1-20. The change in density of KCl with increasing CaCl$_2$ impurity concentration. c = (number of Ca ions)/(number of K ions). Dashed line is the average density of the two salts. After Ref. 22.

crystals, like the mixed crystal $(KCL)_{1-c}(CaCl_2)_c$ with small amounts of CaCl$_2$, one calcium ion replaces two sodium ions. This insures electric neutrality. Naturally, the calcium ion can occupy only one lattice position previously occupied by one potassium ion. The second potassium ion is replaced by a vacancy. Density measurements show that the increase in CaCl$_2$ concentration leads to a decrease in the density, as predicted by this model. However, the observed density decrease is slightly smaller than this model would predict (Fig. 1-20).

A crystal stays electrically neutral if a monovalent metallic ion is replaced by an electron (see Fig. 1-21). Such a defect, vacancy plus electron, is called an "F-center". It leads in some crystals to optical absorption in the visible range. Two adjacent coloring centers, made up of two vacancies and two trapped electrons, give an F$_2$ or M-center (see Fig. 1-22). An F$_3$ or R-center (see Fig. 1-23) consists of three adjacent anion vacancies replaced by three electrons. Since the energy state of an electron in an anion vacancy depends on the interaction with adjacent cations, an impurity cation will lead to a change in the energy states of the electron. Such a center is called an F$_A$-center. If a cation is removed from an ionic lattice, then electrical neutrality can be obtained if one of the neighboring anions loses the same number of electrons as the cation takes out. This can lead to a system with a vacancy and an electrically neutral atom. For the case of KCl, one can describe the defect

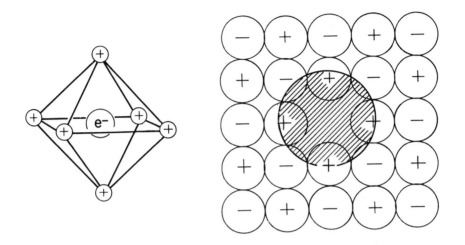

FIG. 1-21. Atomic structure of F-center. After Ref. 11, with permission.

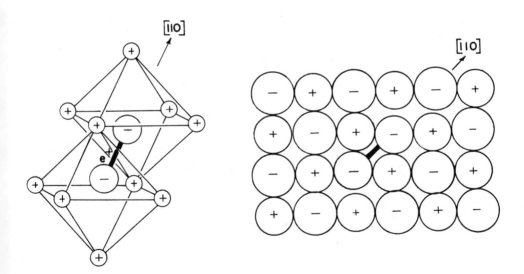

FIG. 1-22. Atomic structure of F_2-center. After Ref. 11, with permission.

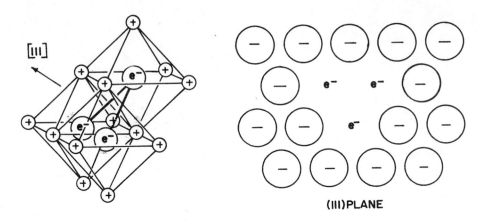

FIG. 1-23. Atomic structure of F_3-center. After Ref. 11, with permission.

as a 'hole' attached to a negatively charged molecule of Cl_2. This is a V_K-center (see Fig. 1-24).

Single crystals usually have linear or "line defects". A simple line defect is an "edge dislocation", which may be visualized as the edge of a half-plane inserted into a perfect crystal. This defect is shown in Fig. 1-25a. The crystal is imperfect

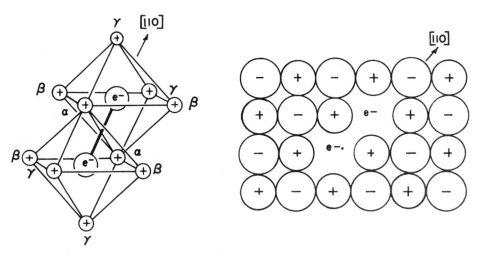

FIG. 1-24. Atomic structure of V_K-center. After Ref. 11, with permission.

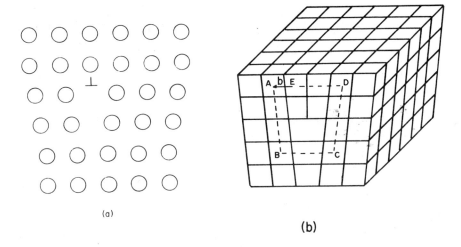

(a)

(b)

FIG. 1-25. The atomic structure of an edge dislocation is given in (a). It is characterized by the symbol ⊥. One can visualize the edge dislocation as the lower edge of the half plane inserted from the top. This edge is the dislocation line. Burger's circuit in a crystal with an edge dislocation is given in (b) as a dashed line and the vector b. The distance in lattice units A-B is equal to the distance C-D. The distance B-C is equal to D-E (both are three lattice units long). In a crystal with a dislocation, the vector EA has to be added to close the Burger's circuit. EA is the Burger's vector. It is perpendicular to the edge dislocation line.

only at the edge of the inserted half-plane. It can be characterized with the "Burger's vector" $\vec{b}$, obtained by constructing a "Burger's circuit". Figure 1-25b shows how it can be obtained. One starts at an arbitrary lattice point A, takes n steps down to B, takes m steps to the right (C), takes then again n steps up to D, and m steps to the left (E). The vector from E to A is the Burger's vector of the dislocation. E would coincide with A if the crystal is perfect, without the extra half plane. For an edge dislocation, the Burger's vector stands perpendicular to the dislocation line, which is the edge of the inserted half-plane.

A second type of dislocation is obtained if the crystal is cut along a half-plane, and if atoms opposite to each other in this half-plane are shifted by one step parallel to the lower end of the section (Fig. 1-26). This is a screw dislocation. The dislocation

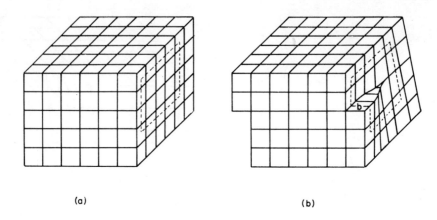

(a)   (b)

FIG. 1-26. Perfect crystal (a) and crystal with screw dislocation (b). The screw dislocation can be produced by moving all atoms
of the two top rows of atoms by one lattice parameter to the left,
up to the middle of the crystal. The Burger's circuit is given in
(a) and (b). The Burger's vector is defined in the same way as in
Fig. 1-25. In this figure, the screw dislocation line emerges from
the center of the Burger's circuit. The dislocation line of a screw
dislocation is parallel to its Burger's vector.

line is the lower end of the cut, and the Burger's vector b of

this defect is lined up parallel to the dislocation line (Fig. 1-26b).

This is a "screw dislocation".

Both types of dislocations may split up into "partial disloca-

tions". The crystal structure of the plane between these two

partial dislocations is different than in the perfect crystal. If

the original lattice is fcc, the nearest and next nearest neighbor

arrangement in the plane between the two partial dislocations (a

"stacking fault") is equal to the hcp structure. Its energy differs

only slightly from the energy in the fcc lattice, since these two

lattices have the same nearest neighbor configurations. Only the

next nearest neighbor configurations are different in hcp and fcc

structures.

The plastic deformation mechanism in metals depends on the dis-

location structure. Therefore, a detailed description of the dis-

location arrangement, the dislocation multiplication and movement,

and the dislocation interaction is given in books on plastic properties of metals. The effect of dislocations on the electric,
magnetic, and thermal properties of metals and compounds is usually
small.

III. ELASTIC PROPERTIES OF CRYSTALS

a. Static Deformation

Atoms in an ideal crystal occupy lattice points. Atoms are
displaced from these positions in real lattices. The displacement
may be due to thermal oscillation, elastic deformations, or the
interaction of atoms with defects. The average displacement of atoms
from the ideal position can be quite large. Some metallic crystals,
so-called "whiskers" since they are hairlike, can be strained by more
than 10% of their length under tension. This elongation is perfectly
elastic, since the whisker snaps back to its original shape as soon
as the load is released. Thermal oscillations of atoms near the
melting point of the crystal are again of the order of 10% of the
distance between nearest neighbors.

One can produce macroscopically visible displacements of atoms
by straining, shearing, or compressing a solid. This leads to elastic deformations of the sample if the deformation is a unique function of the applied forces. The elastic deformation for metals is
usually a linear function of the applied force. This is "Hooke's
Law":

$$\Delta \ell / \ell = \varepsilon = (\text{Force/Area})/E = \sigma/E \qquad\qquad (1-7)$$

where ℓ is the sample length, $\Delta \ell$ the length change of the sample due
to an applied force, $\Delta \ell / \ell = \varepsilon = (\text{Force/Area})/E = \sigma/E$, σ the stress,
defined by "normal force per cross sectional area", and the constant
E is the "modulus of elasticity". If Eq. 1-7 is independent of the

orientation between applied force and crystal sample orientation, the material is elastically "isotropic".

The stress on the sample in a given direction will not only change the sample length in the direction of the applied load but it will usually change the sample length in the directions perpendicular to the applied load. A stress in the x_1 direction, σ_1, will be association with length changes $\Delta\ell_2$ and $\Delta\ell_3$ in the x_2 and x_3 directions, respectively. One finds in isotropic materials that

$$\varepsilon_2 = \varepsilon_3 = -\nu\varepsilon_1 \qquad\qquad (1\text{-}8)$$

where ν is the "Poisson's ratio" and $\varepsilon_1 = \Delta\ell_1/\ell_1$. Equations 1-7 and 1-8 are applicable to both tensile and compressive stresses. Since one can usually superimpose elastic stresses and deformations, it is easy to correlate the compressibility, β, defined by

$$\beta = -\,(dV/Vdp) \qquad\qquad (1\text{-}9)$$

with ν and E. p is the hydrostatic pressure. In this equation, dV is positive for compression. For hydrostatic pressure $\sigma_1=\sigma_2=\sigma_3=-p$. This means that the volume of the sample at pressure p_1, V_1, divided by the original volume V_o at pressure $p = 0$, is $V_1/V_o = (V_o + \Delta V)/V_o =$

$$1 + (\Delta V/V) = (1 + \varepsilon_1)(1 + \varepsilon_2)(1 + \varepsilon_3) = (1 + \frac{\sigma_1}{E} - \frac{\nu\sigma_2}{E} - \frac{\nu\sigma_3}{E})^3 \approx$$

$$1 + 3\varepsilon \approx 1 + 3\sigma_1(1 - 2\nu)/E \quad , \qquad\qquad (1\text{-}10a)$$

if one neglects higher terms. σ_i is a stress in the x_i direction. ΔV in this equation is negative for compression. Equations 1-9 and 1-10a for small deformations give a correlation between compressibility, Poisson's ratio and modulus of elasticity:

$$\beta = 3(1 - 2\nu)/E. \qquad\qquad (1\text{-}10b)$$

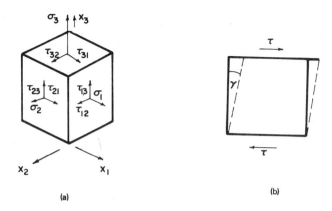

FIG. 1-27. Stresses σ and shear stresses τ acting on a cube (a). Deformation by shear γ due to a shear stress τ (b).

σ_i is a stress applied perpendicular to a surface which has a normal parallel to the x_i-direction. A stress applied parallel to the surface is a shear stress τ_{ij}. The first index indicates the plane on which it acts, the second gives its direction. Equilibrium conditions for static deformations require that $\tau_{ij} = \tau_{ji}$. A shear stress τ leads to a shear strain γ (1-27b):

$$\tau = \gamma G \qquad\qquad\qquad (1\text{-}11)$$

where G is the "shear modulus". The general correlations between normal stresses σ_i, shear stresses τ_{ij}, strains ε_i, and shear strains γ_{ij} for small deformations of anisotropic systems can be given by:

$$\varepsilon_1 = S_{11}\sigma_1 + S_{12}\sigma_2 + S_{13}\sigma_3 + S_{14}\tau_{23} + S_{15}\tau_{31} + S_{16}\tau_{12}$$
$$\qquad\qquad\qquad\qquad\qquad\qquad\qquad\qquad (1\text{-}12a)$$
$$\gamma_{23} = S_{41}\sigma_1 + S_{42}\sigma_2 + S_{43}\sigma_3 + S_{44}\tau_{23} + S_{45}\tau_{31} + S_{46}\tau_{12}$$

and similar equations for ε_2, ε_3, γ_{31}, and γ_{12}. S_{ij} are the "elastic compliance constants". It is also possible to write the stresses and shear stresses as linear functions of strains and shear strains.

This gives:

$$\sigma_1 = C_{11}\epsilon_1 + C_{12}\epsilon_2 + C_{13}\epsilon_3 + C_{14}\gamma_{23} + C_{15}\gamma_{31} + C_{16}\gamma_{12}$$

$$\text{(1-12b)}$$

$$\tau_{23} = C_{41}\epsilon_1 + C_{42}\epsilon_2 + C_{43}\epsilon_3 + C_{44}\gamma_{23} + C_{45}\gamma_{31} + C_{46}\gamma_{12}$$

and similar equations for σ_3, σ_2, τ_{31}, and τ_{12}. A more complete set of equations may be found in Ref. 3. C_{ij} are the "elastic stiffness constants" or the "moduli of elasticity". These two sets of equations with 36 constants each are applicable to arbitrary anisotropic crystals. The elastic energy is a quadratic function of the strains if Hooke's law can be applied. This means that the elastic stiffness constants are symmetrical. The number of independent elastic stiffness constants is therefore only 21. The symmetry of a cubic system reduces this number of independent constants to three.

B. Stress Fields of Point and Line Defects

One would expect near a vacancy a relaxation effect, in which neighboring atoms would be shifted from the equilibrium position of the perfect lattice to a new equilibrium position closer to the center of the vacancy. A similar, but smaller, effect would be obtained if an impurity in a substitutional position were smaller than the matrix atom it replaces. The difficulty in determining theoretically the composition dependence of the lattice parameter, and deviations from Vegard's and Zen's law show clearly that the interaction of vacancies and of point defects cannot be handled quantitatively with sufficient accuracy.

Similarly, the energy of the center of a line defect (core energy) is difficult to estimate. However, it is possible to calculate the stress field surrounding this center with simplifying assumptions for the case of a screw dislocation. One assumes that the center of the dislocation is a hollow tube, that the material is isotropic, and that the linear theory of elasticity can be used. The calculation for elastic energy of a screw dislocation of unit length gives: $E = (Gb^2/4\pi)\ln(R/r)$, with R the upper limit of the stress field, and r the radius of the hollow tube. This equation shows that the elastic energy of a screw dislocation is infinite for an infinitely large sample. In typical materials, one would assume R to be the average distance between dislocations. The energy per unit length of an edge dislocation is $E = Gb^2\ln(R/r)/4(1 - \nu)$.

C. Elastic Waves in Cubic Crystals

Elastic constants can be measured in static experiments. It is, however, frequently easier to determine elastic constants from dynamic tests in which elastic waves travel through the crystal. Displacements of atoms in these tests are small and the deformation is elastic. Elastic forces acting on a small volume element are equal to the forces of inertia. One obtains for the equation of motion of atoms in the x_1 direction [3]:

$$\rho(\partial^2 u_1/\partial t^2) = \partial\sigma_1/\partial x_1 + \partial\tau_{21}/\partial x_2 + \partial\tau_{31}/\partial x_3, \tag{1-13}$$

where ρ is the density of the material, and u_1 the displacement of atoms in the x_1 direction. Similar equations can be set up for the x_2 and x_3 direction. One obtains from Eq. (1-13), replacing σ and τ by ε and γ:

$$\rho(\partial^2 u_1/\partial t^2) = C_{11}(\partial\varepsilon_1/\partial x_1) + C_{12}(\partial\varepsilon_2/\partial x_1 + \partial\varepsilon_3/\partial x_1)$$
$$+ C_{44}(\partial\gamma_{12}/\partial x_2 + \partial\gamma_{31}/\partial x_3). \tag{1-14a}$$

ε_i and γ_{ij} are related to u_i by:

$$\partial u_i/\partial x_i = \varepsilon_i \ , \qquad \partial u_i/\partial x_j + \partial u_j/\partial x_i = \gamma_{ij} \ , \tag{1-14b}$$

This means that Eq. (1-14a) can be changed to an equation which contains only second derivatives in time or space of u_i:

$$\rho(\partial^2 u_1/\partial t^2) = C_{11}(\partial^2 u_1/\partial x_1^2) + C_{44}(\partial^2 u_1/\partial x_2^2 + \partial^2 u_1/\partial x_3^2)$$

$$+ (C_{12} + C_{44})(\partial^2 u_2/\partial x_1 \partial x_2 + \partial^2 u_3/\partial x_1 \partial x_3). \quad (1\text{-}14c)$$

A solution of this equation for a longitudinal wave in the [100]
direction can have the form:

$$u_1 = u_{1,0} \exp(iq_1 x_1 - i\omega t), \quad\quad\quad\quad\quad\quad\quad (1\text{-}15)$$

where $q_1 = 2\pi/\lambda$ is the wave vector and ω the angular frequency of
the elastic wave. Inserting this equation into (1-14) gives for a
longitudinal wave in the [100] direction:

$$\omega^2 \rho = C_{11} q_1^2, \quad v_1(1) = v\lambda = \omega/q_1 = (C_{11}/\rho)^{1/2}. \quad (1\text{-}16)$$

$v_1(1)$ is the sound velocity of the longitudinal wave. A shear wave
moving in the x_1 direction, with atoms displaced along the x_2 or
x_3 directions, gives again a periodic displacement of atoms:

$$u_2 = u_{2,0} \exp(iq_1 x_1 - i\omega t). \quad\quad\quad\quad\quad\quad (1\text{-}17)$$

This assumption gives:

$$\omega^2 \rho = C_{44} q_1^2, \quad v_1(t) = (C_{44}/\rho)^{1/2}. \quad\quad\quad\quad (1\text{-}18)$$

$v_1(t)$ is the sound velocity of the transverse wave. These equations
give relationships between the wave vector $\vec{q}$, the angular frequency
of the sound wave and elastic constants. They make it possible to
determine elastic constants from sound wave propagation measurements.
Figures 1-28 and 1-29 give experimental results for pure metals.

In a typical experiment in which one determines the correlation

between wave length and frequency of sound waves, the wave is generated at the sample surface and moves through the crystal, is back-
reflected from the rear of the sample, travels again through the
sample, and produces in the transducer an electrical signal. The
time interval between emission and absorption of the sound wave can
be used to determine the velocity of sound, since this time interval is equal to twice the sample length divided by the sound velocity.
One can also produce standing sound waves in a crystal. The frequency
of the sound wave is then equal to the frequency of the transducer
signal. The wavelength of the standing wave is a fraction of twice
the sample length. A plot of the frequency of the sound wave as a
function of the wave number, $\vec{q}$, $\nu = f(\vec{q})$, is the "dispersion curve"
of the material.

The longest possible wave length in a solid is twice the length
of the free sample. The ends of the sample move in opposite
directions, and the center of the sample is stationary. The shortest
possible wave length is twice the distance of adjacent low index
atomic planes, d_o. These adjacent planes then move in opposite directions. In other words, they are 180° out of phase. Shorter wave
lengths are physically not meaningful. One obtains all possible oscillations in a one dimensional system by summing up all wave lengths
between $\lambda = 2a$ (a is the distance between atoms in this one dimensional lattice) and $\lambda = 2L$ (L is the length of the sample). Generally, $\lambda = 2L/n$, with n an integer. Wave numbers range therefore
over values given by:

$$-\pi/a \leq q = n\pi/L \leq \pi/a. \tag{1-19}$$

Negative wave numbers correspond to elastic waves moving in opposite
directions to waves with positive wave numbers.

$\vec{q}$ values of stationary elastic waves in a two or three dimensional crystal are contained in the first Brillouin zone in reciprocal
space. A Brillouin zone is constructed in reciprocal space similar
to the Wigner-Seitz cell in real space. One connects lattice points

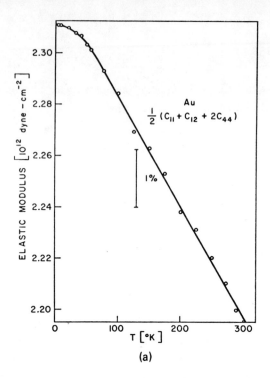

FIG. 1-28a. The elastic stiffness constant $(C_{11} + C_{12} + 2C_{44})/2$ for gold. After Ref. 13, with permission.

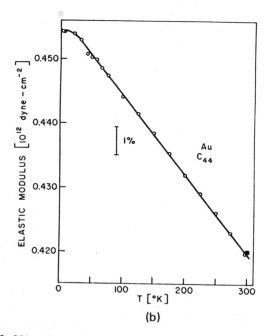

FIG. 1-28b. The elastic stiffness constant C_{44} for gold. After Ref. 13, with permission.

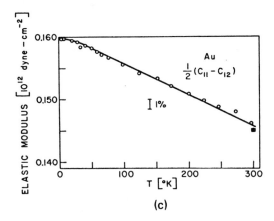

FIG. 1-28c. The adiabatic elastic shear constant $(C_{11} - C_{12})/2$ for gold. After Ref. 13, with permission.

in reciprocal space given by Eq. 1-6b with the origin of the reciprocal lattice. Then one bisects these lines, and constructs planes normal to these lines at the point of bisection. The set of planes closest to the origin is the boundary of the first Brillouin zone. A Brillouin zone is shown schematically in Fig. 1-30. Lines of constant frequency in the (100) plane of the Brillouin zone of aluminum are given in Fig. 1-31. These lines should be spheres for isotropic materials. This is found in Fig. 1-31 for $q/(\pi/a) \leq 0.5$. Larger q values, which correspond to a wave length of the order of the lattice parameter, do depend markedly on the orientation. This is not surprising, since isotropic behavior in a crystal is only to be expected if the wave length of the sound wave is much larger than atomic distances. In such a case, elastic constants are averaged over volume elements which are large compared with the unit cell.

The first Brillouin zone is sometimes replaced by the "Debye sphere" which has the same volume as the first Brillouin zone. This sphere was named after Debye, who proposed that the number of elastic waves should be equal to the number of degrees of freedom, 3N, where N is the number of atoms in the crystal [15]. A simpler model for oscillations in a crystal was used by Einstein [16]. He assumed in

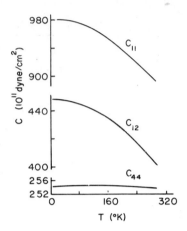

FIG. 1-29. The elastic stiffness constants of BaF_2. After Ref. 14, with permission.

the calculation of the lattice specific heat that the atoms are independent harmonic oscillators. Each atom can oscillate in three directions. Its vibrations are independent of the movement of other atoms,
and the energy levels are quantized, E_n = nhν. The frequency of oscillation can be estimated from the weight of the atom and the elastic constants of the material. The force on one individual atom is
σA, where A is the cross section of the atom. Further, σ = EΔa/a,
where Δa/a is the relative displacement of an atom from its equilibrium position. The elastic force on the vibrating atom is $M\partial^2 a/\partial t^2$
= $Ea^2 \Delta a/a$. This is the equation for a harmonic oscillator with mass
M. Its Eigenfrequency is of the order of 10^{14}Hz for a typical metal.

     It is sometimes convenient to describe atomic oscillations in a
crystal not as lattice waves, but as particles which are called "phonons". Instead of describing the material by the number of elastic
waves one gives the number of phonons. These phonons have properties
typical for particles.

     It is not possible to visualize at the same time the particle
and the wave model to describe a physical phenomena. However, the
duality of two mutually exclusive concepts for the description of
elastic properties of crystals does not disturb scientists very much.

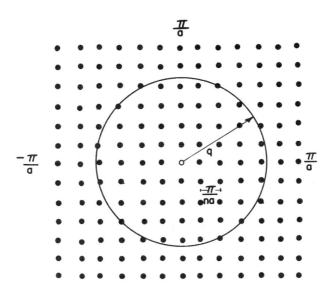

FIG. 1-30. First Brillouin zone for a simple cubic lattice. The circle represents a contour of constant energy for an isotropic system.

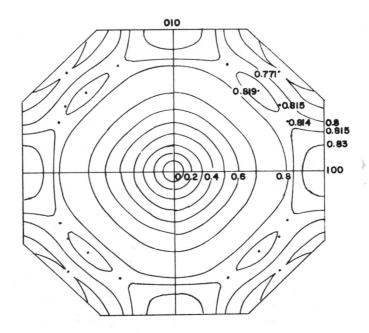

FIG. 1-31. Surfaces of constant frequency in aluminum for longitudinal waves. After Ref. 17, with permission.

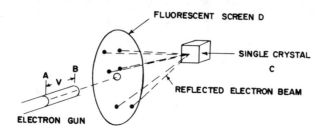

FIG. 1-32. Diffraction of an electron bean on a single crystal.

They are conditioned to accept such duality since they realized at
the beginning of this century that properties of light should be
described by a wave model if one studies light propagation properties,
and by a particle model if one studies the energy transfer from light
to electrons.

If a light beam strikes a metal surface, electrons may be emitted if the frequency of light is above a critical value, ν_o. ν_o is
independent of the light intensity. $W = h\nu_o$ is the "workfunction"
of the material. According to Einstein the energy transfer from
the light beam to the electron is:

$$E = h\nu \qquad\qquad\qquad (1-20)$$

One can describe the physical process of electron emission in the
following way. A light particle, which is called a "photon" and
which has an energy $h\nu$, interacts with one electron which absorbs
the energy $h\nu$. This electron leaves the metal with the energy
$h\nu - W$, provided the electron suffers no additional energy loss.

Again, two different models have to be used to describe an experiment in which electrons are accelerated by an electric field
and then interact with a crystal. Figure 1-32 shows the experimental
arrangement. Electrons at rest at A are accelerated from point A
to B by a voltage V. Their energy at B is eV, where e is the
charge of the electron. The electron velocity can be calculated from

the fact that the kinetic energy $mv^2/2$ is equal to the potential energy eV. These electrons interact with the crystal C in Fig. 1-32 and are diffracted on the fluorescent screen D. The electrons produce on this screen a pattern as if they were waves which interacted with the crystal C. The wave length of the electrons is inversely proportional to the electron momentum mv:

$$mv = p = h/\lambda = \hbar k \tag{1-21}$$

h is the Planck constant, $\hbar = h/2\pi$, and $k = 2\pi/\lambda$ is the wave number of the electron. This equation shows both the particle and wave characters of electrons. It was proposed first by DeBroglie in 1923. This equation is not only used for electron diffraction experiments where electrons are scattered on periodic lattices, but also for other wave-particle correlations.

It is found in neutron diffraction and X-ray measurements that neutrons or X-rays are scattered in a crystal either elastically or inelastically. The momentum of the neutron or photon before the interaction with the lattice, $m\vec{v} = \vec{p} = \hbar\vec{k}$, will change to $\hbar\vec{k}'$. One obtains for the elastic case:

$$\vec{k}' = \vec{k} + \vec{q}_n \tag{1-22}$$

where $\hbar\vec{q}_n$ is the momentum transmitted to the crystal. $\vec{q}_n$ is a vector of a lattice point in reciprocal space. For inelastic interactions the equation is:

$$\vec{k}' \pm \vec{q} = \vec{k} + \vec{q}_n \tag{1-23}$$

where $\vec{q}$ is the wave number of the elastic wave (or "phonon") created or annihilated during the interaction process. Multi-phonon processes are also possible.

It is difficult to determine $\vec{q}$ in X-ray experiments, since the frequency shift of X-rays due to scattering is small. This makes

$\vec{q}$ very small. However, the frequency shift can be measured in neutron diffraction experiments. It is also possible to determine the
energy transfer from the neutron to the lattice. The conservation
of energy requires:

$$\hbar^2 k^2/2m_n = p^2/2m_n = \hbar^2 k'^2/2m_n \pm h\nu = p'^2/2m_n \pm h\nu \qquad (1-24)$$

where m_n is the mass of the neutron, k and k' are its wave numbers before and after its interaction with the lattice, and $h\nu$ is
the energy of the created (+) or annihilated (−) phonon. These
equations make it possible to determine the wave number and the
energy of a phonon.

It should be mentioned that a phonon with $q \neq 0$ has no momentum in the real sense, because for each atom which moves at a given
moment with the velocity $\vec{v}$, there exists another atom which moves
with the velocity $-\vec{v}$. The real momentum of the total number of atoms
is zero. All atoms move at the same time in the same direction only
for q = 0. These philosophical subtleties are, however, usually
neglected. The $\vec{q}$ value calculated from Equations (1-23) and (1-24),
multiplied by $\hbar$, is taken as the momentum of the phonon.

It is frequently important to know how many "phonon states", dn,
exist between ν and $\nu + d\nu$ in reciprocal or ν space. One defines
therefore as "density of phonon states":

$$dn/d\nu = \mathcal{D}(\nu) \qquad (1-25)$$

One defines similarly the number of phonon states in an angular
frequency interval from ω to $\omega + d\omega$:

$$dn/d\omega = g(\omega) \qquad (1-26)$$

One can also use as definition of the density of phonon states:

$$dn/dE = h(E) \qquad (1-27)$$

It is possible to calculate $g(\omega)$ for the approximation of an

isotropic medium with no dispersion. This means that the sound
wave velocity is independent of the frequency. For this case,
$\partial q/\partial \omega = q/\omega$. Further, $\omega/q = v$. The number of states in a given
dq interval is equal to the area of the surface of the sphere with
the radius q times dq. This gives:

$$dn = 4\pi q^2 dq (2\pi/L)^{-3} \qquad\qquad (1-28)$$

Therefore, $dn = g(\omega)d\omega = 4\pi q^2 d\omega/v(2\pi/L)^3$ and one obtains

$$g(\omega) = 4\pi\omega^2/v^3(2\pi/L)^3 \ , \qquad \omega < \omega_D \ . \qquad\qquad (1-29)$$

The concept of an elastic medium cannot be used if one wants
to investigate the elastic properties of solids on an atomic scale.
This can be done with the Born-von Karman model [19], in which one
assumes that the lattice consists of individual atoms, tied together
by elastic springs. The forces on an atom in the simple system of a
linear lattice with atoms of equal mass M and springs with equal
spring constants α would be given, assuming only nearest neighbor
interaction, by:

$$M(\partial^2 u_n/\partial t^2) = -\alpha(u_n - u_{n-1}) - \alpha(u_n - u_{n+1}), \qquad\qquad (1-30)$$

where u_n would be the displacement of atom n from its equilibrium
position in the otherwise undisturbed crystal. An example of the
solution for this equation is:

$$u_n = A\exp(i\omega t - iqx). \qquad\qquad (1-31)$$

The lattice is defined only for $x = na$, with n an integer and a
the lattice constant. Therefore:

$$u_n = A\exp(i\omega t - iqna). \qquad\qquad (1-32)$$

If one considers the symmetry of the system one obtains finally:

$$\nu = \nu_0 \sin(aq/2) \qquad\qquad (1-33)$$

with $\nu_0 = (1/\pi)(\alpha/M)^{1/2}$ as correlation between ν and q [18].
The $\nu = \nu(q)$ curves are called "dispersion curves". The longest
wave length in this system would again be $\lambda = 2L = 2Na$, with N
the number of atoms in the chain. Results of a calculation of the
density of states for such a chain is given in (1-33). Whereas the
continuum model predicts a constant $\mathcal{D}(\nu)$ value for a linear lattice,
$\mathcal{D}(\nu)$ of the Born-von Karman model increases rapidly with ν. Further,
the group velocity $\partial\omega/\partial q$ of the elastic oscillations of the linear
chain is a function of frequency.

The density of phonon states can also be calculated for the case
of a linear lattice with two different types of atoms with masses M'
and M'', where adjacent atoms are connected by identical springs [3].
Similar results are obtained for a system in which all atoms are
equal, but where the types of springs with spring constants α_1 and
α_2 alternate (1-34). Figure 1-35 gives the results of calculations
for the frequency as a function of the wave number, and the density

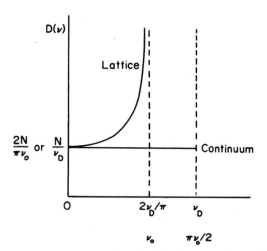

FIG. 1-33. Schematic diagram of the density of phonon states in
a 1-dimensional Born-von Karman lattice.

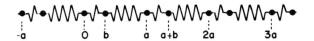

FIG. 1-34. A 1-dimensional lattice with base. All atoms have the same mass.

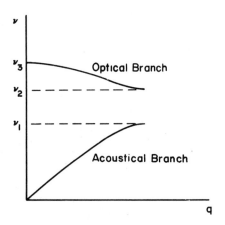

FIG. 1-35. Dispersion curve of the optical and acoustical branch of a 1-dimensional lattice with base.

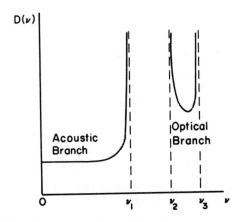

FIG. 1-36. Density of phonon states for a 1-dimensional lattice with base.

of phonon states as a function of frequency (1-36). The low frequency branch of the dispersion curve represents the "acoustical mode"
for the system which contains light and heavy atoms or ions. The
acoustical mode is due to the relatively slow movement of heavy atoms or ions in respect to each other. The high frequency section is
the "optical branch". Essentially, it is due to the vibrations of
light atoms or ions. The light ion and heavy ion represent an electric dipole. The frequency of oscillation of such a dipole is very
high. It can be of the same order of magnitude as the frequency of
light in the infrared range. Typical examples of crystals with an
optical branch in the frequency spectrum are ionic crystals like
NaCl, where both atoms are strongly ionized so that the dipole moment of adjacent Na- and Cl-ions is large. The low mass of the Na-
ion is responsible for the high frequency of the dipole. The low
frequency range contains the combined movement of the adjacent ions
of the dipole. Figure 1-37 gives the dispersion curves obtained for

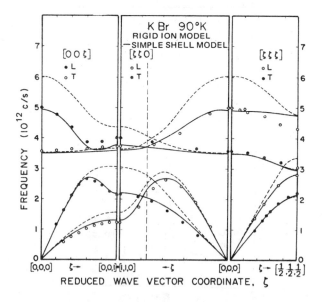

FIG. 1-37. Dispersion curve of KBr at 90°K. Points give experimental results, lines give calculations. Reproduced from Ref. 20,
with permission.

KBr from neutron diffraction experiments. Its shows the acoustic and optical branch.

The linear chain model has only central forces acting between atoms. A second type of force is a function of the angle which the line joining the moving atoms makes with the equilibrium position of the line. This is the "angular force". Another term of the elastic energy is due to the electron gas. This is in first approximation only volume dependent. One may call it a "volume force" [21].

$g(\omega)$ and $\mathcal{D}(\nu)$ have been calculated for different crystal structures. The approach follows essentially the same line as discussed for the linear chain. One knows from the boundary condition the wave length of possible planar waves. The corresponding q-values occupy points in the first Brillouin zone in reciprocal space. Then one solves the equation of motion of atoms in such elastic waves from the elastic parameters and masses of the atoms of the crystal and obtains for each u_n value the frequency of both longitudinal and transverse waves. To obtain $g(\omega)$, one determines contours of constant frequency in reciprocal space and integrates over all occupied states. One obtains:

$$\int g(\omega) d\omega = \int w(q) d^3 q = (L/2\pi)^3 \int d^3 q. \tag{1-34}$$

This gives with

$$d\omega = |d\vec{q} \cdot \nabla_q \omega| = |\nabla_q \omega| dq_\perp: \tag{1-35}$$

$$\int g(\omega) d\omega = (L/2\pi)^3 \int dA_\omega dq_\perp = (L/2\pi)^3 \int (dA_\omega / \nabla_q \omega) d\omega, \tag{1-36}$$

where the subscript ω in dA implies that dA is a surface element in q-space on contours of constant angular frequency. $\nabla_q \omega$ is equal to the group velocity of the elastic waves. One has to add both transverse and longitudinal lattice vibrations.

It is possible to obtain an analytical solution for the case of an elastically isotropic medium. There $g(\omega) \propto \omega^2$ for $\omega \leq \omega_D$, and $g(\omega) = 0$ for $\omega > \omega_D$. In this calculation one replaces the Brillouin zone by the Debye sphere of the same volume.

D. Frequency Spectrum of Elements and Compounds

Computer calculations of $g(\omega)$ of ideal systems and of some elements and compounds are available. Figure 1-38 gives results of a calculation by Fine [15] for tungsten which is a bcc element with elastic constants very close to that of an isotropic system. Fine solved a system of 140 equations to determine $\mathcal{D}(\nu)$. One may describe the results of the calculation in first approximation as the super-position of two curves. One is due to longitudinal, the other to transverse oscillations. Figure 1-39 gives results of a calculation on aluminum. In this diagram, the transverse and longitudinal fre-quency spectra are given separately.

Neutron scattering experiments make it possible to measure the dispersion curves in crystals. Figure 1-40 shows experimentally de-termined phonon energies. Calculations are given as full lines. The calculation was based on the Born–von Karman model. The model used had twelve adjustable parameters. Generally, models which gave a good fit with neutron diffraction experiments gave also good agree-ment with experimentally determined elastic constants. The tempera-

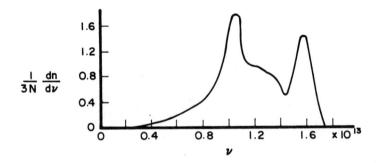

FIG. 1- 38. Density of phonon states for tungsten after a cal-culation by Fine.

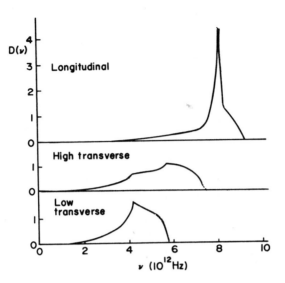

FIG. 1-39. Density of phonon states for aluminum. After Ref. 40, with permission.

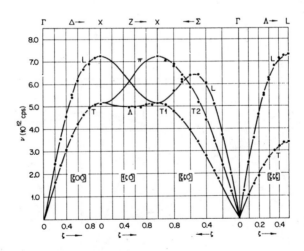

FIG. 1-40. Phonon dispersion curves for copper at 49°K. Lines represent a calculation with a sixth nearest-neighbor model. Dots give experimental data. Reproduced with permission from Ref. 25.

ture dependence of the frequency spectrum is rather small. Changes
in the phonon spectrum were only of the order of a few percent for
wave numbers smaller than $0.2\pi/a$. This is expected from the temperature dependence of the elastic constants which change by 3 to 4%
in this temperature range.

The frequency spectrum of diamond, an insulator with covalent
bonding, is given in Fig. 1-41. It shows that recent calculations
agree well with theoretical curves. One would expect that the dispersion curves for other group IV semiconductors are similar, since
bonding in germanium and silicon is also covalent. Bonding is partly
covalent and partly ionic in III-V and II-VI semiconducting compounds.
The ionic contributions should lead to an electric dipole moment. The
shape of the dispersion curves depends only on the spring constants
and the masses of the atoms. Therefore, the dispersion curves of GaAs
are not very different from that of germanium, since the masses do
not differ very much for these three elements. Only the longitudinal
optical branch of the dispersion curve of GaAs is shifted to slightly
higher values. The width of this branch in GaAs is of the same order
of magnitude as the energy associated with the absorption of infrared radiation.

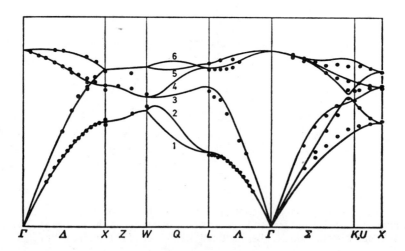

FIG. 1-41. Dispersion curve of diamond. After Ref. 26, with
permission.

E. Frequency Spectra of Disordered Alloys

Substitutional or interstitial impurity atoms can be easily visualized as small particles tied with springs to the rest of the
crystal. Such an atom will usually oscillate with a frequency which
is different from the frequency of the host atoms of the lattice, because either its mass or its spring constant is different from what
is found in the rest of the sample.

The effect of an impurity atom on the frequency spectrum of an
Einstein model can be easily visualized. In this model, all atoms of
the host lattice oscillate with one frequency ν. The impurity would
add one additional frequency ν' if it is an interstitial atom. As
for a substitutional atom, one naturally has to subtract one frequency from the number of host-frequencies.

A slightly more sophisticated model for impurity atoms consists
of a linear chain with atoms of mass M, where one atom is replaced
by another atom with mass $M' < M$. The spring constants between atoms
are assumed to be identical, and springs act again only between nearest neighbors. The equation of motion for atoms in this chain is :

$$M'(\partial^2 u_o/\partial t^2) = \alpha(u_1 + u_{-1} - 2u_o). \qquad (1-37)$$

$i = 0$ is the position of the impurity atom. For all other atoms one
obtains:

$$M(\partial^2 u_i/\partial t^2) = \alpha(u_{i+1} + u_{i-1} - 2u_i), \quad i \neq 0. \qquad (1-38)$$

These equations lead to:

$$\omega = \omega_{max} M^2/(2MM' - M'^2) \qquad (1-39)$$

as angular frequency of the light impurity atom. $\omega_{max} = (4\alpha/M)^{1/2}$ is
the cutoff frequency of the lattice with $M = M'$.

One finds similarly that a heavy atom will oscillate with a
different frequency than most other atoms in the lattice. However,
whereas a light atom usually adds a high frequency above the regular cutoff frequency of the lattice without impurity, the heavy atom oscillates more slowly than the host atoms if the spring constants are the same. The frequency spectrum will show a small additional peak below the cutoff frequency.

Computers make it possible to calculate the frequency distribution of ordered and disordered models. Figure 1-42 gives an example of a calculation where the crystal is three dimensional. The
ratio of masses is three. This would be similar to the ratio of
masses in the copper-gold system. The addition of 5% of light atoms
to the matrix leads to some additional high frequency terms.

The dispersion curves of several niobium-molybdenum alloys as
obtained from neutron diffraction experiments are given in Fig. 1-43.
In spite of the fact that the masses of these two elements are very
similar, that their melting points differ by less than 10% (this indicates bonds of similar strength), and that they form a complete
range of solid solutions (this indicates that their electronic structure is not too dissimilar), their disperion curves differ noticably.
For instance, the frequency versus wave number curve of niobium (1-
43a) has a minimum between the center point [000] and the point H
at the Brillouin zone for $q = 0.8 \cdot [001]$. This minimum shifts to the
H point for pure molybdenum (1-43c). Other characteristic points of
the dispersion curve such as F_1 between H and P shift in a systematic way with composition. It is presently not possible to explain
these shifts or the detailed shapes of the dispersion curves from
first principles. These curves differ significantly if one compares
the effect of adding niobium to molybdenum or molybdenum to niobium.
The level of frequencies increases in the second case, but there is
practically no effect on the frequency distribution. The opposite
effect is found if small amounts of niobium are added to molybdenum.
In that case, the level of frequencies is practically constant, but
the shape of the dispersion curve changes markedly with composition.

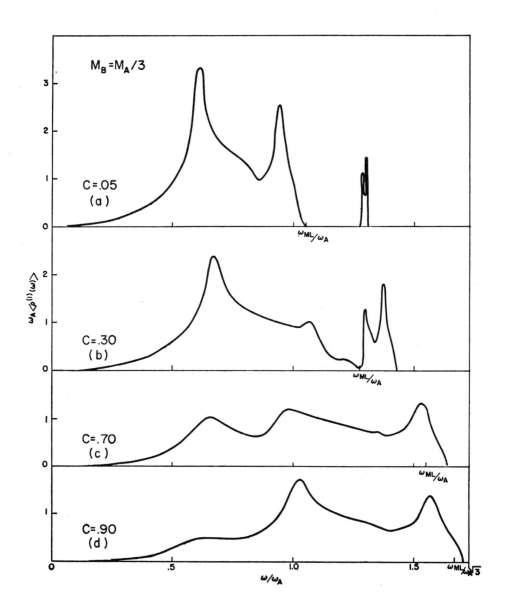

FIG. 1-42. Calculated frequency spectra of a binary disordered alloy, where the mass ratio of atoms A and B is equal to three. c is the concentration of B-atoms. After Ref. 27, with permission.

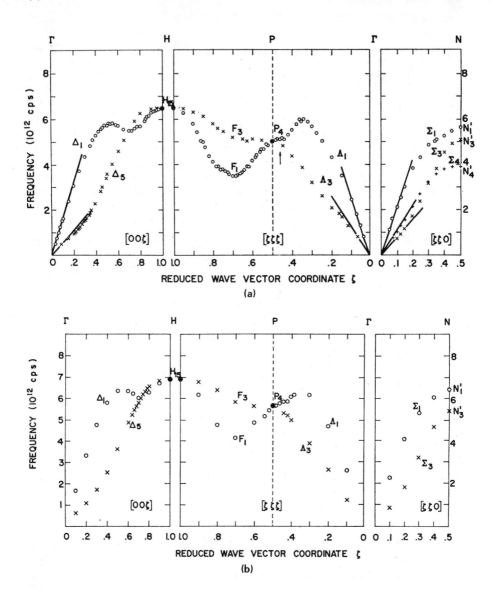

(a)

(b)

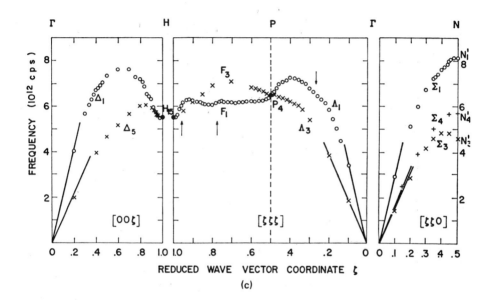

FIG. 1-43. Dispersion curves of niobium (a, opposite page), a niobium-molybdenum alloy with 15 at.% Mo (b, opposite page), and molybdenum (c). Reproduced with permission from Ref. 28.

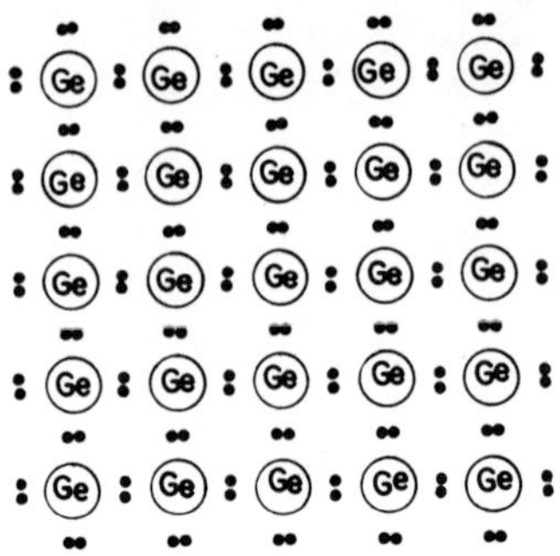

FIG. 1-44.   Spacial distribution of outer electrons in pure germanium.

IV. ELECTRON ENERGY STATES

A. Localized Electrons

The discovery of individual electrons around 1900 led to the development of some models in which it was assumed that a crystal consists of a system of regularly arranged ions and, in the case of metals, of freely movable electrons which have thermal energies of the order of $k_B T$. In insulators all electrons were assumed to be tightly bound. Now one would assume for a semiconductor like germanium that some of these bound electrons can break away from their atoms.

Chemists have developed for diamond, germanium and silicon the concept of "covalent bonding", in which adjacent atoms share pairs of electrons. They are responsible for cohesion. It seems natural to use such a model to explain other properties of semiconductors. The picture of bonds consisting of electron pairs is vivid. One can remember it easily, and one can imagine how minor modifications like substitutional impurities affect the electron distribution. Figure 1-44 gives a schematic diagram of the bonds in a semiconductor. One sees how the four outer electrons are shared. One can visualize how

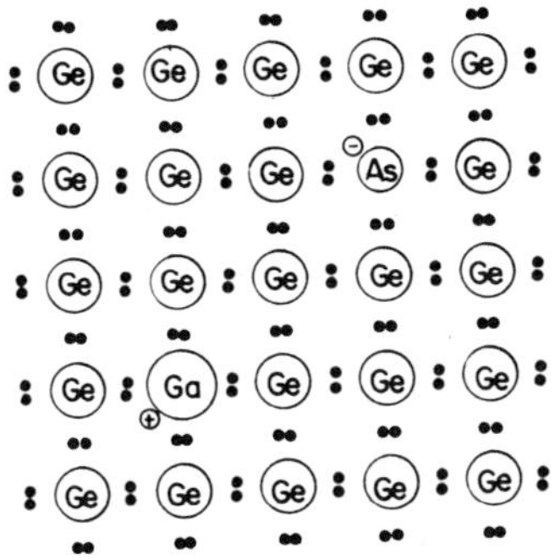

FIG. 1-45. Spacial distribution of outer electrons in germanium with gallium and arsenic impurities. The extra electron near
arsenic would tend to contribute an extra outer electron; the hole
at gallium would tend to absorb an electron.

an electron breaks away from its normal position, and floats through
the crystal due to the effect of an applied electric field. Figure
1-45 shows the electronic structure of such a system if a germanium
atom is replaced by an arsenic or gallium atom. Arsenic has one electron more than germanium. This extra electron may break loose more
easily than the electrons shared by neighboring atoms. It is bound
only to one atom and not two. Therefore, one would expect that the
electrical conductivity of germanium doped with arsenic should be
larger than in pure germanium. Arsenic is an "electron donor". Replacing a germanium atom by gallium leads to a deficit of electrons,
since gallium has one electron less than germanium. There is space
for another electron. Gallium would therefore be an "acceptor".
An electron may jump into this first hole, creating a hole in its
original position. It looks as if a positively charged hole is
moving through the crystal.

    If an electron in a regular bond breaks away from an atom it
leaves a hole behind. Both electron and hole are bound together if
not too far apart. Since the hole can move through the lattice, both
hole and electron may move together. They form an "exciton". An
electron and a hole are weakly bound together, and together carry no
electric charge. However they carry energy through the lattice.

A conduction electron will not only interact with holes, it will
attract neighboring ions. This leads to an elastic stress field around
an electron, with an associated stress energy. Electron and stress
field (or strain field) form a "polaron". The moving electron drags
this stress field along. The energy required to move both stress field
and electron is larger than the energy required to move the electron
alone. This leads to the impression that the polaron is heavier than
a free electron. The mass of the polaron is $M_{pol.} = m(1 + \alpha/6)$, if
the coupling constant between lattice and electron α is much smaller than 1. α is defined by $\alpha = 2(lattice deformation energy)/(op-
tical phonon energy)$. One may view $\alpha/2$ as the number of phonons
which surround a slowly moving electron in a crystal. α is small
only for semiconductors. One obtains values above one for ionic crystals. Typical values for α are: 5.6 for KCl; 1.9 for AgCl; 1.6 for
AgBr; 0.85 for ZnO; 0.16 for PbS; 0.06 for GaAs; 0.014 for InSb.

The highly mobile conduction electrons in metals interact with
each other via long range Coulomb interaction. Since electrons repel
each other, one would expect a cloud around each condution electron
in which the conduction electron concentration is smaller than the
average. This cloud is equivalent to a positive charge. The combination of this "electron plus positive cloud" acts as a unit, just as
the polaron, except that it can move much faster through the lattice.
It is a "quasi particle". It can have a mass different than that of
a free electron. Its electrostatic field decays much more rapidly
with distance than the Coulomb field of an individual electron; this
may partly justify the frequently made assumption that the interaction of conduction electrons can be neglected in the calculation of
electron energies.

As was said before, models of individual electrons can be
easily visualized. However, a quantitative description of electron
energies can be obtained more easily if one treats conduction electrons in a crystal as "waves". This makes it possible to calculate
their energies. One can even describe the position of an individual
electron with the wave model.

B. Energy Bands

DeBroglie proposed in 1923 that momentum and wavelength of electrons are correlated by $\lambda = h/p$. A few years later, Schrödinger suggested that quantized electron energy levels in hydrogen are equivalent to standing electron waves. Electron energies and the spacial electron distribution can be determined from the Schrödinger equation, which has for the time independent one electron system the form:

$$(\hbar^2/2m)(\partial^2\psi/\partial x_1^2 + \partial^2\psi/\partial x_2^2 + \partial^2\psi/\partial x_3^2) + (E - V)\psi = 0. \quad (1\text{-}40)$$

This is also written as:

$$(\hbar^2/2m)\nabla^2\psi + (E - V)\psi = 0. \qquad (1\text{-}41)$$

V is the potential energy of the electron, E is the total energy of the electron, and ψ is the electron wave function. Its physical meaning can be visualized if one uses the interpretation that $|\psi(\vec{r})|^2$ is proportional to the probability of finding the electron at point $\vec{r}$. ψ is normalized by setting $\int|\psi|^2 dx_1 dx_2 dx_3 = 1$.

The Schrödinger equation is sometimes written as:

$$H\psi = E\psi \qquad (1\text{-}42a)$$

where H is:

$$H \equiv -(\hbar^2/2m)\nabla^2 + V = (p^2/2m) + V, \qquad (1\text{-}42b)$$

This form of the Schrödinger equation indicates that H represents the energy of an electron or electron system with the kinetic energy term $-\hbar^2\nabla^2/2m$ and a potential energy term V. This concept makes it

possible to write down the Schrödinger equation for systems which
have not only kinetic and electrostatic potential energies, but also
the vector potential of electrons moving in a magnetic field.

Individual atoms far apart from each other should have electron
distributions of spherical symmetry, if no outside electric or magnetic
fields are applied. In such a case, one changes the Schrödinger
equation to spherical coordinates (1-46) and sets:

$$\psi(\vec{r}) = R(r)\Phi(\phi)\Theta(\theta). \tag{1-43}$$

Inserting this expression into the Schrödinger equation for the hydrogen
atom gives a solution with: $E_n = -(13.6/n^2)eV$, where n= 1, 2,
3, 4, ... is the principal quantum number. $\psi = \psi_1 = C_1\exp(-C_2 r)$ for
n = 1, and $\psi = \psi_2 = (C_3 - C_4)\exp(-C_2 r)$. Figure 1-47a shows these
functions. Solutions of the Schrödinger equation show that the angular
momentum of the electron orbiting around the nucleus has a value
G, which is a multiple of the Planck constant divided by 2π. $G = \hbar\ell$,
where ℓ is the angular momentum number with $0 \le \ell < n$. Solutions
of the Schrödinger equation with $\ell = 0$ have spherical symmetry. The
electrons associated with this wave function are s-electrons. The
wave function ψ is zero for certain r values. In other words, ψ
has nodal spheres. Electrons with the angular momentum $\hbar$ are

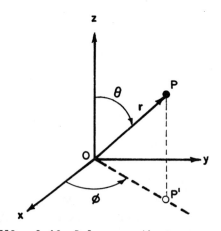

FIG. 1-46. Polar coordinates

p-electrons. $\Phi(\phi)\Theta(\theta)$ may have values of $\cos\theta = x_3/r$, $\sin\theta\cos\phi =$ x_1/r, or $\sin\theta\sin\phi = x_2/r$. This means that p-electron wave functions have nodal planes where the wave function is zero. $\ell = 2$ corresponds to d-electron states which have nodal planes going through the origin. The d-electron functions have the form $(x_1^2 - x_2^2)/r^2$, $(x_3^2 - x_1^2)/r^2$, $x_2 x_3/r^2$, $x_3 x_1/r^2$ and $x_1 x_2/r^2$.

The energy and the spin orientation of an orbiting electron depends also on the applied magnetic field H. Only specific orientations between $\hbar\ell$ and H are allowed. The projection of ℓ into

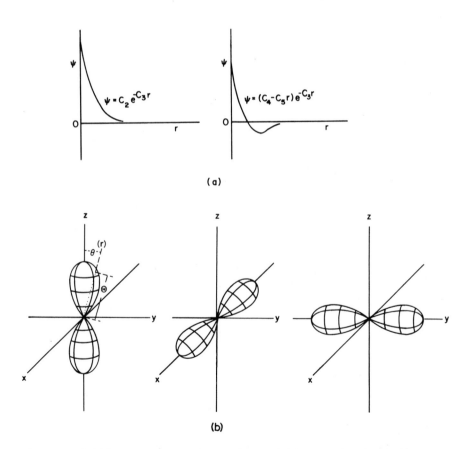

(a)

(b)

FIG. 1-47a + b. Electronic wave functions for single electrons in the 1s and 2s state are given in (a). The angular wave function dependence of the 2p-electron is given in (b).

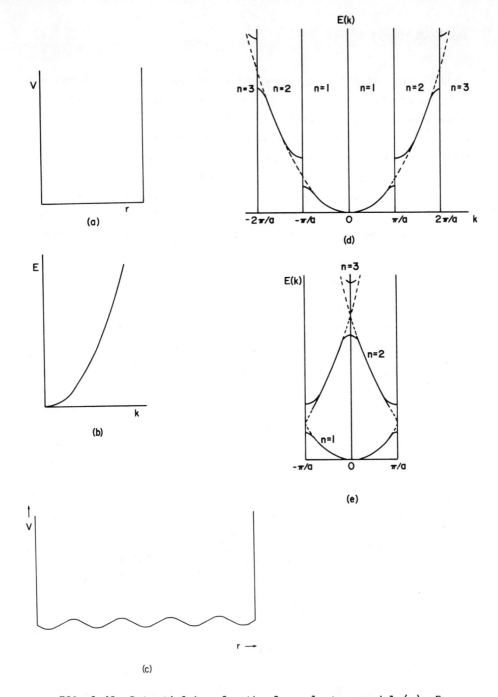

FIG. 1-48. Potential box for the free electron model (a). Energy of free electron (b). Potential of nearly free electrons (c). E = E(k) in the one-dimensional nearly free electron model for the extended zone scheme (d), and for the reduced zone scheme, in which sections of the E(k) curves are shifted by ± nπ/a (e).

the direction of H has to be an integer, m. m is the magnetic
quantum number of the electron and can have values given by $- \ell \leq m$
$\leq + \ell$. Each electron has, independent of its orbital motion, a
magnetic moment μ_s. This moment is close to the Bohr magneton μ_B.
Each electron also has a spin moment of $s\hbar$, with $s = \pm 1/2$. s is
the spin quantum number. Each electron is uniquely defined by n,
ℓ, m, and s. The Pauli principle states that only one electron may
occupy one electron state characterized by these four quantum numbers.

At present, the Schrödinger equation for a crystal with a large
number of electrons and ions cannot be solved. Simplifying assumptions
have to be made. One frequently uses the "one electron model". There
one neglects the interaction of one conduction electron with the
other conduction electrons. One assumes further that the potential
inside the metal is constant and that the potential walls at the
boundary of the metal prevent electrons from escaping. This is the
"potential box model" (1-48a). Electrons move freely inside this
box. Solving the Schrödinger equation for this case of the "free
electron model", one obtains for a cubic specimen with sides equal
to L_1, L_2, L_3 wave functions of the form:

$$\psi = A \sin(\pi \ell_1 x_1 / L_1) \sin(\pi \ell_2 x_2 / L_2) \sin(\pi \ell_3 x_3 / L_3) \qquad (1-44)$$

ℓ_i are integers and A is a constant which can be determined from:

$$\int |\psi|^2 dx_1 dx_2 dx_3 = 1 \qquad (1-45)$$

The energy values for the electrons in the free electron model are:

$$E = (\hbar^2 \pi^2 / 2mL^2)(\ell_1^2 + \ell_2^2 + \ell_3^2) \qquad (1-46)$$

if $L_1 = L_2 = L_3 = L$. One sets:

$$\pi^2(\ell_1^2 + \ell_2^2 + \ell_3^2)/L^2 = k_1^2 + k_2^2 + k_3^2 = k^2 \qquad (1-47)$$

where k is the wave number of the electron. This corresponds to the same usage as given in the previous sections on lattice waves. Again, $\vec{k}$ is a vector in three dimensions, occupying points in the reciprocal space, given by a set of numbers:

$$k_i = \pi \ell_i / L \tag{1-48}$$

with ℓ_i = 0, 1, 2, 3, 4 .... The expressions for ψ and E in Eqs. 1-44 and 1-46 may be written as exponential functions:

$$\psi = A \exp(i\vec{k} \cdot \vec{r}) \tag{1-49}$$

and

$$E = \hbar^2 k^2 / 2m \tag{1-50}$$

Figure 1-48b shows this quadratic wave number dependence of the electron energy. Each electron state, characterized by a k-vector in reciprocal space, can be occupied by two electrons, one with spin up and one with spin down. Similar to the density of phonon states, one defines the density of electron states N(E) as the number of electron states dn per energy interval dE:

$$dn = N(E)dE \tag{1-51}$$

The number of electrons in a given energy interval should be equal to the volume of the sample Ω times the number of states in the wave number space $4\pi k^2 dk / (2\pi)^3$. Therefore, according to [29], one obtains dn = Ω $4\pi k^2 dk / (2\pi)^3$. This gives, together with Eqs. 1-50 and 1-51:

$$N(E) = \Omega \ (2m)^{3/2} E^{1/2} / 4\pi^2 \hbar^3 \tag{1-52}$$

for free electrons in a potential box. The number of electrons as a function of energy can be calculated from the last two equations. One obtains as maximum energy for an electron [29]:

$$E_{max} = 36.1 \ (n_o/\Omega)^{2/3} (eV) \tag{1-53}$$

if one assumes that two electrons can occupy one energy state. n_o is the number of electrons per atom, and Ω in this equation is the volume of one atom in Angstrom units. This energy would be the "Fermi energy" of a metal or alloy at absolute zero temperature. Equation 1-53 may be used only for monovalent atoms or for simple alloys with an outer electron concentration close to one. E_{max} is equal to 3.16(eV) for sodium and 7.1(eV) for copper. These values are high compared with the thermal energy of atoms at room temperature. The free electron model takes no account of the crystal structure. Therefore, this model may be applied to alloys as well as to pure metals.

One can obtain experimentally N(E) values from low temperature specific heat and susceptibility measurements. It was found in a number of these experiments that N(E) had about the same values for several alloy systems, if N(E) was plotted as a function of the outer electron per atom ratio, el./at. One describes this result with the concept of the "rigid band model": "N(E) of alloy systems is a function of el./at. only". It is easy in this model to change a plot of N(E) versus el./at. into a plot of N(E) versus E, since Δ(el./at.) $\propto$ Δn = N(E)ΔE. This gives $\Delta E \propto \int d(el./at.)/N(E)$.

C. The Nearly Free Electron Model

To bridge the gap between the free electron model with a constant potential and the real metal with a periodic potential which has singularities at the ions, one uses the "nearly free electron model" (N.F.E.M.). There the potential of the ions is periodic, but

the deviations from the average potential value is small. Figure
1-48c gives a schematic diagram of such a potential. The "Bloch theorem" states that solutions of the Schrödinger equation with a periodic potential are of the form [29]:

$$\psi_k(\vec{r}) = u_k(\vec{r}) \exp(i\vec{k}\cdot\vec{r}) \tag{1-54}$$

where $u_k(\vec{r})$ is a function with the periodicity of the lattice. Typically, one assumes that the wave function for the one dimensional
case is given by $\psi = u(x) \exp(ikx)$ with $u(x) = \Sigma A_n \exp(-2\pi inx/a)$.
a is the lattice parameter of the lattice, and one forms the sum
from $n = -\infty$ to $n = +\infty$. The mean value of the potential V should
be zero. $V(x) \propto \exp(i2\pi x/a)$ would be such a function. Inserting
these expressions for $V(x)$ and ψ into the Schrödinger equation
and assuming that the absolute values of A_n is much smaller than
A_o for $n \neq 0$, gives as first approximation again [29] $E = \hbar^2 k^2/2m$.
In second approximation, one obtains:

$$E = E_o + \Sigma |V_n|^2/(E_o - E_n) \tag{1-55}$$

with $E_o = \hbar^2 k^2/2m$, $E_n = \hbar^2(k - 2\pi n/a)^2/2m$, and $V_n = (1/a)\int V(x)\cdot$
$\exp(2\pi inx/a)dx$. The result given in (1-55) is similar to the free
electron result, if k is not too close to $\pi n/a$. However, at
$k = k_n = \pi n/a$ the slope of the $E = E(k)$ curve is horizontal,
and E is proportional to $(k - k_n)^2$. The major result of this calculation is that deviations from the free electron behavior are expected, if a multiple wavelength of the electron wave is equal to 2L.

This result is not unexpected. For $\lambda = 2L/n$, the 1/2 wavelength
of the electron wave is a simple multiple of the lattice parameter.
It is "in phase" with the lattice. Therefore, the interaction of the
electron wave and the lattice in one part of the lattice is the same
as in any other part of the sample. This should lead to strong interactions between crystal and electron wave, and to a marked deviation
from the free electron model, which neglects these interactions. If

electron wave and lattice are not in phase, the interaction of the lattice and electron wave in one part of the sample would be cancelled by an opposite interaction in another part of the sample.

One can expand this concept to explain the interaction of electron waves with a three dimensional lattice. It is required that the projection of the electron wave on the lattice plane leads to an "in phase" correlation. This means that strong lattice-electron wave interaction is possible only if the projection of the k-vector on the x_i-direction should be equal to a multiple of π/a_i in an orthorhombic lattice. These k-vectors lie in a plane perpendicular to the x_i-direction, intersecting the x_i-axis at $\pi n/a_i$. This construction is the same as that of the Wigner-Seitz cell, where the lattice vectors are given by $n2\pi/a_i$. The Wigner-Seitz cell in this reciprocal lattice is called "Brillouin zone". Its boundary contains all k-vectors for which the $E = E(k)$ curve can have a discontinuity. The n-th Brillouin zone would have as outer boundaries planes which intersect the x_i-axes at $n\pi/a_i$. The inner boundaries would intersect these directions at $(n-1)\pi/a_i$. It is frequently convenient to rearrange the individual sections of higher Brillouin zones in this "extended zone scheme" (1-48d) by shifting them by multiples of $2\pi/a_i$. This gives the "reduced zone scheme" (1-48e).

Random deviations of the real lattice from the ideal perfect lattice should not affect the lattice-electron wave interaction. The "in phase" relationships should not be influenced markedly by such disturbances. Therefore, the nearly free electron model can describe properties of substitutional alloys, where the different types of atoms are randomly distributed. Likewise, the N.F.E.M. applies also to crystals where atoms oscillate randomly around equilibrium positions. Only atomic ordering in alloys could lead to major changes in the Brillouin zone structure. This has been shown in the $E(q)$ curves of the elastic frequency spectrum of a lattice with base. It can also be demonstrated in a simple planar cubic system, in which the composition is AB. If atoms are randomly

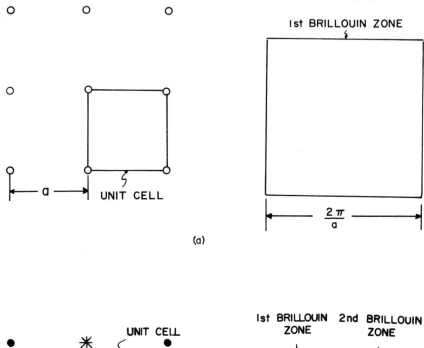

(a)

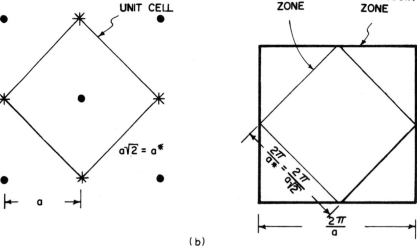

(b)

FIG. 1-49. 1st Brillouin zone in a 2-dimensional disordered
alloy (a), and in an ordered AB alloy (b).

distributed, the first Brillouin zone intersects the axis at b = π/a. If ordering takes place (1-49), the lattice parameter of the new unit cell is $a\sqrt{2}$, and tilted by $45°$ in respect to the original lattice. Its cubic Brillouin zone is also tilted by the same amount, and the distance from the Brillouin zone boundary to the center is $\sqrt{2}\pi/a$.

In the later sections $E = E(k)$ diagrams of several elements will be shown. They may seem much more complex than what is shown in Fig. 1-48. This is not necessarily due to strong lattice-electron interaction but only to the way the data are presented. For instance, Fig. 1-50 shows in a typical representation the $E(k)$ curves of the diamond lattice. This diagram gives results for the "empty lattice". In this case, no interaction between lattice and electron gas is assumed. Therefore, the $E(k)$ curves show no energy gaps at the Brillouin zone boundaries. The complexity of these $E(k)$ diagrams is due to the fact that the intercepts of the $E = \hbar^2 k^2/2m$ curves with second and third Brillouin zone boundaries are given.

The $\vec{k}$ values of electrons with the highest energy at absolute zero temperature in k-space form a surface which is named the "Fermi surface". The electron energy at this surface is the "Fermi energy". The Fermi surface of the free electron gas is a sphere. In the nearly free electron model, the interaction between lattice and electron waves leads to distortions from this spherical shape. The definition $N(E) = dn/dE$ leads to:

$$dn = N(E)dE = (L/2\pi)^3 \int dk^3 \qquad (1-56)$$

One can also integrate over shells in k-space between contours of constant energy. This gives:

$$dn = (L/2\pi)^3 \int dA dk_\perp \qquad (1-57)$$

where dA is a small surface section on surfaces of constant energy, and $dk_\perp$ is the wave vector perpendicular to dA. The integration

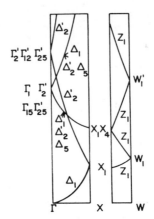

FIG. 1-50. Partial representation of free electron energy levels for the diamond lattice.

takes place over occupied states. Since $dE_\perp = \mathrm{grad}_k E \cdot dk_\perp$, one obtains:

$$N(E)dE = dn = (L/2\pi)^3 \int (1/\mathrm{grad}_k E)dAdE \qquad (1\text{-}58)$$

$$N(E) = (L/2\pi)^3 \int (1/\mathrm{grad}_k E)dA. \qquad (1\text{-}59)$$

One tries to visualize properties of metals with simple models. The idea of freely moving electrons is very convenient since it appeals to the intuition. The concept of the density of states, or of $\mathrm{grad}_k E$, is more difficult to handle. Fortunately, one can incorporate $\mathrm{grad}_k E$ into a simple electron model. One has to keep in mind that $\mathrm{grad}_k E = \hbar v$, where v is the electron velocity. In other words, the spacing and shape of contours of constant energy in k-space will be reflected in the velocity distribution of electrons. Since the velocity of particles depends on their mass and the acceleration, one may suspect that the $E = E(k)$ relationship could be reflected in the mass of electrons. Naturally, this mass would be only fictious. One calls it the "effective mass", m*. It is defined as:

$$m^* = \hbar^2/(\partial^2 E/\partial k^2) \qquad (1\text{-}60a)$$

for a 1-dimensional system. It is defined as:

$$m^*_{ij} = \hbar^2/(\partial^2 E/\partial k_i \partial k_j) \tag{1-60b}$$

for a 3-dimensional system. This seems to be a reasonable definition, since for the free electron model with $E = \hbar^2 k^2/2m$ it leads to $m^* = m$. m^* can become negative near the Brillouin zone boundary. The band is nearly filled and $N(E)$ in the nearly free electron model is proportional to $(E' - E)^{1/2}$ where E' is the electron energy at the boundary.

Electrons in states where k touches the Brillouin zone boundary cannot contribute to the electrical current or the electronic specific heat. Therefore, the density of states $N(E)$ is proportional only to the area of the Fermi surface which does not touch the Brillouin zone boundary. Fig. 1-51 shows the correlation between $N(E)$ and the filling of the Brillouin zone with electrons for a simple cubic lattice. The peak in $N(E)$ corresponds to the first contact between the Fermi surface and the Brillouin zone boundary at E_1. Further additions of electrons will decrease the number of electron states. At E_2, this zone is filled, and the electrons in this zone or band do not contribute to the electric current, or to the electronic specific heat. It is possible that electrons can be added to the second Brillouin zone before the first Brillouin zone is filled (1-51b). For such an energy band system the sample has for any Fermi energy level free electrons. It is a metal. However, if the bands are separated and are either full or empty, then the crystal would be an insulator. If the distance between these bands is small compared with thermal energies, then it is possible that a noticeable number of the electrons of the lower band (valency band) will be excited into the next empty band (conduction band). Such a crystal is a semiconductor. In some elements, the valency and the conduction bands overlap slightly. These materials are semimetals.

The $N(E)$ curves for the fcc crystal would have two peaks (1-52) instead of the one found in the simple cubic lattice. These peaks

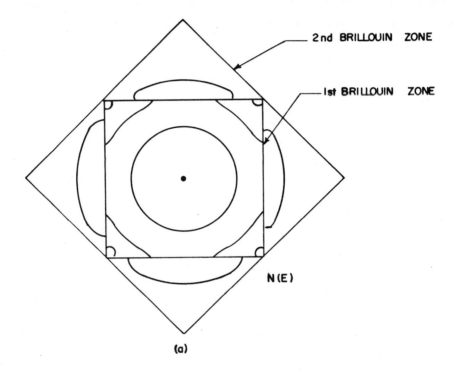

(a)

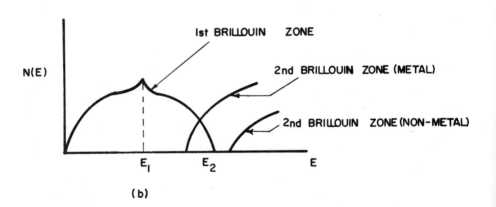

(b)

FIG. 1-51. Brillouin zones (a) and density of states curve (b) for a simple cubic lattice.

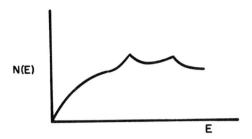

FIG. 1-52. Density of states curve for a fcc lattice.

correspond to the first contact of the Fermi surface with Brillouin
zone boundaries in the [111] and [100] directions, respectively. It was initially proposed that the electron per atom ratio of
the first contact point between a spherical Fermi surface and the
Brillouin zone boundary would give the limit of the α-brass stability range. A further increase of the electron per atom ratio
would lead to a rapid increase of the total electron energy, since
$dE = dn/N(E)$, and $N(E)$ decreases rapidly beyond the first contact
point. Another phase, the bcc phase, would become more stable, since
the contact point with the Brillouin zone occurs at a higher electron concentration.

The calculated electron per atom ratio for the first contact
between a spherical Fermi surface and the Brillouin zone is 1.35.
This is very close to the limit of the stability range of noble
metals with higher valency second components. Experiments show however that the Fermi surface of copper already touches the first
Brillouin zone. Therefore, the explanation for the stability range
of α-brass has to be more complicated.

Electrons in different bands, but with the same wave vector
and the same energy will interact. This will modify the $E(k)$ curves.
Figure 1-53 shows possible $E(k)$ curves for the conduction band and
d-band. The conduction band contains wave functions of 4s, 4p, and
4d symmetry, which are orthogonal to 3d-wave functions. The two
$E(k)$ functions without interaction would overlap along specific
symmetry lines in k-space. This interaction will lead to

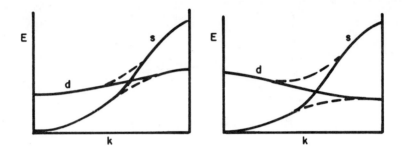

FIG. 1-53. Modifications of E(k) curves by "hybridization
(dashed lines). After Ref. 30, with permission.

"hybridization" which changes E = E(k) as indicated by the dashed
line (1-53). It means that wave functions may change their character
and be in one section in the Brillouin zone of predominantly d-char-
acter, and in another of predominantly s-character. This may also
occur on different sections of the Fermi surface. It is possible that
the sign of the radius of curvature in the E(k) curve changes.
Therefore, some sections of a Fermi surface may contain holes, and
others may contain electrons, if the sign of the radius of curvature
changes. This is found in copper.

D. The Tight Binding Approximation

Instead of assuming that electrons are kept inside a box by a
potential where the deep potential wells of individual ions are represented by small periodic deviations from the average potential,
one may start with a model in which the atoms are far apart and where
the wave functions of electrons are the wave functions of the free
atom. Moving the atoms closer together will lead to overlapping outer
electron orbits. Electrons will interact. Since each electron state
in interacting systems can be occupied only by one electron, the
identical outer electron levels of individual atoms will change their

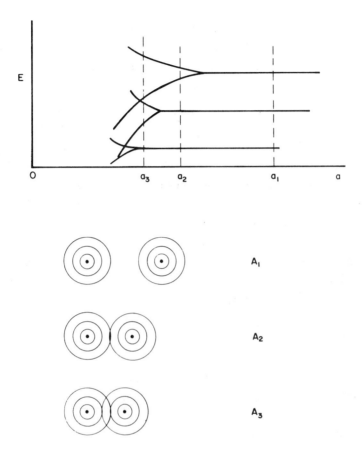

FIG. 1-54. Atomic energy levels in the tight binding approximation. The energy levels are sharp if the atoms are widely separated.
(A_1: separation between atoms is a_1). Reducing the separation to
a_2 or a_3 with overlapping electron orbits (A_2, A_3) leads to broad
energy levels.

positions. Figure 1-54 gives schematically the rearrangement of energy levels as a function of the interatomic distance. The three
energy levels in Fig. 1-54 are sharp lines at the interatomic distance a_1, since electron wavefunctions shown schematically in A_1
as circles do not overlap. For $a = a_2$, outer electron orbits overlap (A_2). Electrons with the highest energy will form an energy band.
The electrons in lower orbits will still have sharp energy levels.
For $a = a_3$, lower lying electron orbits will overlap (A_3), and the

energy of electrons in the second orbit spread out.

   The starting point in this model is a system of non-interacting
atoms. Each electron is associated with only one atom. The interaction of electrons with other atoms, after atoms are brought more
closely together, is treated as a perturbation. Therefore, this model
is called the "tight binding model". It is surprising that the tight
binding model and the nearly free electron model lead to similar results, in spite of the fact that the nearly free electron model
assumes that ion cores are only responsible for small perturbations
of free electron wave functions and energies, whereas the tight
binding approximation assumes that atomic orbits are modified only
due to perturbations by the interaction of electrons of neighboring
atoms. The tight binding model shows that metallic bonding has some
similarity to "covalent bonding". There electrons of different atoms
are shared between neighboring atoms because this leads to a reduction
in the total energy. An electron between two positive ions is at a
lower energy than if it interacted with only one ion. The covalent
bond is of short range. Only nearest neighbors are important. With
the tight binding model it is easy to study the effect of impurities
which will interact only with nearest neighbors.

   One would expect that wave functions in the tight binding model
would be of the form [31]:

$$\psi_k = \Sigma \, \exp(i\vec{k}\cdot\vec{r}_n)\beta_m\phi_n^{(m)} \, (\vec{r} - \vec{r}_n) \qquad\qquad (1\text{-}61)$$

where $\exp(i\vec{k}\cdot\vec{r}_n)$ is a phase factor, β_m is an adjustable parameter,
$\phi_n^{(m)}$ is a set of atomic orbits of the atom at point $\vec{r}_n$, and the
summation is over m and n. This trial function is a linear combination of atomic orbits (L.C.A.O.) with adjustable parameters. The
L.C.A.O. method has the disadvantage that it only gives good wave
functions near the core of the atoms. It cannot easily represent
the wave functions in the regions between atoms. On the other hand,
this method has the advantage that the E = E(k) function automati-
cally is periodic in reciprocal space. This method is very useful
for the calculation of molecular orbits.

E. Cellular Methods, Orthogonal Plane Waves and Pseudopotentials

It seems plausible to assume that the electron distribution in
a metal will be described with wave functions which are similar to
those of the free atom near the nucleus and approach plane wave
character between ions. One of the oldest methods of obtaining
such wave functions is the Wigner-Seitz method in which one assumes
that the potential and the wave functions inside a sphere surrounding
the nucleus are of spherical symmetry. The radius of this sphere is
equal to the atomic radius. Matching wave functions from one cell to
the next are obtained by assuming that the electron wave functions
are joined smoothly at the boundaries. Therefore, one sets:

$$(\partial\psi/\partial r)_{r=r_a} = 0 \qquad\qquad\qquad (1-62)$$

Again one uses the Bloch theorem to obtain solutions of the
Schrödinger equation, and one can calculate cohesive energies of simple solids with reasonable accuracy. Figure 1-55 gives examples of
calculations for sodium, copper and silver. Sections between nodal
spheres (between $\psi = 0$ sections) correspond to electrons with
different quantum numbers. The lowest section would give the 1s,
the next the 2s state, and so on. The calculated wave functions
will depend only on the number of occupied electron shells of the
metal and on the atomic volume. It is insensitive to all other physical parameters. It is therefore surprising that calculations with
the Wigner-Seitz method give reasonable values for the cohesive energies of simple metals.

One assumes in the "augmented plane wave technique" (A.P.W.T.),
another example of the cellular method, that the potential is constant between ions, and spherical around the nucleus [31]. This
gives the "muffin tin" potential (1-56). One solves the Schrödinger
equation within each sphere and matches it to plane waves in the

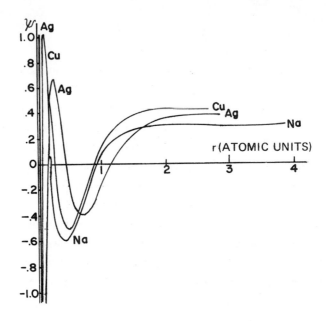

FIG. 1-55. Wave functions of electrons in their lowest states. Functions are normalized. 1 atomic unit is equal to 0.529 Angstroms. After Ref. 29, with permission.

FIG. 1-56. Muffin tin potential. After Ref. 31, with permission.

constant potential section. The solution of this approach is not a
solution of the Schrödinger equation of the crystal, because no special relationship between E and k is assumed. New trial functions are set up with the help of the Bloch theorem. It turns out
that a single augmented plane wave with the kinetic energy of a free
electron is frequently a good approximation of the wave function.
Electrons behave in most parts of the crystal as if they were free.
Mixing with other states is important only near the zone boundary.
This is a rather surprising result. It means essentially that one
can neglect the deep potential wells of ions.

The cellular methods are very useful if a potential of spherical symmetry can be assumed. A more general approach, used frequently
for transition elements, is the "orthogonal plane wave technique".
In this approach, one subtracts from the plane wave function the
atomic core functions, which represent the rapid electron density
variations near the nucleus. Figure 1-57 shows schematically the
plane wave function, the core function, and the combined plane wave
core function.

The "pseudopotential theory" has been used in recent years to
calculate the band structure of several noble metals. This computational technique explains why the nearly free electron model frequently gives good results, in spite of the fact that the perturbation potential in the free electron model is certainly much too weak
to be realistic.

It is shown with the pseudopotential theory that one can replace
the large potentials of ion cores by much smaller potentials.
Electron waves interacting with this potential will undergo a phase
shift. The theory replaces the actual potential by a pseudopotential
which leads to the same phase shift of the electron wave. The calculation follows to a certain degree an approach similar to the augmented plane wave method.

Since the potential in the nearly free electron model (and, in
principle, in all models with unrealistically weak potentials) is not
the real potential, one calls it also a pseudopotential. Figure

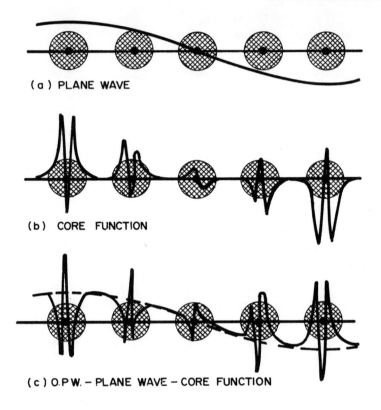

(a) PLANE WAVE

(b)   CORE   FUNCTION

(c) O.P W. – PLANE WAVE – CORE FUNCTION

FIG. 1-57.   Synthesis of orthogonal wave.   After Ref. 31, with permission.

1-58a gives an example of such a pseudo potential for sodium.

For a review of these and other computational techniques to determine the wave functions and energies of electrons in crystals, the reader is referred to more advanced treatises by Ziman [2, 33] and Dimmock [34]. The calculations require fast computers, and the investigator has to use his judgment in the selection of the strength of the method, if he wants to obtain specific results. For instance, the augmented plane wave method concentrates on the plane waves in the interstitial regions between atoms. It is difficult with this approach to obtain a good fit with atomic wave functions. The augmented plane wave method fits more naturally into the nearly free electron scheme. If the investigator is interested in structural and

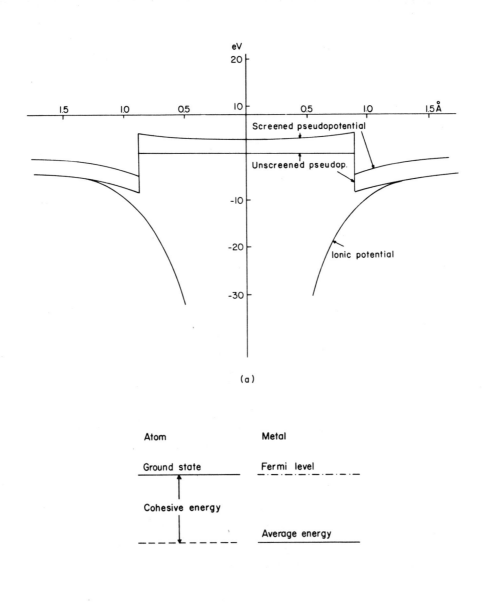

FIG. 1-58. Schematic diagram of the pseudopotential of sodium (a). Energy levels of sodium in its atomic and metallic state (b).

atomic properties, he may use the Green's function method, also known
as the KKR (Korringa, Kohn, and Rostoker) method, in which both of
these terms can be separately evaluated. We are not so much interested in the details of these computations, but rather with the question of how the calculated energy-wave functions and Fermi surfaces
can be connected with experimentally determined electron transport
and optical properties of crystals.

F.   Energy Bands and Fermi Surfaces of Elements,
Ordered Alloys and Compounds

The first Brillouin zone of sodium, a bcc crystal, is 50% occupied by electrons. Measurements show that its Fermi surface departs
from spheroidicity by less than 0.1%. E = E(k) of sodium is given
in Fig. 1-59. Figure 1-60 gives results of calculations for copper.
An augmented plane wave method was used in this calculation by Snow
[37]. Symmetry points in reciprocal space are given by letters;
the same designation is used in these diagrams. Γ corresponds to the
center of the Brillouin zone, and points between Γ and X or Γ
and K would correspond to k-vectors between these point pairs.
For the free electron model, the E(k) curves for these lines should
be parabolic. $\partial^2 E/\partial k^2$ of these curves is inversely proportional to
the effective electron mass. Therefore, one would expect that electrons close to Γ, 0.6Ry below the Fermi energy, or electrons between
Γ and X at the Fermi energy level have masses similar to free
electrons. One can easily see that the bands between −0.4 and −0.2
Ry with nearly horizontal slopes should have heavy electrons. m*
has to be large, since $m* \approx \hbar^2 \pi^2 / 2a^2 \Delta E$, with ΔE the change of the
electron energy from the bottom to the top in one band. Therefore
small values of ΔE lead to large values of m*. Small ΔE values
should give also high N(E) values, since $\Delta E = \int dn/N(E)$. One would
therefore expect from E(k) curves given in (1-60a), that the density of states curve for copper has very high values between −0.4

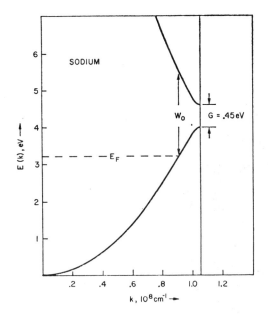

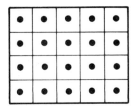

FIG. 1-59. E(k) of sodium in the nearly free electron model (left figure). Volume of metallic ion cores with deep potential wells is frequently small compared with the total volume. These cores are schematically given by full circles in the right figure. Reproduced with permission from Ref. 36.

and $-0.2\mathrm{Ry}$ below the Fermi energy level, and since it contains five d-electron subbands with ten electrons. Figure 1-60b gives the calculated $N(E)$ curves derived from Figure 1-60a.

Optical, electron emission, and magnetic properties depend on electron states in large energy intervals. Electron transport properties like the Hall effect or the electrical resistivity are affected only by electrons at the Fermi energy level. To understand their properties, the electron energy structure near the Fermi energy level has been studied in detail. Whereas the $E(k)$ curves are usually determined only in specific symmetry directions in k-space, Fermi surfaces are determined in three dimensions. In theoretical investigations one would calculate $E(k)$ curves and inspect their characteristics near E_F, the Fermi energy. For instance, the $E(k)$ curves would intersect E_F in (1-60a) between Γ and X on an $E(k)$

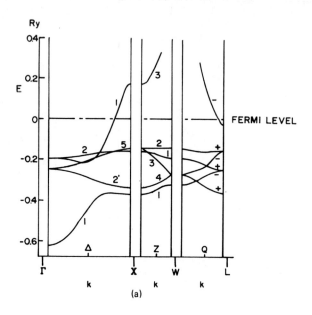

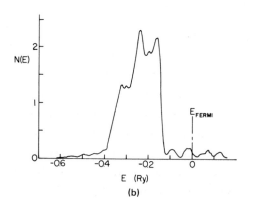

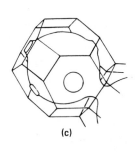

FIG. 1-60. Calculated E(k) curves (a) and density of states curves (b) of copper. After Ref. 37, with permission. Fermi surface of copper (c).

curve which has a shape expected for free electrons. In three dimensions, this intersection would correspond to a large sphere inside
the first Brillouin zone. However, the high negative curvature in
other sections close to the Brillouin zone boundary indicates that
the Fermi surface has small sections where the carriers are heavy
holes. A complete model of the Fermi surface of copper is given in
Fig. 1-60c. It shows clearly the large spherical belly sections,
small sections with negative curvatures, and the contact areas with
the first Brillouin zone boundary. A discussion of the different experimental methods to determine the Fermi surface is outside the
scope of this book. Such measurements can be made only under rather
stringent requirements. Temperatures have to be low and samples have
to be of high purity. However, the shape of the Fermi surface explains
also room temperature measurements in alloys. A two electron band
model for copper may be explained by assuming that electrons in the
belly section corresponds to one band, and the holes near the neck
to a second band.

The E(k) diagrams and Fermi surfaces of silver and gold are
similar to those of copper. Even solid solutions of noble metals with
each other have Fermi surfaces similar to the Fermi surfaces of the
individual elements. A major change is expected only if atomic ordering takes place, regardless of how small the energy of ordering
is. Atomic ordering will modify the Brillouin zone drastically because new Brillouin zone boundaries will form. They will intersect
the Fermi surface and change it. Figure 1-61 shows the Fermi surface
of ordered Cu_3Au as obtained by Harrison. He constructed it by
folding and shifting sections of the Fermi surface of pure copper.
Results similar to those obtained by Harrison were obtained in a
calculation by Gray and Brown [38], who used a modified plane wave
method. These calculations, in which a muffin tin potential was used
with the atomic potentials of gold or copper ions, yielded E(k)
curves which were similar to those of copper. Near the Fermi energy
level, differences calculated for selected points varied by only
0.04Ry. Some d-like points, well below the Fermi energy, showed

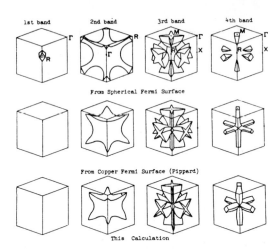

FIG. 1-61. Comparison of the calculated Cu_3Au Fermi surface
with Harrison's folded surface. Reproduced with permission from
Ref. 38.

deviations of as much as 0.1Ry from Harrison's work.

Again using a muffin tin type potential and the augmented plane
wave method, Arlinghaus [39] calculated the E(k) curves for β-brass.
This is an ordered compound with 3/2 electrons per atom. Its composition is CuZn. This system should be especially interesting for an
electron energy calculation, since it is one of the Hume-Rothery
phases, where the stability is determined from the electron concentration. In this E(k) calculation (1-62a), the potential for the
spherical part of the potential near the nucleus of zinc and copper
was again assumed to be that of ions. However, the constant potential between the atoms was not selected just by matching it with
the periodical potentials at the interface. Instead, the potential
between spheres was the average potential produced by the copper
or zinc ions. This leads at the interface to a discontinuity in the
potential. It is 0.2435Ry at the copper ion sphere, and 0.2396Ry at
the zinc ion sphere. The E(k) curves calculated with this model
show conduction bands near the Fermi energy level, and two separate
d-bands below the Fermi energy. The copper d-band extends from E_F-
3.3eV to E_F- 5eV, and the zinc 3d-band lies about 11eV below the

Fermi energy.   The Fermi surface of this compound is given in Figs.
1-62b and c.

Metals with two outer electrons crystallize preferably in the
hcp structure.   Different energy bands overlap.   The two conduction
electrons per atom nearly fill the first Brillouin zone.   However,
one finds electrons not only in the next zone, but also small segments in the third and fourth Brillouin zones filled with electrons.
These small electron pockets and the hole pockets in the first
Brillouin zone have been measured with high accuracy.   In the case
of magnesium [40], the determination of the Fermi surface is sensitive to variation of 0.1% of the Fermi energy.   The calculation of
the energy band structure for this element with the single ortho-
gonalized plane wave methods is in good agreement with the experiments.   The accuracy of such calculations is of the order of 5%
[42].   To improve agreement between experiments and theory, Kimball
et al. [40] used a semi-emperical model Hamiltonian based on the
pseudopotential theory.   This approach has the dual advantage that
it acts as an interpolation procedure where limited experimental
information on the Fermi surface can be used to generate the entire
Fermi surface, and where one can then also obtain E(k) bands above
and below the Fermi energy level.

The calculated $E = E(k)$ curves are given in Fig. 1-63a.   It
shows the typical parabolic energy versus wave number curves.   The
various sections of the Fermi surface for this element are shown in
Fig. 1-63b.   They were obtained with the single O.P.W. model calculation of the Fermi surface.   Since these sections are very close
to the Brillouin zone boundary, the effective mass of the electrons
may differ markedly from the free electron value.   Surprisingly,
the density of states values obtained for these elements from electronic specific heat measurements are very close to the free electron model value.   This means that the large values of m* which
lead to small electron velocity values $v = \mathrm{grad}_k E / \hbar$ in Eq. 1-59
compensates for the smaller area of the Fermi surface [2].

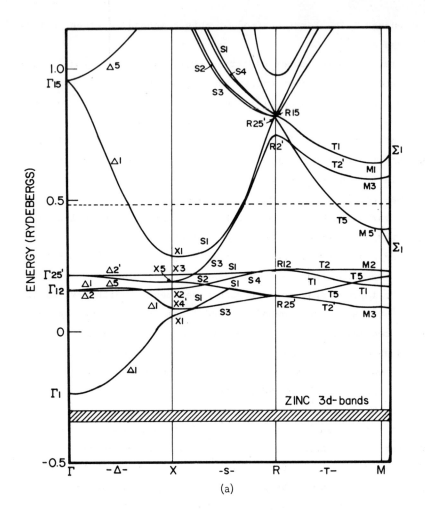

(a)

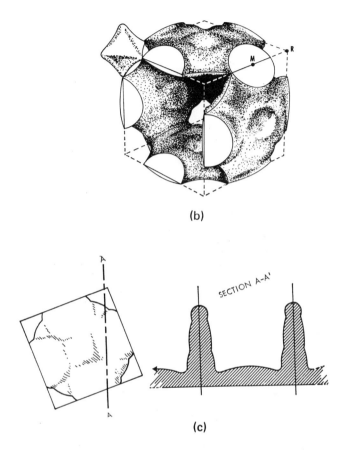

(b)

(c)

FIG. 1-62. E(k) of ordered β-brass (a, opposite page). Fermi surface of β-brass (b). Open orbit in β-brass (c). After Ref. 39, with permission.

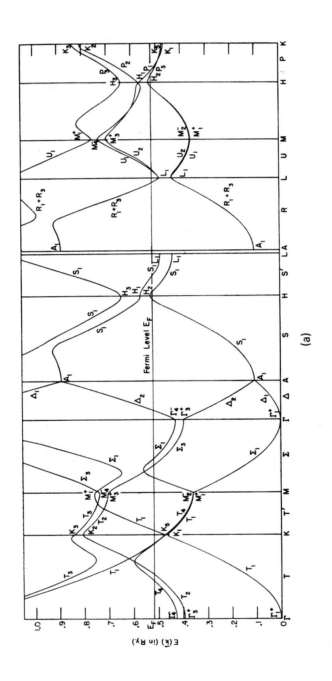

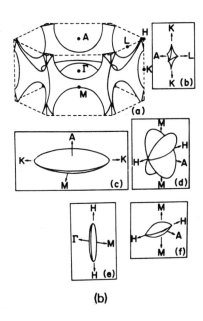

(b)

FIG. 1-63. E(k) of magnesium for lowest energy bands. The calculation uses a non-local pseudopotential model (a, opposite page). After Ref. 40, with permission. Fermi surface of magnesium after an orthogonal plane wave model calculation (b). After Ref. 41, with permission.

Several energy band studies of aluminum, an fcc element, are available. The Brillouin zone is given in Figure 1-64 [43]. Aluminum has three outer electrons. The first Brillouin zone cannot accept more than two. If one would use the free electron approximation, the Fermi sphere would intercept the first Brillouin zone boundary, leave small hole sections at corners, and would bulge out in the hexagonal and on the square faces. The tight binding calculation would give no bulging at the square surfaces. The experimental data obtained from DeHaas-van Alphen experiments are somewhere in between these two theoretical predictions. The main result of these investigations is that the Fermi surface of aluminum is rather close to the result predicted by the free electron model [2]. Figure 1-64b shows the E(k) diagram of aluminum. A recent calculation of E(k) for energies more than 11 Ry above the Fermi energy level by Hoffstein

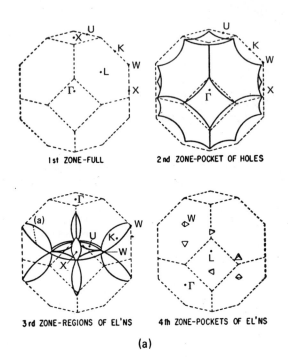

1st ZONE-FULL   2nd ZONE-POCKET OF HOLES

3rd ZONE-REGIONS OF EL'NS   4th ZONE-POCKETS OF EL'NS

(a)

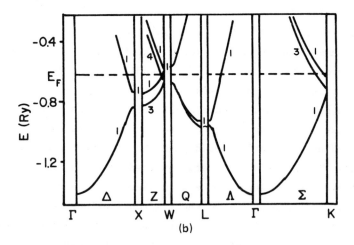

(b)

FIG. 1-64. Fermi surface of aluminum according to a single O.P.W. approximation (a). Calculated E(k) curves of aluminum (b). After Ref. 43, with permission.

and Boudreaux [45] shows that the E(k) curves become more complex
at higher energies and deviate from the free electron model. Studies of electron transport phenomena or optical properties do not
require such high energy calculations. However, if one is concerned with electron emission experiments, electron diffraction, or
with soft X-rays, information on electron energies to such high
energy values is required.

Transition elements have a rather complex electronic structure since not only conduction electrons (s- and p-electrons) but
also d-electrons occupy partly filled bands.

Some of the features of the electronic structure of the fcc
transition elements with 10 outer electrons are similar to the
structure of noble metals. This is not surprising, since palladium and silver or platinum and gold differ by only one electron
and have the same crystal structure. Qualitatively, one could
just reduce the Fermi energy level in silver or gold until the
electron concentration per atom is reduced by one. This should
move the Fermi energy level so far downward that d-bands intersect the Fermi energy level, making the material a transition
element (1-65).

The nickel energy band structure is even more complex since
nickel is ferromagnetic at low temperatures. In this system, one
would initially decrease the Fermi energy level. Then one subdivides the bands into sub-bands with spin-up and spin-down, and
shifts these two (symmetrical) sections with respect to each other
in such a way that one has 0.6 electrons per atom more than the
other. This would give at absolute zero temperature just the
right magnetization.

The chromium group elements form bcc crystals. Since tungsten has a very high melting point, and low vapor pressure, it was
one of the first transition elements which could be prepared in
the form of high purity single crystals. This made early Fermi
surface studies on this material possible. The Fermi surfaces of
tungsten, vanadium and paramagnetic chromium, which belong to the

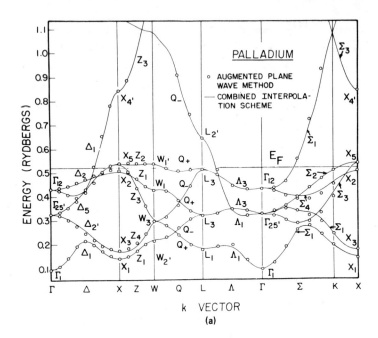

(a)

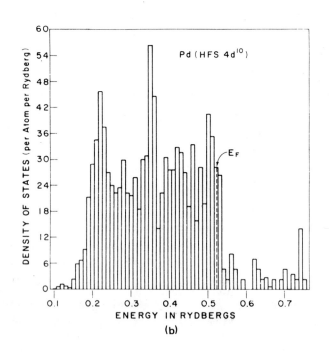

(b)

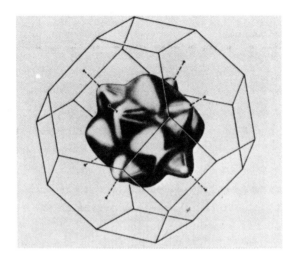

(c)

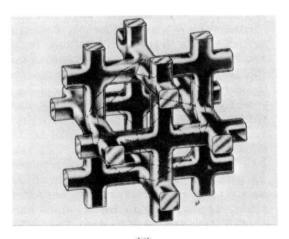

(d)

FIG. 1-65. Calculated E(k) curves (a, opposite page), N(E) curves (b, opposite page), Fermi electron surface (c), and Fermi hole surfaces (d) for palladium. Reproduced with permission from Ref. 46.

same group in the periodic system are similar. The Fermi surface for
molybdenum is given in Fig. 1-66a. It consists of two main closed
surfaces [48]. One is the 'electron jack' centered around the middle
of the Brillouin zone at Γ, the the other is the hole 'octahedra' at
the symmetry point H. The small independent ellipsoids surrounding
the 'jack' also contain holes. The 'jack' consists of an octahedral
center connected by balls. The octahedral center is, in the case of
chromium, very similar in size and shape to the octahedral holes at
H. This is a quirk of nature. It leads to a peculiar type of antiferromagnetism. It makes it possible that the electron octahedra at
Γ and a hole octahedra at H may combine and cancel if chromium
becomes antiferromagnetic. Since all these electron states are
associated with electrons or holes in partially filled state, this
antiferromagnetism is associated with non-localized electrons.

Mattheiss [47] used an A.P.W. method to calculate the energy
band of tungsten. The muffin tin potential in this calculation was
obtained as a superposition of atomic potentials of Pd. The
$(5d)^5(6s)^1$ configuration of free tungsten was selected, since it is
probably close to the potential in the solid. The E(k) curves of
tungsten are given in Fig. 1-66b. It shows that the width of the d-
band is of the order of one Rydberg. The N(E) curve has a sharp
minimum at the position of the Fermi energy level. The chromium group
elements have six outer electrons, and there are ten d-states and two
s-states available. The six outer electrons could fill the two d-
subbands and the conduction band. However, electrons spill into
higher bands, leaving the same number of holes behind. Therefore,
these elements are compensated just as zinc. Fig. 1-66b shows the
Fermi energy level of molybdenum and of niobium. These elements have
the same crystal structure. Niobium has one outer electron less than
molybdenum. It is assumed that the energy bands of these two elements
are the same (rigid band model) and that the removal of one electron
just lowers the Fermi energy level.

Elements with the diamond structure have four outer electrons.
One should be an s-electron, and three should be p-electrons.
Naively one would expect two different types of bonding electrons,

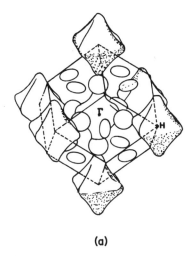

(a)

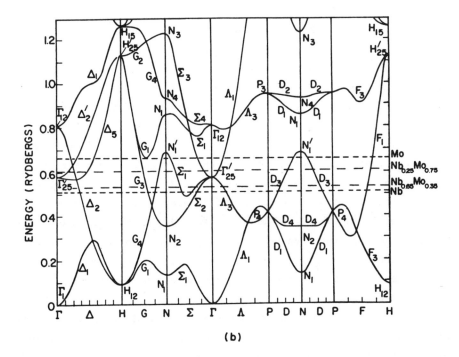

(b)

FIG. 1-66. Fermi surface of chromium group elements (a). After Ref. 47, with permission. E(k) curves for tungsten with Fermi energy levels of niobium, molybdenum, and some of their alloys. After Ref. 48, with permission.

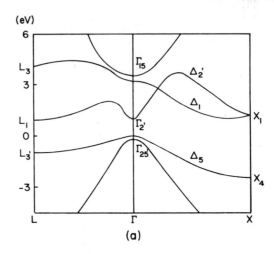

FIG. 1-67a. Schematic diagram of energy levels in germanium.

since the one   s-electron has radial symmetry, and the three   p-electrons contribute to directional bonding. However, one finds that bonding in diamond, silicon, or germanium leads to a tetrahedral nearest neighbor configuration, in which all four nearest neighbors are identical in their electron configuration. Therefore, one should expect that bonding has the same symmetry. One describes bonding in such a system as sp^3-bonding. The four electrons oscillate between different states in such a way that the average electron bonding configuration is the same between one atom and all its four neighbors. Small sections of the calculated   E(k)   diagram for germanium are given in Fig. 1-67.   The Fermi energy level lies in between the filled valency and the empty conduction band. Properly speaking, germanium and silicon have no Fermi surface. However, one can draw contours of constant energy just above and just below the Fermi energy. Figure 1-67b and c   shows these surfaces.

Energy band details near band edges for semiconductors are very important for the calculation of effective masses and energy gaps. Therefore, some special techniques have been developed to calculate E(k) curves near maxima in the valency band, and minima in the conduction band. A frequently used type of perturbation calculation,

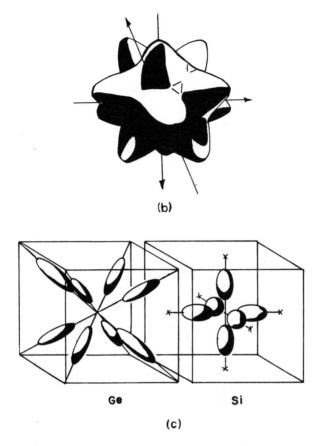

(b)

(c)

FIG. 1-67b + c. Constant energy surfaces for heavy holes in silicon (b), and for electrons in germanium and silicon (c). After Ref. 2, with permission.

proposed by Kane [49], has been used for several semiconductor elements and compounds. One essential feature of this calculation is that one takes coupling between the spin of the electron and the orbital movement of this electron (spin-orbit coupling) into account. The Schrödinger equation for such a model has the following form:

$$[p^2/2m + V + (k/2m^2c^2)(\text{grad}\vec{V} \times \vec{p}) \cdot \vec{\sigma}]\psi_k = E\psi_k. \qquad (1\text{-}63a)$$

Here $\vec{p}$ is the momentum operator and $\vec{\sigma}$ the spin operator. The last

term in the bracket, neglected in previous sections, is due to energies between the spin of an electron and its orbital movement. This
theory gives for the constant energy surfaces functions of the form:

$$E(k) = -\Delta E + (\hbar^2/2m)(Ak^2 \pm [B^2k^4 + C^2(k_1^2k_2^2 + k_2^2k_3^2 + k_3^2k_1^2]^{1/2}) \tag{1-63b}$$

assuming that the energy at the valency edge is zero [50]. $k^2 = k_1^2 + k_2^2 + k_3^2$. The function is the same for both silicon and germanium.
Values of the parameters are given in Table 1-5. However, it is fre-
quently sufficient to use

$$E(k) = \hbar^2(k_1^2 + k_2^2)/2m_t + \hbar^2k_3^2/2m_\ell \tag{1-63c}$$

where t stands for transverse, and ℓ for longitudinal mass.

Silicon and germanium form a complete series of solid solutions.
One would expect that the $E(k)$ curves are also smooth functions of
composition. However, the energy gap between valency and conduction
band is not a linear function of composition. Experiments indicate
that there is a change in slope of the energy gap versus composition at 79% silicon.

Only a few examples of the electronic band structure of ordered
alloys and compounds can be given. The band structure of semiconducting compounds of the zinc-blende type is frequently determined with
a semi-quantitative model proposed by Herman [51]. The zinc-blende
type lattice of these compounds is very similar to the diamond lattice of germanium and silicon. Therefore, Herman proposed that one
can develop the $E(k)$ curve of GaAs from the $E(k)$ curves of
germanium which lie just between these two elements in the

TABLE 1-5

Ge: A = -13.3; $|B|$ = 8.6; $|C|$ = 12.5
Si: A = - 4.0; $|B|$ = 1.1; $|C|$ = 4.1

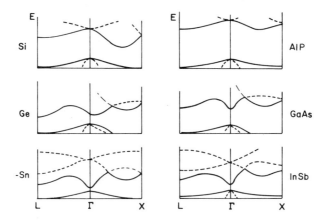

.FIG. 1-68.   Qualitative E(k) diagrams of several III-V semiconductors.   These curves were obtained by perturbation methods from
Group IV semiconductors.   After Ref. 52.

periodic system.   The main difference between the zinc-blende and
the diamond lattice is that the former lacks a center of inversion.
A change from $\vec{r}$ to $-\vec{r}$ can lead to a different potential.   The potential of the compound will differ from that of germanium.   This
difference is treated as a perturbation in Herman's calculation.
The results of such a calculation are given in Fig. 1-68.

In these systems, the valency band consists of two subbands
which degenerate at k = 0 (in other words, the electrons with
k = 0 in both subbands have the same energy).   Another band is split
by spin orbit coupling and found for k = 0 below both these
degenerate bands.   The spin-degeneracy is lifted in this lower lying
band.   This results in a slight shift of the maximum of this band
away from the [111] orientation.   A quantitative description of
the energy band system of III-V semiconductors is obtained from
these calculations if one specifies values for energy gaps and
electron masses, etc.   These data are usually obtained from experiments.

In recent years, self-consistent calculations for semiconducting compounds have become available.   They give essentially

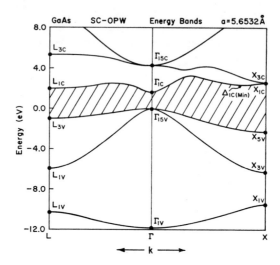

FIG. 1-69.   Calculated band structure of GaAs.   Reproduced with
permission from Ref. 53.

the same results as the semiquantitative models developed by Herman.
These new calculations are, naturally, not completely without adjustable parameters.   For instance, the electron interaction is
difficult to treat.   The Hamiltonian contains electron-electron
interaction terms.   The results of such calculations are shown in
Fig. 1-69.   The exchange energy (see Chapter 2) is included in
this calculation.   Agreement between calculated and measured energy
gaps for states with the same k-values is good.   "Off axis" energy
differences are probably incorrect.

Pentavalent metals like arsenic, antimony, and bismuth crystallize in the rhombohedral structure.   The change of the sc lattice system to the rhombohedral system requires a shearing opera-
tion of only a few degrees (the angle $\alpha = 60°$ between the (111)
and (100) directions changes to 57° 14.2') and the shift of one
atom by only a few percent of the lattice parameter.   This small
change in the atomic configuration explains why the Brillouin zone
of this crystal structure resembles very much the Brillouin zone
of the sc lattice.   Rhombohedral shear makes the L and T point in

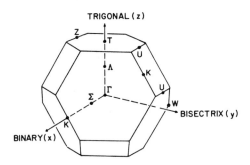

FIG. 1-70. Brillouin zone of the A7 crystal structure. After
Ref. 54, with permission.

the Brillouin zone non-equivalent. It lowers the s band at the L
point and raises it at the T-point. Therefore, one has electron
states at L, and hole states at T. These points are therefore of
special interest. One distinguishes between 'principal L-points',
and 'non-principal L-points'.

The Brillouin zone of this system is given in Fig. 1-70. It
can contain ten electrons for the two atoms. The E = E(k) curves
show that the overlap between the first Brillouin zone and the
second Brillouin zone is rather small. Therefore, the first Brillouin zone is nearly filled. The next zone has the same small
number of electrons. Only a small band shift is required to turn
these elements into semiconductors.

The small energy gaps are associated with high carrier mobilities, small effective masses and non-parabolic E = E(k) relationships. The small energy gap makes accurate calculations of the
E(k) diagram near the energy gap very difficult. These calculations can give only the broad outline of the energy dependence of
electrons in these elements, since the accuracy of such first principal calculations is only of the order of 0.2 to 0.3 eV presently.
Since most of the experimental studies are concerned with electron
states at L and T, one frequently uses a phenomenological two-band
to explain experimental data. Then the band parameter can be determined with a high accuracy.

G. Impurity States

A zinc atom has one electron more than copper. Substituting
zinc into the copper matrix will lead to an extra charge at the
nucleus which has to be compensated for in the crystal. Let us
assume that one places the extra conduction electron into the 'sea'
of conduction electrons of the sample. To compensate for the excess
positive charge of the zinc ion, one has to take this extra electron from the 'sea' of conduction electrons and move it close to
the zinc ion. This will reduce the energy of the electron gas, U,
by [31]:

$$dn(r)/N(E) = - dU(r) \qquad\qquad (1-64)$$

where r is the position of the extra electron. The change in the
electron density $dn(r)$ corresponds to a change in the local charge
density near the zinc ion which leads in turn to a potential $d\Phi$.
It is shown with the potential theory that $d\Phi$ must satisfy the
equation:

$$\nabla^2 d\Phi = -4\pi q^2 dn(r) = 4\pi q^2 N(E) dU(r) = \lambda^2 dU \qquad\qquad (1-65)$$

One obtains as the solution for the potential of the extra electron,
if one assumes that $dU = d\Phi$ except near the zinc ion:

$$dU = (q^2/r) \exp(-r/r_o) \ ,$$

where r is the distance from ion core to charge and $1/\lambda = r_o$ is the
screening radius. The potential dU is the "screened Coulomb potential", since the normal Coulomb potential, q^2/r, is multiplied with
a function which decreases exponentially with the distance from the
impurity. The screening radius in metals is of the order of 10^{-10} m.
The impurity will interact therefore only with electrons in this
neighborhood.

Naturally, the interaction of transition element impurities with electrons in a non-transition element matrix is more complicated [55,56]. The transition element has both conduction electrons and d-electrons near the Fermi energy level. The d-electrons will keep their wave character even in the noble metal matrix since the d-wave functions will not extend far away from the transition metal ion into the noble metal matrix. However, the d-electron will interact with the conduction band since the d-electron has energies similar to the conduction electrons.

The analysis of this interaction process shows that it has some similarity to the problem of two coupled oscillators. This coupling leads to new modes of vibrations, and to new energy states. The new energy states which are created due to the interaction of the d-electrons with the s-band are different from the original energy states of the pure matrix and the free impurities; these new states are "resonance states" with resonance energies ϵ_r. These states are characterized by a band width Γ. The energy level ϵ_r can be below, at, or above the Fermi energy level. Figure 1-71a shows a typical example of such a density of state curve. The spacial distribution of the virtual bound state is given in Fig. 1-71b.

The interaction between transition element impurities and a non-transition element matrix like aluminum can be quite strong. The interaction may lead to a splitting of the impurity states. States with spin-up or spin-down have different energies. Figure 1-71c to e shows schematically such a system. Figure 1-71c gives atomic d-levels. Intra-atomic forces lead to a line up of spins. Electrons with different spins will have different energy levels (1-71d). Interaction of these d-electrons with s-electrons will further modify energy levels. This interaction is the "hybridization" by resonance (1-71e).

Impurity states can be represented, just as pure periodic crystals, by pseudopotentials. Figure 1-71f shows the unscreened Coulomb potential. The pseudopotential at the impurity is given in Fig. 1-71g. It again is modified by screening (1-71h). This

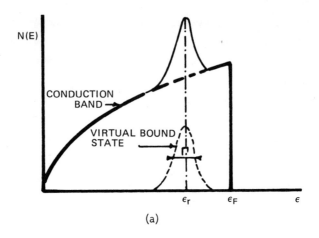

(a)

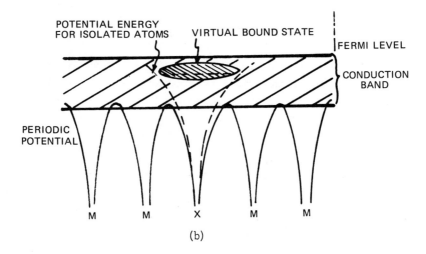

(b)

potential is much smaller and of shorter range than the pure Coulomb field.

Impurities in semiconductors can occupy both interstitial and substitutional positions. Lithium is an example of an interstitial impurity in germanium and silicon. Most impurities are, however, found in substitutional positions. Arsenic would add one electron to the germanium crystal, and gallium would subtract one. Since

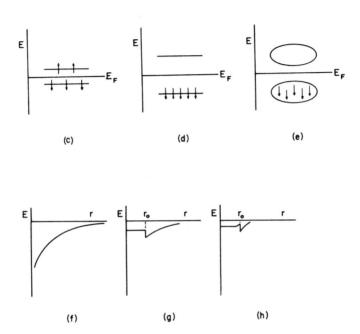

FIG. 1-71. Virtual bound state of an impurity in a conduction band in the N(E) diagram (a, opposite page). Spacial distribution of virtual bound state (b, opposite page). Polarized atomic d-levels (c). These levels are split by d-d interaction (d). Broadening of states given in (d) by hybridization with s-electrons (e). Bare ion potential (f). Bare ion pseudopotential (g). Screened pseudopotential (h). After Refs. 31 and 57, with permission.

the semiconductors have such a low concentration of conduction electrons or holes, one should not expect that screening charges form around the impurity atom. The potential of the extra electron would have the Coulomb form. The energy of the electron can be estimated if one compares it with the case of the hydrogen atom. There the Bohr theory gives as energy $E_n = - e^4 m*/2\varepsilon^2 \hbar^2 n^2$, where ε is the dielectric constant. The Bohr radius would be given by $r_n = \varepsilon n^2 \hbar^2 / e^2 m*$.

The dielectric constant is the most important correction. ε = 15.8 for germanium and ε = 11.7 for silicon. The effective mass for germanium is m* = 0.1m; for silicon m* = 0.2m. This gives as ionization energy 0.006 eV for germanium, and 0.02 eV for silicon. The value for the first Bohr radius is about 80 $\overset{o}{A}$ for germanium, and about 30 $\overset{o}{A}$ for silicon. These values are rather large compared with the smallest Bohr radius for a free hydrogen atom: 0.53 $\overset{o}{A}$.

Group V element impurity atoms are called "donors" since they add electrons to the crystal. Impurities which replace a matrix atom with higher valency would lead to a deficit in electrons in the crystal. They introduce "acceptor states" which can absorb electrons or add "holes". The modified Bohr model applies just as well to such acceptor states as to the previously discussed "donor states". Table 1-6 shows the energy levels of acceptor and donor states in germanium.

H.   Energy Band Models and Fermi Surface of Disordered Alloys

The simplest model for the energy bands of alloys is the "rigid band model". This name implies that the "energy band", which may be the E = E(k) curve for one given "n" value in Fig. 1-48d and 1-48e, the corresponding N(E) curve, or even the combination of such curves for different "n" values, is independent of the way electrons are added to the crystal. The Fermi energy depends in

TABLE 1-6
Donor and Acceptor Ionization Energies in Germanium and Silicon

	P	As	Sb	B	Al	Ga
Ge:	0.0120eV,	0.0127eV,	0.0096eV,	0.0104eV,	0.0102eV,	0.0108eV
Si:	0.045 eV,	0.049 eV,	0.039 eV,	0.045 eV,	0.057 eV,	0.065 eV

this case only on the number of electrons per volume, and N(E) is
a unique function of the electron concentration. The free electron
model is an example of the rigid band model. One would expect from
the rigid band model that the addition of gold to copper will lead
to a minor change in the density of states at the Fermi energy level,
because the only effect would be a small reduction in the number of
conduction electrons per unit volume, since the atomic volume of gold
is slightly larger than the atomic volume of copper.

The volume change due to alloying in extended solid solutions
is usually small, because the atomic radii of participating elements
cannot differ by more than 15%. Therefore, the effect of volume
changes is usually neglected in the discussion of the rigid band
model. Only the average concentration of conduction electrons is
regarded as the important parameter. Typically, one is concerned with
the question, does the addition of 2 at.% zinc to copper have the same
effect as adding 1 at.% gallium, since in both cases one adds the same
number of electrons to the system. Copper with the addition of both
1 at.% zinc and 1 at.% nickel has the same electron concentration as
pure copper, since nickel has one electron less and zinc one electron more than copper. This ternary alloy is therefore called "pseudo copper". It should have the same N(E) value as copper according
to the rigid band model.

The density of states in an extended alloy system can be a complicated function of alloy concentration. The change of energy with
alloying can be obtained for such a system, if the rigid band model
can be used, with the equation:

$$dE = dc(z_1 - z_2)/LN(E),\qquad\qquad(1\text{-}66)$$

where c is the concentration of the second component of the alloy,
$z_1 - z_2$ is the difference of the number of electrons in outer shells
of the two elements, and L is the number of atoms per unit volume.
No assumption on the shape of the density of states curve has to be
made. N(E) curves obtained from N(E) = f(c) have been used to correlate results from a number of experiments in alloy systems.

A typical example for the use of the rigid band model is the
Nb-V alloy system as shown in Fig. 1-66b, where the number of electrons filling the energy band is assumed to be just the sum of the
electrons of the constituents. The Fermi energy of two alloys obtained in this way are indicated by dashed lines in this figure.
The density of states curve obtained in such alloy studies may then
be regarded as representative of the density of states of an element above and below its Fermi energy level.

One sometimes modifies this rigid band model with the concept
of the "soft rigid band model". This was proposed by Cohen and
Heine [58]. They assumed that alloying modifies the energy gap at
the Brillouin zone. For instance, if energy gaps of copper at L in
the first Brillouin zone decrease with the addition of zinc, then
the Fermi surface of CuZn may become more spherical with increasing
zinc concentration.

The discussion of the screened impurity charge in the previous
section already showed that the rigid band model is a simplification.
The screening has the effect that the extra charge of an impurity
atom will be 'seen' by the conduction electrons only in the neighborhood of the impurity. Several lattice parameters away from the
impurity, the crystal should be essentially in the undisturbed
state with no change in the Fermi energy. One would expect that
the Fermi energy shifts with alloying only if the screening clouds
of impurity atoms interact. This problem has been studied by Friedel
[59]. He showed for finite solute concentration that part of
the excess electrons will not be used for screening, because the
potential fields between solutes will interact with each other. To
accomodate the same number of electrons as the rigid band model, the
Friedel model requires that some states have to be added below the
Fermi energy level. Since the ions which contribute the extra
screening electrons have naturally also the same number of extra
positive charges in the ions, they can push the potential down.

A disordered alloy is a non-periodic structure, in spite of the
fact that it is a perfect crystal. It is very difficult to solve the
Schrödinger equation for such a system. Therefore, one tries in one

approach to solve the Schrödinger equation for a model in which the
potential for each lattice site is the same, namely the average potential of the different atoms in the crystal. This is the "virtual
crystal approximation"; The potential of each lattice site of the
alloy is assumed to be of the form:

$$U(\text{alloy}) = c_A U(A) + c_B U(B).\tag{1-67}$$

This is a weighted average of the potentials of the individual atoms.
The potential given in Eq. (1-67) may not correspond to a potential
found for any atom of the crystal. However, the virtual crystal approximation potential may be reasonable even for this case. If an
electron moves through the lattice, it interacts with host and solute atoms. The likelihood of interaction is proportional to the concentration of these atoms. Therefore, the use of an average interaction potential seems to be a reasonable first approximation.

Figure 1-72 gives results obtained with the virtual crystal
approximation. Figures 1-72a and 1-72b show two possible examples of
solutions. The two components of the alloy have different energy
bands in their pure state. The important parameters which characterizes the interaction of electrons of different atoms in the alloy is
the band width in each pure metal, E_w^i, and the distance between the
centers of the two bands, E_d. One would naively expect that the combined energy band is in first approximation just the sum of the density of states of the two bands. This is found only if both bands
are widely separated. The virtual crystal approximation shows for
the case of a band overlap (Fig. 1-72c + d give the individual bands)
that the density of states curve is not the average N(E) of both
components. The density of states curve of the alloy could exhibit
a deep minimum (1-72e) or be split. The sharp minimum is due to an
oversimplification in the calculation. The real alloy should show
only a minimum in the density of states curve.

The general features of the virtual crystal approximation for
transition element alloys have been described by Beeby [60] in the
following way. The d-band electrons in these metals interact with

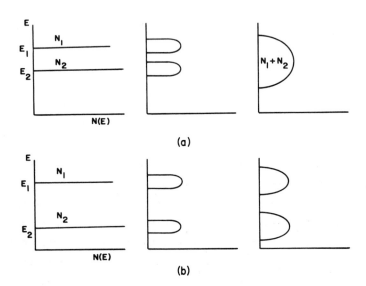

FIG. 1-72. Energy band models for alloys after Beeby [60]. Density of levels formed by N_1 ions at levels E_1 and N_2 ions at level E_2. Separation of bands is so small that bands will finally overlapp (a). Band separation is to large for formation of overlapping bands (b). Ideal representation of N(E) of two pure metals (c + d, opposite page) and for their alloys (e + f, opposite page). Suggested band structure for $\underline{Ni}Zn$ (g, opposite page), $\underline{Ni}Fe$ (h, opposite page), $\underline{Ni}Cr$ (i, opposite page), and $\underline{Ni}Mn$ (j, opposite page).

the electrons of the s- and p-band. This interaction may have little effect on the d-band levels. Therefore one neglects the interaction between d-band and conduction band. It is then possible to discuss qualitatively the d-bands, without using any specific calculations, but keeping some of the physical meaning in the discussion. Naturally, one cannot correlate all physical properties with such a model. Ferromagnetic ordering effects are usually neglected. Typically, only correlations of the unsymmetrical position of bands with spin up or spin down is taken into consideration as a magnetic ordering effect.

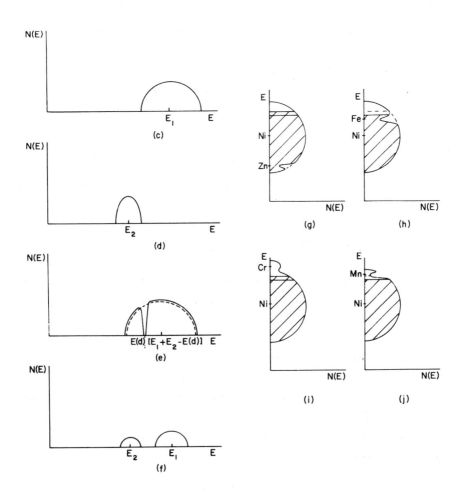

In the examples given in Fig. 1-72g to j, zinc, chromium, and manganese are added to nickel. The separation of the energy levels of the host and the impurity atom should be small for the case of zinc impurities in nickel. The density of states curve of Fig. 1-72g with the minimum of N(E) well below the Fermi energy could therefore be expected. The filling of the bands closely follows the rigid energy band model for NiCu or NiZn. Similarly, iron (or cobalt) impurities will keep the band shape essentially unchanged. The gap will be found in this case close to the Fermi energy level. The magnetic

moment should increase with decreasing electron concentration in the
Ni-Fe system, whereas it increases in the Ni-Cu or Ni-Zn system for
decreasing electron concentration. Chromium or vanadium have valen-
cies considerably different from nickel. One would therefore expect
that the energy band splitting is large. The rigid band model is
not applicable any more. The situation should be similar to the case
given in Fig. 1-72e. The two bands will split, the lower band being
completely filled by impurity d-electrons. The filling of the d-band
accounts for the decrease in the magnetic moment, as found for these
alloys experimentally, whereas the rigid band model would suggest
an increase in the magnetization.

It can be shown that the virtual crystal approximation can be
treated as a special case of the "coherent potential approximation"
[61]. In this approach, the movement of an electron wave through the
lattice is regarded as a succession of scattering events. The scattering processes are then averaged over the crystal sites. One individual scattering site would be investigated. It would be assumed
that the ion would be imbedded in an arbitrary medium selected in
such a way that the calculation can be made self-consistent.

Calculations of density of states curves for some of these models have been conducted for simple energy band structures. The
starting point is the assumption of a band for the pure elements (i)
which is of the form:

$$N^i(E) = (2/\pi W_i^2)(W_i^2 - \{E - E^i\}^2)^{1/2}, \quad N(E) \geq 0. \tag{1-68}$$

These bands correspond to half-ellipses with the center at $E = E^i$,
and a band width of $2W_i$. The relative separation of the two bands
is given by:

$$\delta = (E^A - E^B)/W, \tag{1-69}$$

where E^A and E^B are the energy of the center of the two bands. The
energy W is a scaling parameter (see Ref. 61 for details).

Since the system is symmetrical for both types of atoms, it is sufficient to calculate the density of states curves for concentrations
smaller than 1/2. Figure 1-73 gives results of calculations where
the concentration of impurity atoms is c = 0.15, and where the relative separation of the host and impurity band is 0.4, 1, and 2.
The center of the energy band of the crystal containing only B-atoms
is above the center of the energy level of the host A-atom matrix.
Figure 1-73 shows that the virtual crystal approximation (1-73a)
leads to quite different results from the coherent potential model
(1-73b) for these energy bands and their separations. Whereas the
impurities could lead in the virtual crystal approximation only to a
large broadening of the energy band of the alloy, even for the largest δ values used, the coherent potential approximation leads to a
broad, slightly unsymmetrical N(E) curve only for the δ = 0.4 value.
$\delta \geq 1$ leads to two separate bands. The composition dependence of
the band shape is shown in Fig. 1-74 for several δ-values.

    The coherent potential approximation has been used by Levin and
Ehrenreich [62] to calculate density of states curves for the complete range of composition of Au-Ag alloys. The model used in their
calculation contains one s-band and one d-band which hybridizes with
the s-band. This leads to band splitting. Calculated density of
states values are given in Fig. 1-75. The band splitting is found
in this calculation for both pure elements and for the alloys.
This represents a rather unphysical result. One may naturally wonder how meaningful the other parts of the calculated N(E) curves
are. It turns out that one can obtain useful correlations, e.g.
with optical measurements, by selecting reasonable values for the
width of the s- and d-bands of the two elements, their relative
positions, and the position of the resonance energy. One controversial result of this calculation is that it gives acceptable results only if one assumes a charge transfer from gold to silver in
AuAg alloys. The concept of a charge flow has been used before,
but usually a flow of charges from silver to gold has been assumed.

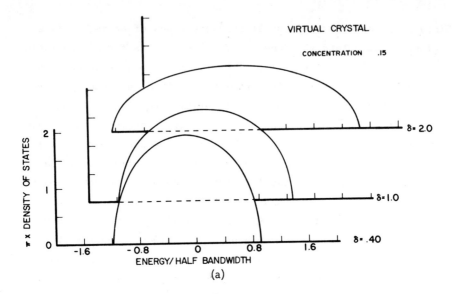

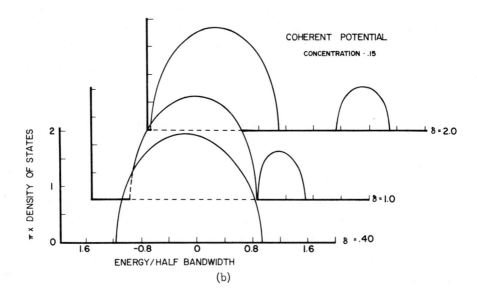

FIG. 1-73. N(E) as calculated in a self consistent virtual crystal approximation (a), and in a coherent potential approximation (b). The impurity concentration is 0.15. After Ref. 61, with permission.

The energy band structure of copper rich Cu-Zn alloys has been determined with the Kohn-Rostocker method, in a combination with a "pseudo-periodic" potential [63]. The potential in the calculation is of the form $V(r) = (1 - c)V_A(r) + cV_B(r)$ where $V_A(r)$ and $V_B(r)$ are the appropriate "muffin-tin" potentials for the individual atoms. The energy bands have been calculated for several zinc concentrations near the symmetry points Γ, X, and L. The maximum zinc concentration of 30 at.% used in this calculation gives E = E(k) curves shown in Fig. 1-75 as dashed lines. The full lines correspond to the E(k) curves of pure copper [64]. The bands were adjusted in such a way, that they agreed for zero zinc concentration with Segall's results [64]. The energy Eigen values are given in Table 1-7 as a function of composition. The error in the graphical evaluation of these data is estimated to be $\pm$ 0.13 eV. Alloying leads to a displacement of the conduction bands. This displacement is not uniform (1-75). Shifts at Γ_1, X_4', X_1 above E_F and L_2', are of the same order of magnitude. The shift of L_1 above E_F is slightly larger, so that the energy gap $L_2 - L_1$ is reduced. Further, a narrow d-band of zinc is found about 2.6 eV below the Γ_1 level.

The relatively "rigid" displacement of some sections of the energy bands with alloying is partly due to the initial assumptions.

TABLE 1-7

Conduction Band Energies (in Ry) for CuZn

	0%	10%	20%	30%Zn
Γ_1	-0.836	-0.87	-0.88	-0.89
X_4'	-0.029	-0.05	-0.06	-0.08
X_1	+0.389	+0.37	+0.35	+0.33
L_2'	-0.247	-0.26	-0.28	-0.29
L_1	+0.189	+0.12	+0.10	+0.05

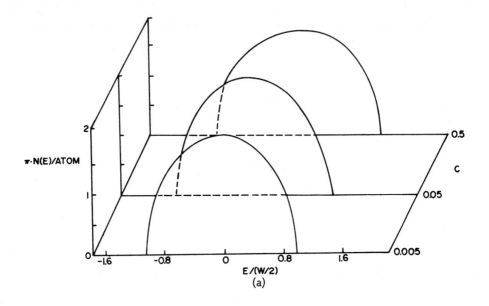

(a)

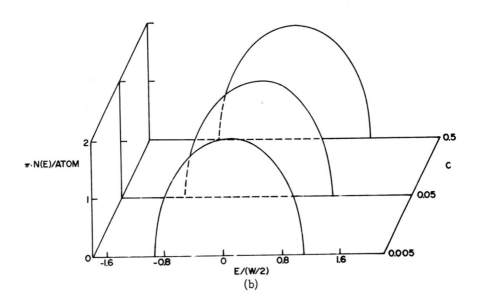

(b)

FIG. 1-74a + b. Model calculation of N(E) in the coherent potential approximation as a function of the impurity concentration for
$\delta = 0.25$ (a) and $\delta = 0.5$. After Ref. 61, with permission.

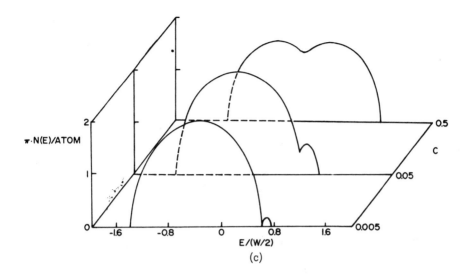

FIG. 1-74c. Model calculation of N(E) in the coherent potential approximation as a function of the impurity concentration for
$\delta = 0.75$. After Ref. 61, with permission.

If one uses the neutral free atom potentials as a starting point,
then one implies that the excess conduction electrons of zinc should
be found most of the time near the Zn^{2+} ion. This is consistant with
Friedel's concept of screening. The reduction of the L_2, $- L_1$ energy gap leads to a modification of the Fermi surface. It should
lead to a more spherical Fermi surface, since a reduction in energy gaps corresponds to a smaller interaction between lattice and
electron gas.

The modification of the energy band structure, as obtained
from this calculation, is in good agreement with Friedel's band
model. The addition of zinc ions to the matrix shifts the energy
band of the conduction electrons to a lower position. However, the
number of states moved down is less than the number of electrons
added to the system. The net effect is, therefore, that the Fermi
energy level increases with the addition of zinc atoms, but not as
much as expected from the rigid band model.

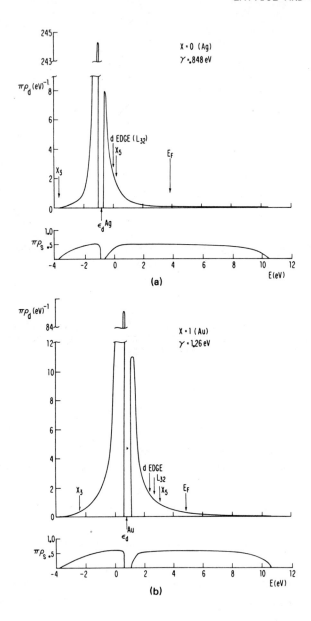

FIG. 1-75a + b. Calculated N(E) curves for s- and d-bands of silver (a) and gold (b). Reproduced with permission from Ref. 61.

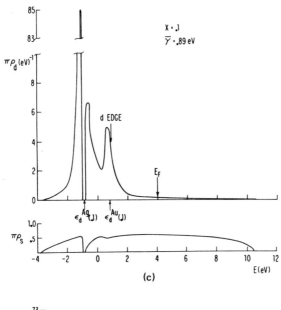

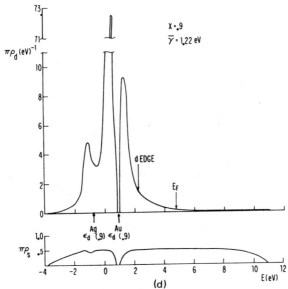

FIG. 1-75c + d. Calculated N(E) curves for s- and d-bands of
Au$_{.1}$Ag$_{.9}$ (c) and Au$_{.9}$Ag$_{.1}$ (d). Reproduced with permission from Ref. 62.

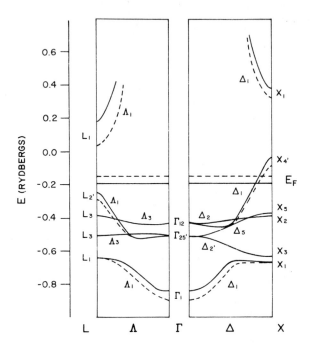

FIG. 1-76. Energy bands of α-Cu$_{70}$Zn$_{30}$ (dotted lines) and
pure copper (solid curves). Reproduced with permission from
Ref. 63.

Kirkpatrick et al. [65] used the coherent crystal calculation
to determine the density of states of copper nickel alloys. They
neglected the s-band, and determined first the N(E) curve for a
model, in which the d-bands of both elements were of identical
shape, with a sharp peak at the top, but with both bands shifted
in respect of each other. The calculation shows that the peaks
of N(E) of the pure elements are also found in the alloys. The
50-50 alloy has two minor rounded peaks at the same positions,
where the pure elements have their sharp peaks. The high energy
peak of the bands nearly disappears if the corresponding band is
less than 30% concentration. In the case of the alloy, where the
major component has an energy band which is lower than that of
the minor component, the alloy band is split into two sections.
Application of this model calculation to a more realistic

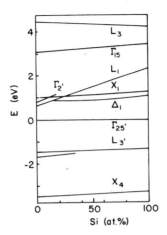

FIG. 1-77. Band structure of Ge-Si alloys. After Ref. 67.

model for the copper-nickel system shows again that the position
of peaks will not change much with alloying.

The energy band structure of a semiconductor alloy can be obtained more easily than the energy band structure of metallic conductors, since there is no "sea" of conduction electrons where the
energy and the spacial distribution may be affected markedly by
changes in the potential of ions. In semiconductors like germanium and silicon, where the electronic bonding is predominately covalent, the energy-wave number relationship changes gradually from one
pure element to the other in a series of solid solutions. Figure
1-77 shows the change in the energy levels as a function of silicon
concentration.

An example of a virtual crystal calculation for this system of
isoelectric elements has been given by Stuckel [67]. He calculated
the energy band structure of the disordered $Ge_{.50}Si_{.50}$ alloy. He
used the virtual crystal approximation. In other words, he used the
average potential of germanium and silicon for his calculation.
Not surprisingly, the calculated energy band form of his Ge-Si
alloy is very similar to that of the two elements. There is hardly

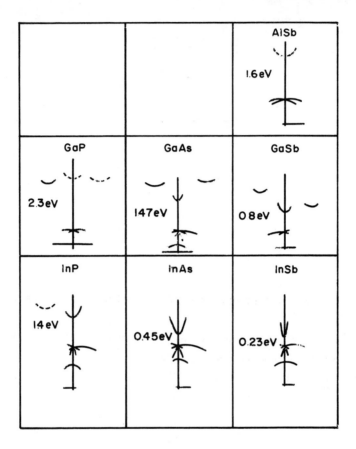

FIG. 1-78. Qualitative survey of the band structure of III-V compounds. Numbers give energy gap estimates. After Ref. 52.

a difference in the self consistent OPW calculations. However, the author finds in his calculation that energy differences at symmetry points for the 50-50 alloy differ from a value obtained by a linear interpolation of the pure germanium and silicon values.

Even the combination of a partially covalent, partially ionic bonding does not lead to major changes in the energy band structure of semiconductors. Figure 1-78 gives the hypothetical band structure of III-V compounds which possess such a bonding mixture. The energy gaps of these structures are given in Fig. 1-79.

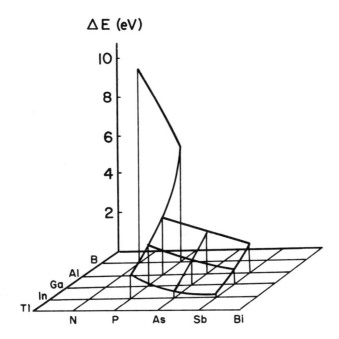

FIG. 1-79. Energy gaps of III-V compounds. After Ref. 52.

Semimetals are characterized by the very small overlap of
nearly filled and nearly empty energy bands near the Fermi energy
level. They are similar to semiconductors where one has a small
energy gap between the highest filled energy band, the valency
band, and the lowest empty conduction band. The transition from
the semimetal to the semiconductor band structure can be continuous,
as in the case of the solid solution series of bismuth antimony
alloys. Some alloys in the solid solution binary composition
range alloyed with $0.07 < c_{Sb} < 0.22$ are semiconductors.

Figure 1-80 shows schematically the energy band structure of
these elements and their alloys. Conduction band minima of bismuth center at the L symmetry points of the Brillouin zone. The
Brillouin zone of this element, which has a rhombohedral structure,
is given in Fig. 1-70. The conduction bands consist of three equivalent pockets, which have nearly ellipsoidal non-parabolic surfaces.

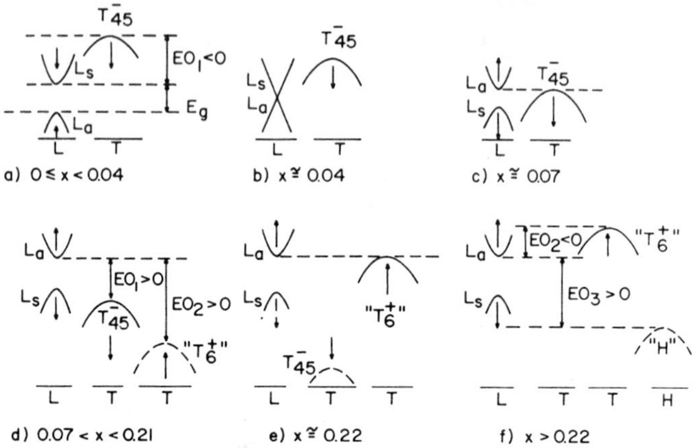

FIG. 1-80. Energy band model of $Bi_{1-x}Sb_x$ alloys. The arrows give direction of shift with increasing Sb-concentration. After Ref. 69, with permission.

The electron states associated with these surfaces are highly anisotropic. The energy gap between the valency band and conduction band at L is rather small. The valency band at T reaches higher up than the bottom of the L-conduction band. Therefore, bismuth is a semimetal.

The energy band structure of $Bi_{1-c}Sb_c$ bands as shown in Fig. 1-80 has been determined from a combination of experiments and calculations. The essential parts of the bismuth energy band for transport properties is one hole band at L_a, one hole band at T_{45}^- and one electron band at L_s. The bands at L are inverted in antimony. The L_a band contains electron pockets and L_s is associated with holes. The band at T_{45}^- is now moved far below the Fermi energy. Instead, a band at H_6^+ is nearly filled. These two bands at T_{45}^- in bismuth and the band at H_6^+ are responsible for the semimetal character of the two elements, since their tops are slightly above the Fermi energy level. This means that the nearly filled and nearly empty bands of these two elements overlap. Since alloying decreases the position of the top of these two bands, it is

possible that the alloys may become semiconductors with a finite
gap between filled and empty states. Experiments on this alloy system can be explained if one assumes that the inversion of the two
bands at L occurs for the alloy with about 4 at.% Sb. The second
semimetal-semiconductor transition seems to occur at 22 at.% Sb.

The proposed energy band structure for pure bismuth and for
alloys with 5, 10 and 22 at.% Sb shows the inversion point at 8 to
9 at.% Sb close to the semimetal-semiconductor transition. Beyond
22 at.% Sb, the pure element Sb has been studied in detail. The
Fermi surfaces of Si-Bi alloys have been investigated in the composition range from pure bismuth to the alloy with c = 0.12. Electrons and holes at L retain their highly prolate ellipsoidal-like
shape, whereas the holes at T are essentially ellipsoidal.

Group V elements, such as bismuth, antimony and arsenic, are typical examples of semimetals. They are characterized by a very small
overlap of the last nearly filled valency band and the first nearly
empty conduction band. Such small energy gaps influence strongly
the electronic properties of these elements and lead to important
similarities between group V semimetals and group IV semiconductors,
since in both cases the energy gaps are of the order of the thermal
energy of electrons at room temperature or smaller. One expects
therefore in both cases strong coupling between electrons or holes
to phonons. Both semiconductors and semimetals are characterized
by low carrier concentrations, high carrier mobilities and highly
anisotropic non-parabolic energy momentum correlations.

The calculation of energy states of alloys, discussed in the
previous section, should be complemented with measurements of their
Fermi surface. Unfortunately, the techniques developed for the
measurement of Fermi surfaces are best applied to systems of perfect periodicity, or in other words to super pure elements at very
low temperatures, where even deviations from the periodicity due to
thermal oscillations are small. Frequently used for dilute alloy
studies is the deHaas-vanAlphen effect, in which conditions on
purity are not as critical as with most other techniques. In this

type of measurement, the sample is subjected to a magnetic field.
This leads to a Lorentz force on the electrons. It will not change
the absolute value of the velocity of the electron, only its direction since the Lorentz force acts perpendicular to both field and
electron velocity direction. The wave vector of the electron k =
mv/$\hbar$ rotates along contours of constant energy. This allows the
measurement of the cross sectional area surrounded by the moving
k-vector on the Fermi surface.

There exist some new methods for Fermi surface determination,
which are not affected by lattice defects and should, therefore, be
very useful in the study of non-dilute alloys. One is the "positron annihilation" technique. If a positron, which has the same
mass as an electron but the opposite charge, is injected into a
metal (either by radiation of a reactor, or, as in the case of copper, by radioactive decay of copper isotopes), it will usually
reach thermal equilibrium and then interact with an electron. Both
combine, annihilate each other, and the energy released in this
process leads to the creation of two photons with the same energy.
The photons would leave the sample in opposite directions if the
electron has zero momentum (neglecting the small momentum of the
electron and positron due to their thermal kinetic energy). However, if the electron has a momentum then the two photons are not
emitted in opposite direction. If $[\pi - \alpha]$ is the angle between the
two photon flight directions, then $mc\alpha = \hbar k$ would be the momentum
of the electron and positron in the direction perpendicular to the
photon direction before the interaction. It is possible to determine the momentum of the electron before annihilation since the
momentum of the positron can be estimated. One would expect that
this technique should be able to give the Fermi surfaces of alloys,
since no requirements on the mean free path of electrons, or on the
periodicity of the lattice have to be given. Unfortunately, correlation effects between electrons and the effect of core electrons
etc. are presently difficult to separate out, and quite a bit
more work is required to use this technique in the study of alloys.

Another method, which is based on the Kohn-effect, shows potential for the investigations of alloys. Here one uses the interaction of the electron gas with elastic waves to determine the Fermi surface. Elastic waves distort the lattice. Ions are periodically displaced from their equilibrium position. Conduction electrons rearrange themselves to these fluctuations of the ion distribution, since the electrons screen the ion. One can show that this changes the screening parameter $\lambda = 1/r_{screen}$ which is now a function of the wavelength of the elastic wave. One can show that the screening parameter has a logarithmic term which is singular at $\vec{q} = 2k_F$ where $\vec{q}$ is the wave vector of the elastic wave, and $\vec{k}_F$ is the electron wave vector at the Fermi surface. The Kohn effect has been used for measurements of the Fermi surface of pure elements. In principle it should be possible to apply it to the study of non-dilute alloys.

Parameters which characterize the Fermi surface have also been obtained from Faraday rotation measurements. They make it possible to determine the neck radius not only of noble metal Fermi surfaces at the contact with the 1st Brillouin zone, but also of their alloys [69,70].

Most of the alloy studies of the Fermi surface of alloys have been made to date only with the deHaas-vanAlphen effect, in spite of the fact that this technique gives only extreme cross sections of Fermi surfaces and alloys with up to about one percent impurities can be investigated. This leads in metals with simple Fermi surfaces, as in copper base copper zinc alloys, only to small changes in cross sectional areas. One should expect more noticeable effects in metals with a more complex electronic structure such as lead, aluminum, or zinc.

Anderson and Hines [71] measured the deHaas-vanAlphen effect in dilute alloys of bismuth in lead. They determined the oscillation in a few symmetry directions like [1 1 1] and [1 0 0] and calculated the shift of the Fermi energy level from the change in the dH-vA frequencies. They assumed in the interpretation of the

data that the alloys form substitutional solid solutions. They
assumed further that the alloy is disordered, and that the lattice
parameter will not change with alloying. Then they calculated the
shift of the Fermi energy level assuming a rigid band. One considers a metal with n conduction electrons per atom, and replaces a
fraction of these atoms, c, with impurity atoms which have n + z
electrons per atom. This gives an average electron per atom ratio
of n + cZ. Expanding the energy of the metal near the Fermi level
in a Taylor series, they obtained

$$\Delta E = [\Delta N/N(E_F)] - [(\Delta N)^2/2N(E_F)^3][\partial N(E_F)/\partial E] \qquad (1-70)$$

where $N(E_F)$ is the density of states at the Fermi level, and $\Delta N = cz$.
The comparison of this calculation with the dH-vA measurements requires that one knows the effective mass of the electrons, and the
curvature of the Fermi surface which affects the density of states.
Since both effects are correlated, enhancement effects will not
strongly affect the calculation. Much more important may be the
effect of screening of extra electrons around impurities. Mott [72]
pointed out that the Fermi energy level a few lattice parameters
away from the impurity should have the potential of the pure element.
Friedel [59] suggested that, even with no shift in the Fermi
energy level of the matrix, all energy levels are shifted so that
the distance from the bottom of the band to the Fermi level is moved
by the same amount as in the rigid band model. Brailsworth [73]
showed that this is valid even if the energy level shift is a function of energy.

Figure 1-81 shows the calculated shift of the dH-vA oscillation
with impurity concentration and experimental results. It shows that
the rigid band model is a good first approximation to calculate
energy shifts in these alloys, if impurity concentrations are not
too high.

A similar investigation was executed by Shepherd and Gordon
[74] on aluminum base alloys with zinc, silicon, germanium, magnesium

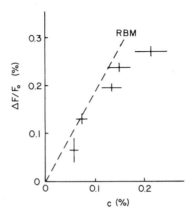

FIG. 1-81. Relative shift of de Haas and van Alphen frequency $\Delta F/F_0$ as a function of Bi-concentration. Crosses give experimental results. The dashed line represents the Rigid Band Model calculation. After Ref. 72.

and silver impurities. The results were again interpreted with the rigid band model, and good agreement between measurements and calculations were obtained.

A more detailed study of the de Haas – van Alphen effect in zinc alloys showed [75] that the rigid band model can only partly explain the effect of alloying in some segments of the Fermi surface. Zinc has a rather complex Fermi surface. Some segments are quite small. Alloying will have a large effect on the size and shape of these segments. This change in shape is due to the fact that zinc is an hcp crystal, where the c/a ratio frequently changes with alloying. This complicates the calculation if one uses the nearly free electron model, and it is necessary to devise a special geometrical formula for each segment of the Fermi surface and its change with alloying. In no case can the sign of the observed change on the periodicity of the de Haas – van Alphen frequency be related to the electron or hole character of the orbit. However, it is possible to obtain with such an approach the area

change of the needle-like sections of the Fermi surface and compare
it with the "nearly free electron" calculations by Harrison. Again,
agreement between de Haas – van Alphen measurements and calculations
based on the nearly free electron model is good. The main success
of this approach is that it demonstrates the importance of the c/a
ratio for the interpretation of the data. This explains why the
addition of copper, which is supposed to subtract electrons from
the alloy since copper has one electron less than zinc, leads to an
increase in the size of the electron orbit rather than to a decrease. The change in the c/a ratio in alloying with aluminum leads
to a change in the needle orbit, which is compensated by the change
in the increase in the electron concentration due to the addition
of an extra conduction electron with each aluminum atom. It is not
possible to interprete all the features of the measured parameters
of the Fermi surface with results from the nearly free electron
model. This is partly due to the fact that one does not know the
energy gaps between bands. The assumption of zero energy gap models
underestimates the change in the needle area by a factor of two.
The model also underestimates the magnitude of the increase of the
area near P_2 by a factor of ten, if one assumes that the energy gap
is zero, or it underestimates it by a factor of three if one raises
k_F by 3.5%. This 3.5% change would shrink the Fermi area section
of pure zinc in such a way that one obtains agreement between theory and dH–vA measurements for the pure element.

Chapter 2

MAGNETISM

I. INTRODUCTION

The effect of the earth's magnetic field on loadstones was used
more than 3000 years ago by the Chinese, and it was mentioned in
Greek writings around 600 B.C. Gilbert, appointed physician to Queen
Elizabeth I of England, prepared a survey on magnetism. The mechanical forces between magnetic poles were investigated quantitatively
during the 18th and 19th century, culminating in the Coulomb law.
This law is now frequently used as an introduction to magnetostatics,
but it is really the end product of long years of studies. Maxwell's
equations show the similarity of magnetic and electric phenomena.
The main difference between these two phenomena is that there exists
no magnetic equivalent to an individual electric charge in the Maxwell equations. Magnetic poles have always been found in pairs with
opposite signs. The search for magnetic monopoles is still going on.

Magnetic properties of materials were measured in static experiments in the 19th century. An understanding of these properties was
only possible with the discovery of the electron and the measurement
of its electric charge, spin, and magnetic moment. The investigation
of the population of different energy levels and relaxation times
became of increasing interest only in this century. Progress in this
field was helped very much by advances in the construction of high
frequency electronic equipment.

The magnetic properties of materials can be defined by forces
in magnetic fields, or by electric fields induced by changes of magnetization with time. A long rod, magnetized by applied fields or
spontaneously parallel to its long axis, is characterized by its pole
strength p and its geometric dimensions. It has poles of opposite
signs at each end. The mechanical forces $\vec{f}$ between two different
poles, p_1 and p_2, are given by the Coulomb equation:

$$\vec{f} = (1/4\pi\mu\mu_o)(p_1p_2/r^2)(\vec{r}/r) \ , \tag{2-1}$$

where μ is the relative permeability (equal to one in vacuum) and
μ_o is the magnetic permeability in free space.

Two major measuring systems are presently used for magnetic
units. The context makes it clear which system is used. One is the
Gaussian or cgs system, the other the rationalized MKS or SI system.
In the Gaussian system, μ_o is set equal to $1/4\pi$; in the SI system
$\mu_o = 4\pi \cdot 10^{-7}$ Henry/meter. The distance r between poles in the
Coulomb equation in the cgs system is given in centimeters, and forces
in dynes. Therefore, only p in Eq. (2-1) is unknown if poles of
equal strength interact. This equation can be used to define the
pole strength.

One can rewrite Eq. (2-1) in the following way in the cgs system:

$$\vec{f} = (p_1p_2/r^2)(\vec{r}/r) = p_1\vec{H} \qquad \text{(cgs)}, \tag{2-2}$$

where $\vec{H}$ is the magnetic field. $\vec{H}$ is defined as the force on a
unit pole in a magnetic field. Its unit is the Oersted (Oe).

A rod with a + pole on one end and a   - pole at the other end
is a magnetic dipole with the dipole moment $p\vec{\ell}$, where ℓ is the
length of the rod. The dipole moment per unit volume is:

$$p\vec{\ell}/V = \vec{M} \qquad \text{(cgs)} \tag{2-3}$$

the magnetization of the sample, $\vec{M}$. The pole strength per unit cross sectional area of the sample, p/A, is equal to M. M is given in Gauss (G) in the Gaussian system. Since the magnetization of a volume element in vacuum is equal to H, one frequently finds magnetic fields given in Gauss instead of Oersted.

The Gaussian system is now in the process of being replaced by the SI system, in which one can define the magnetization polarization $J = \mu_o M$, or, more correctly, the magnetic polarization change with time $\partial J / \partial t = \mu_o \partial M / \partial t$ in the following experiment (see Fig. 2-1a). A wire of length s surrounds a homogeneous sample with cross section A. The voltage induced in this wire, V, is:

$$V = \int F ds = \mu_o \int (\partial M / \partial t) dA \qquad \text{(SI)}. \qquad (2\text{-}4)$$

The magnetic field in the center of an infinitely long solenoid is defined by (2-1b):

$$H = (n/\ell) I \qquad \text{(SI)} \qquad (2\text{-}5)$$

with I the current in the wire, and n/ℓ the number of turns per unit length of the solenoid. The magnetic field in the SI system is given in ampere-turns/meter (A/m), the magnetic polarization in tesla, weber/meter^2 or volt-second/meter^2 ($T = Wb/m^2 = Vs/m^2$), and the magnetization M in ampere-turns/meter (A/m).

The magnetization of solids which show no spontaneous magnetization is for small fields a linear function of the applied magnetic field H:

$$M = \chi H \qquad \text{(cgs)}$$
$$M = \kappa H \qquad \text{(SI)} \qquad (2\text{-}6)$$

The proportionality constant χ is the susceptibility of the sample. κ is the rationalized volume susceptibility. One also uses the susceptibility per weight ($\chi_{gram} = \chi/\rho$, ρ is the density), or the susceptibility per atom or per mole ($\chi_{mole} = \chi A/\rho$; A is the atomic weight).

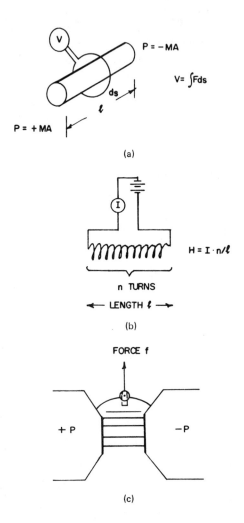

(a)

(b)

(c)

FIG. 2-1. Experimental arrangement to determine the correlation between the magnetization M of a sample with cross sectional area A and the the voltage integral ∫Vdt induced in a wire with length elements ds (a). The voltage is given by V = ∫Fds, with F the electric field in the wire. Correlation between current I and magnetic field of a solenoid of length ℓ with n turns (b). Force f acting on a small sample in an inhomogeneous field (c).

Materials with positive χ values are "paramagnetic", and those with negative χ values are "diamagnetic". One frequently measures susceptibilities with a "force method". For small samples the Faraday method is used (2-1c). The sample is exposed to an inhomogeneous magnetic field. The pole strength of the sample is $\pm$ MA. The force on both poles (1) and (2) is:

$$f = H(2)M(2)A + H(1)M(1)A = \{H(2) - H(1)\}M(1)A \qquad (cgs). \qquad (2-7)$$

Developing H in a Taylor series and neglecting higher terms gives (if ℓ is the distance between both poles):

$$f = \{H(1) + (\partial H/\partial x)\ell - H(1)\}M(1)A = H\chi(\partial H/\partial x)V \qquad (cgs) \qquad (2-8)$$

since $\ell A = V$ is the volume of the sample. Therefore:

$$f/V = \chi H(\partial H/\partial x) = \chi(\partial H^2/\partial x)/2 \qquad (cgs). \qquad (2-9)$$

The magnetization of solids is essentially due to the orbital movements of electrons around ions and the dipole moments associated with each electron spinning around its own axis. The orbiting electron corresponds to an electric current in a loop. Such a current is responsible for a magnetic dipole moment. Therefore, the orbiting electron produces a magnetic dipole moment. Since the angular momentum of the orbiting electron is quantized, the magnetic moment of the electron is quantized. One obtains for an electron orbiting the nucleus of a hydrogen atom on its lowest energy level the dipole moment:

$$\mu_B = e\hbar/2mc \qquad (cgs)$$
$$\mu_B = e\hbar/2m \qquad (SI) \qquad\qquad (2-10)$$

This dipole moment is defined as the "Bohr magneton".

Aside from the angular momentum due to the orbital movement
around the nucleus, each electron spins around its own axis. Its
angular momentum is $\hbar/2$ and its spin quantum number m_s can have
values of $+ 1/2$ and $- 1/2$. This leads to a magnetic dipole moment
equal to:

$$\mu_s = ge\hbar/4mc \qquad \text{(cgs)}$$
$$\mu_s = ge\hbar/4m \qquad \text{(SI)} \tag{2-11}$$

where g is the "gyromagnetic factor". g is equal to 2.002 for the
free electron. In many cases the difference between μ_s and the Bohr
magneton μ_B can be neglected. Therefore, we will in the following
sections frequently use μ_B instead of the more correct μ_s value.
It is sometimes convenient to describe the magnetization of a material with the "number of Bohr magnetons per atom", μ_B/atom. This
usage avoids the problem with magnetic units and makes it easy to
visualize properties on an atomic basis.

The old quantum theory predicted that the angular momentum of
an orbiting electron should be a multiple of $\hbar$. The correct value
is equal to $\{\ell(\ell + 1)\}^{1/2}\hbar$, where ℓ is the angular quantum number
of the electron. It can have values from zero to $n - 1$, with n
the principle quantum number. Its magnetic dipole moment is equal to:

$$\mu = (|e|/2mc)\{\ell(\ell + 1)\}^{1/2}\hbar . \tag{2-12}$$

The projection of ℓ in the direction of the magnetic field (sometimes another specified direction has to be used) has to be an integer, m, the magnetic quantum number, with $-\ell \leq m \leq +\ell$. The interaction of electrons in an atom with each other and with an applied
magnetic field are described with the Russel-Saunders coupling model
for weak fields, and with the Paschen-Back effect for strong fields.
In weak fields one adds the angular momentum vectors of the orbiting
electrons and obtains the resultant vector $\vec{L\hbar}$, and then separately
adds the spin angular momentum vectors S$\vec{\hbar}$. The vector

$(\vec{L} + \vec{S})\hbar$ will then precess slowly around the magnetic field axis
H. This is shown in Fig. 2-2a. The vector in the direction of the
magnetic moment $(\vec{L} + 2\vec{S})\hbar$ will then precess rapidly around the
$\vec{J}\hbar$ vector. The angular momentum for electrons in the Russel-
Saunders coupling is given by:

$$|G| = \{J(J + 1)\}^{1/2}\hbar \qquad (2\text{-}13)$$

with $\vec{J} = \vec{L} + \vec{S}$ the total quantum number of the electrons of an
atom, $J = L - S, L - S + 1, .. L + S$. The gyromagnetic factor is:

$$g = 1 + \{J(J + 1) + S(S + 1) - L(L + 1)\}/2J(J + 1). \qquad (2\text{-}14)$$

The energy of this electron system in a magnetic field is then given
by:

$$E = Mg\mu_B H \qquad (2\text{-}15)$$

where M is the projection of J into the direction of the applied
field. Coupling between $L\hbar$ and $S\hbar$ can be neglected in the limiting case of a strong field (Paschen–Back effect). Each of these two
vectors moves in a precession independently of the other around the
axis of the applied field. The projection of each vector is again
quantized on the direction of the magnetic field. One obtains as
energy of the electrons in the Paschen–Back effect (see Ref. 76):

$$E = h\nu_{Larmor}(L + 2S). \qquad (2\text{-}16)$$

In some materials, electrons associated with different atoms
will interact with each other and their magnetic spins will line up
in preferred orientations. These materials show spontaneous magnetic
ordering. A typical example is iron where on the average at $0°K$
$2.22 \ \mu_B$/atom will line up in the $< 1\ 0\ 0 >$ direction. An iron
single crystal may, however, show no macroscopic noticeable ordering,

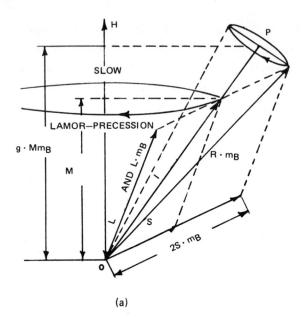

(a)

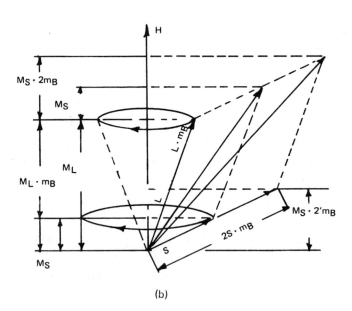

(b)

FIG. 2-2. Coupling of electrons in a small field (a, Russel-
Sounders coupling) and in a large field (b, Paschen-Back effect).

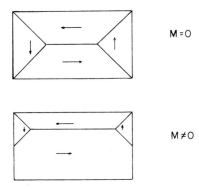

FIG. 2-3. Magnetic domains in a single crystal

because the single crystal consists of a large number of magnetic
domains with parallel aligned spins in each domain, but with different magnetization directions in different domains. The domains are
separated by "Bloch walls" 100 to 1000 A thick. The magnetization
direction changes gradually in these Bloch walls. Each domain will
be magnetized in a < 1 0 0 > direction, the direction of "easy
magnetization". The sum of all magnetic moments may have any arbitrary value between zero and the "saturation magnetization" M_s, as
indicated by some domain configurations in Fig. 2-3. The macroscopically measured magnetization M depends not only on the applied
field but also on the history of the magnetization process.

Applying a field of a few Oersted to an iron single crystal in
the < 1 0 0 > direction would lead firstly, for small fields, to a
line-up of the magnetization $\vec{M}$ into the < 1 0 0 > direction.
The projection of the saturation magnetization $\vec{M}_s$ from this
< 1 0 0 > direction in the < 1 1 0 > will give for low fields the
measured magnetization $M = M_s/\sqrt{2}$. No macroscopically noticeable
magnetization component should be found perpendicular to the applied
field, since the components of the three possible directions for
easy magnetization will cancel each other. The sample will reach
saturation parallel to < 1 1 0 > only if high enough magnetic fields
will rotate the dipole orientation from the easy direction into the

direction of the applied field. Similar arguments can explain the
magnetization curve of nickel and cobalt single crystals (2-4).

The magnetization curve of a polycrystalline sample is shown
schematically in Fig. 2-4d, where an iron sample (initially with
M = 0) is subjected first to a plus field, then a minus field, and
then again to a plus field. The M versus H curve shows a "hysteresis loop". Its area is equal to the energy absorbed during one
magnetization cycle, since:

$$E_{mag.} = V\int HdM \qquad \text{(cgs)}$$
$$E_{mag.} = V\int HdJ \qquad \text{(SI)}. \tag{2-17}$$

The magnetization process is "irreversible" in section (b) of this
curve (2-4d), where magnetization changes are due to jumps of Bloch
walls. Reversible bulging of Bloch walls in section (a), and rotation of spins away from the direction of easy magnetization into the
direction of the applied field in section (c), are responsible for
"reversible" magnetization changes in the sample.

Since frequently a combination of M and H is measured (the
voltage measured in the experiment given in Fig. 2-1a will be due to
changes in M and to the field H which produces the changes in M),
one characterizes the magnetization process of materials by the
"induction":

$$B = H + 4\pi M \qquad \text{(cgs)}$$
$$B = \mu_o(H + M) \qquad \text{(SI)}. \tag{2-18a}$$

The permeability μ is defined as:

$$\mu = B/H \qquad \text{(cgs or SI)}. \tag{2-18b}$$

Since the susceptibility or rationalized susceptibility is defined
as M/H, one obtains:

$$\chi = M/H = (\mu - 1)/4\pi \qquad \text{(cgs)}$$

$$\kappa = M/H = \mu - 1 \qquad \text{(SI)}.$$

(2-18c)

μ depends in the same way as M on the sample history (2-4d).
μ^0 is the permeability measured at the origin of the "virgin"
magnetization curve (magnetization curve which starts at the origin).
$(\Delta B/\Delta H)_{H_i} = \mu_\Delta(H_i)$ is the "incremental" permeability. For its measurement, H oscillates by ΔH around H_i. The field required to
reduce the magnetization from M_s to zero is the "coercive force"
or "coercive field" H_c (H_c is not really a force. The name is
historical.). Changing the field to zero after reaching first
saturation leads to a sample with "remanent magnetism". The magnetization at $H = 0$ is the "remanence magnetization" $4\pi M_r = B_r$.

II. PARA- AND DIAMAGNETISM

A. Paramagnetism of Localized Electrons

Electrons in an ion will interact with an applied field H
through their dipole moment μ_B. They have a potential energy of:

$$U_H = -m_j g \mu_B H$$

(2-19)

where m_j is the magnetic quantum number. The magnetization of N
ions in a unit volume is:

$$M = N\{\Sigma_j m_j \mu_B g \cdot \exp(m_j \mu_B gH/k_B T)\}/\{\Sigma_j \exp(m_j \mu_B gH/k_B T)\}.$$

(2-20a)

$m_j \mu_B$ is of the order of a few Bohr magnetons, H is of the order
of 10^4 Oe, and T may be 10^2 to 10^{4o}K in typical laboratory sit-
uations. This means that $m_j \mu_B H \ll k_B T$. Therefore one obtains from

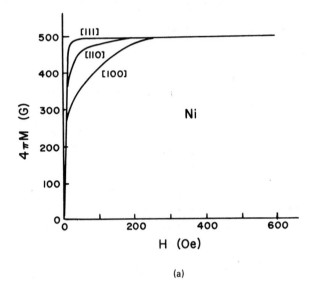

(a)

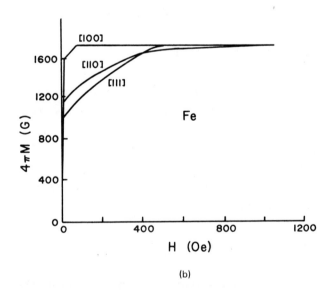

(b)

FIG. 2-4a + b. Magnetization curve of a nickel (a) and an iron
single crystal (b). After Refs. 77 and 78.

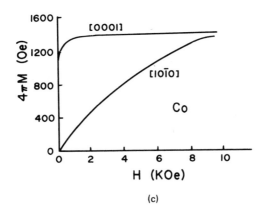

(c)

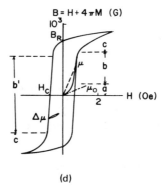

(d)

FIG. 2-4c + d. Magnetization curve of a cobalt single crystal
(c. After Refs. 77 and 78). Magnetization curve of polycrystalline
iron (d). The permeability μ is not a materials constant. It depends on the magnetization process. μ^0 is the initial permeability
of a demagnetized sample. Δμ is the   incremental susceptibility.
Sections a and c are associated with reversible magnetization
changes, b and b' are associated with irreversible magnetization
changes. B_R is the remanence , H_c the coercive field (schematic
diagram).

Eq. 2-20 as first approximation:

$$M = N\mu_B g\{\Sigma m_j (1 + m_j g\mu_B H/k_B T)\}/\{\Sigma(1 + m_j g\mu_B/k_B T)\}. \qquad (2\text{-}20b)$$

Terms of the form $\Sigma_{-j}^{+j} m_j$ are equal to zero. Further, $\Sigma_{-j}^{+j} j = 2j + 1$; $j\Sigma_{-j}^{+j} m_j = j(j + 1)(2j + 1)/3$. This gives for the susceptibility:

$$\chi = Ng^2 j(j + 1)\mu_B^2/3k_B T = Np_{eff.}^2 \mu_B^2/3k_B T \qquad (2\text{-}21a)$$

with

$$P_{eff.} = g(j\{j + 1\})^{1/2} \qquad (2\text{-}21b)$$

as the effective number of Bohr magnetons per ion. Equation 2-21 shows that the paramagnetic susceptibility of ions should be inversely proportional to the absolute temperature. This is the Curie law:

$$\chi = C/T \qquad (2\text{-}22)$$

which gives good results for oxygen molecules in the gaseous state, and for some salts. A rigorous quantum mechanical calculation yields an additional temperature independent term, which in most cases can be neglected.

A more general solution to Eq. 2-20 can be obtained in the following way. One sets $g\mu_B H/k_B T = x$. This gives:

$$M = N\Sigma m_j \mu_B g \cdot \exp(m_j x)/\Sigma\exp(m_j x). \qquad (2\text{-}23)$$

The summation is over m_j. Keeping in mind that $\Sigma\exp(m_j x)$ is a geometric progression, and that $\Sigma m_j \exp(m_j x) = (\partial\{\Sigma\exp(m_j x)\}/\partial x)$, one obtains:

$$M = Ng\mu_B j\{(1 + 1/2j)\coth(1 + 1/2j)y - (1/2j)\coth(y/2j)\}. \qquad (2\text{-}24a)$$

This is identical to:

$$M = Ng\mu_B j B_j(y) \qquad\qquad (2\text{-}24b)$$

where $B_j(y)$ is the Brillouin function, and $y = jg\mu_B H/k_B T$. For
$j \to \infty$, $B_j(y)$ changes to the Langevin function $L(y) = \coth(y) - 1/y$.
Agreement between experimental results and calculated M values on
salts is excellent (2-5).

Magnetic ions are well separated in these salts, so that the
interaction between different magnetic dipoles can be neglected.
Only the applied field is important for the magnetization of localized electrons with magnetic moments. The interaction of spins,
which can lead to spontaneous magnetic ordering, can be important
even in the paramagnetic state. The interaction may be represented
by an effective field, the molecular or Weiss field H_m, which is
assumed to be proportional to the magnetization of the sample:

$$H_m = W_m M \quad . \qquad\qquad (2\text{-}25a)$$

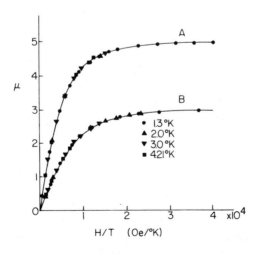

FIG. 2-5. Magnetic moment in Bohr magnetons/ions. A corresponds
to ferric ammonium alum, B to gadolinium sulfate octahydrate.
Lines follow Eq. 2-24b. After Ref. 81, with permission.

W_m is the "molecular field constant". The field acting on a dipole
is then:

$$H = H_{appl.} + H_m = H_{appl.} + W_m M. \qquad (2\text{-}25b)$$

Together with Eq. 2-22, one obtains:

$$\chi = C/T = M/H = M/(H_{appl.} + H_m) = M/(H_{appl.} + W_m M). \qquad (2\text{-}26a)$$

Solving this equation for M gives:

$$M(W_m C/T - 1) = - H_{appl.} C/T$$
$$\chi = M/H_{appl.} = -C/(W_m C - T) = C/(T - T_c') \qquad (2\text{-}26b)$$

where T_c' represents a "critical temperature". Equation 2-26b is the
Curie-Weiss law. $T_c' < 0°K$ is associated with antiparallel coupling
between dipoles, and $T_c' > 0°K$ is associated with parallel coupling.
A plot of $1/\chi$ versus T gives in both cases a linear function (2-6).

B. Diamagnetism of Localized Electrons

The diamagnetic susceptibility of atoms is due to orbiting
electrons which circle the nucleus. The susceptibility of the orbiting electrons of one atom is obtained by adding the contributions
of all electrons of this atom, and the resulting diamagnetism of all
atoms per unit volume is:

$$\chi = - N(e^2/6mc^2) \sum_i \overline{r_i^2} \qquad (cgs). \qquad (2\text{-}27)$$

The sum is formed over all electrons of an atom, i, and r_i is the
radius of the orbiting electron. Since r_i is of the order of 10^{-8}
cm, χ has values of the order of 10^{-6} in cgs units. r_i, and with
it χ, is essentially independent of temperature, in good agreement

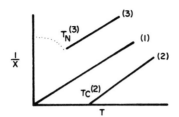

FIG. 2-6. Susceptibility following the Curie law (1), the Curie-Weiss law with positive critical temperature (2), or the Curie-Weiss law with negative critical temperature (3).

with experiments. However, it is not possible to obtain more than qualitative agreement between this equation and most experimental data, since van Vlack showed in a quantum mechanical calculation that other terms have to be added which cannot be accurately calculated. He showed [80]:

$$\chi = -N(e^2/6mc^2)\Sigma \overline{r_i^2}$$
$$+ (2/3)N(e/2mc)^2 \Sigma_{n'}\{\int \psi_n^*, G\psi dx_1 dx_2 dx_3/(E_{n'} - E_n)\}. \qquad (2\text{-}28)$$

In this equation, n' represents the excited state, n the ground state, and G is the angular momentum operator.

C. Paramagnetism of the Free Electron Gas

The previous section describes models of localized electrons and their states due to applied and molecular fields. This section deals with energy states of a degenerate electron gas where one treats energy states as a continuum (energy bands), and where electrons move around freely. Each of the possible energy states for electrons in a metal may be occupied by one electron with spin up or down. Without a magnetic field, both states have the same energy. Bands with spin up or down are equally occupied, and the sample shows no magnetic moment. An outside magnetic field will change the electron distribution as indicated in Fig. 2-7a. This figure gives a plot of the density of states as a function of energy. The energy band is separated into two subsections with spin up or down. Applying a magnetic field in the up-direction will increase the potential energy of the electrons by a magnetic contribution:

$$U_m = \mu_B H. \tag{2-29}$$

For the case of zero temperature, this magnetic potential energy will turn electrons with spin down in the energy range $E_F - \mu_B H$ to E_F around into the spin-up orientation. They will then occupy electron states in the spin-up subband. This means that there are more electrons with the spin parallel to the applied field than opposite to it. The sample is magnetic. The magnetization per unit volume is equal to the excess number of electrons with spin up. This gives for M:

$$M = \mu_B N(E_F) \cdot \mu_B H \tag{2-30a}$$

or, as the susceptibility per unit volume:

$$\chi = M/H = \mu_B^2 N(E_F). \tag{2-30b}$$

Interaction energies $E_{int.}$ between electrons may be taken into account with $E_{int.} = \mu_B H_{int.}$, and if one replaces H by $H_{appl} + H_{int.}$. This gives with $\chi = M/H_{appl.}$:

$$\chi = \mu_B^2 N(E_F)/(1 - I \cdot N(E_F)) \tag{2-30c}$$

where $I = E_{int.} = \mu_B H_{int.}$ is the interaction energy per density of states. For finite temperatures, one replaces the step function for the electron distribution by the Fermi-Dirac function, $f^o(E)$, which gives the probability that an electron state with energy E will be occupied at temperature T (see Chapter III for more details). This

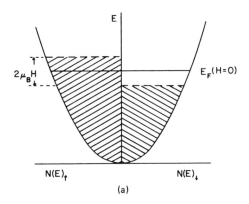

(a)

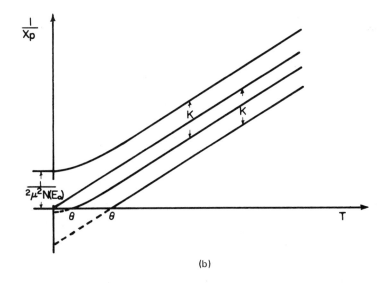

(b)

FIG. 2-7.   Pauli paramagnetism at zero absolute temperature (a).
States are filled up to the Fermi energy level E_F at zero field.
Shaded area gives occupied electron states in a magnetic field.
Temperature dependence of the reciprocal Pauli susceptibility of the
free electron gas without exchange energy (b, top line), the reciprocal Curie susceptibility of localized electrons (b, second line from
top), the reciprocal Pauli susceptibility of the electron gas with
exchange energy (b, third line from top), and the reciprocal Curie-
Weiss susceptibility of localized electrons with exchange energy
(b, bottom line).   θ is the critical temperature for the onset of
ferromagnetism. This is the Curie temperature. After Ref. 83.

gives for M (neglecting exchange energies):

$$M = \mu_B \int_0^\infty \{f^0(E - \mu_B H) - f^0(E + \mu_B H)\} N(E) dE/2. \qquad (2\text{-}31)$$

Developing the Fermi-Dirac function $f^0(E) = [\exp(E - \zeta)/k_B T + 1]^{-1}$, where ζ is the Fermi potential ($\zeta = E_F$ at absolute zero temperature), in a Taylor series gives, since $\mu_B H$ is much smaller than E_F:

$$M = -\mu_B^2 \int_0^\infty \{\partial f^0(E)/\partial E\} N(E) dE \cdot H . \qquad (2\text{-}32)$$

Developing now N(E) in a Taylor series gives [82]:

$$M = -\mu_B^2 H[N(E_F) \int_0^\infty \{\partial f^0(E)/\partial E\} dE + N'(E_F) \int_0^\infty (E - E_F)\{\partial f^0(E)/\partial E\} dE$$

$$+ N''(E_F) \int_0^\infty (E - E_F)^2 \{\partial f^0(E)/\partial E\} dE/2 + \ldots] . \qquad (2\text{-}33)$$

N'(E) and N''(E) indicate first and second derivatives. The first integral in this equation is equal to -1. The second integral is zero, since the first derivative of the Fermi-Dirac function is an even function of $E - E_F$. The third integral is equal to $\pi^2 k_B^2 T^2/3$. Including another term of the form $N'(E)^2/N(E)$, but neglecting higher terms in $E - E_F$ gives then the approximation [83]:

$$\chi_P = \mu_B^2 \{N(E_F) + (\pi^2 k_B^2 T^2/6)[N''(E_F) - N'(E_F)^2/N(E)] + \ldots \}.$$
$$(2\text{-}34a)$$

The susceptibility of the electron gas is the "Pauli susceptibility". Figure 2-7b gives a plot of the normalized reciprocal susceptibility as a function of temperature. It shows that χ_P changes only slightly with temperature for $k_B T \ll E_F$ and approaches the Curie-Weiss law at high temperatures. Equation 2-34a gives for the free electron model with one electron per atom susceptibilities of the order of 10^{-6} in the cgs system. This is in the range of experimentally found values for paramagnetic simple metals and alloys. It is in the same range as the diamagnetism of the ion cores.

Equation (2-34a) for the Pauli susceptibility is applicable
only to systems with no interaction between electrons. This equation would not be applicable to all transition elements. Some of
the transition elements show susceptibility values which can be explained only if one assumes some type of interaction between the
electrons. This was first analyzed with the concept of the Weiss
molecular field which resulted in a shift in the $1/\chi$ versus T curve,
as shown in the Curie-Weiss law. Corrections of this type have been
applied to the equation for the Pauli susceptibility by Stoner and
Wohlfahrt [84 and 87]. These calculations lead to a correction to
the reciprocal susceptibility similar to that of the Weiss molecular field to the reciprocal susceptibility of localized electrons.
In both cases, the $1/\chi$ curve is shifted downwards (Fig. 2-6 and
Fig. 2-7b). Since the Pauli susceptibility is finite at $T = 0^{\circ}K$, a
downward shift will not necessarily lead to infinitely large χ
values. In other words, the molecular field below a critical field
for parallel alignment of spins will, in the Stoner model, not lead
to ferromagnetic ordering at any temperature; whereas the Curie-
Weiss law for such an interaction leads always to a critical temperature T_c, where $1/\chi = 0$. This corresponds to the case of ferromagnetic ordering. The Stoner equation for the susceptibility with
electron interaction is [83]:

$$1/\chi_p' = 1/\chi_p + \text{const} \qquad\qquad (2\text{-}34b)$$

χ_p' is the susceptibility with molecular field, χ_p is the susceptibility without a molecular field. For the case of a parabolic
band, one obtains:

$$\chi_B = (3N\mu_B^2/2E_F)[1 - (\pi^2/12)(Tk_B/E_F)^2 \ldots]. \qquad (2\text{-}34c)$$

D. Diamagnetism of the Free Electron Gas

The susceptibility of an electron gas should be zero according to classical physics, since the energy of a moving electron (without magnetic dipoles) will not change if a magnetic field is applied. Only the direction of the momentum changes, since the Lorentz force $\vec{f}_L$ is acting perpendicular to the velocity of the electron:

$$\vec{f}_L = (e/c) \; \vec{v} \times \vec{H} \qquad (cgs)$$

$$\vec{f}_L = e \; \vec{v} \times \vec{B} \qquad (SI).$$

(2-35)

A non-zero value for the susceptibility of free electron is found in quantum mechanical calculations. One obtains for the Hamiltonian in the magnetic field the expression [82]:

$$H = (1/2m)(p - (e/c)T)^2 + V \qquad (cgs) \qquad (2\text{-}36)$$

where T is the vector potential of the electron, with:

$$mv = p + (e/c)T \qquad (cgs). \qquad (2\text{-}37)$$

The vector potential can be set equal to $A_{x_1} = -Hx_2$, $A_{x_2} = A_{x_3} = 0$. This gives for the Schrödinger equation:

$$(\partial\psi/\partial x_1 - (ie/hc)x_2 H)^2 \psi + \partial^2\psi/\partial x_2^2 + \partial^2\psi/\partial x_3^2 + (2mE/\hbar)\psi = 0. \qquad (2\text{-}38)$$

Assuming:

$$\psi = \exp(i[k_1 x_1 + k_3 x_3]) \; \phi(x_2) \qquad (2\text{-}39)$$

gives a differential equation for $\phi(x_2)$ which is:

$$\partial^2\phi/\partial x_2^2 + (2m/\hbar)[E_\perp + 1/2\ m\omega(x_2 - x_{2,0})] = 0 \ . \qquad (2\text{-}40)$$

This is the equation for a linear harmonic oscillator with the potential energy $U = (1/2)\omega^2 x_2^2$, with $\omega = -\ eH/mc$. The energy values
are:

$$E_n = (2n + 1)\mu_B\ H = (n + 1/2)\hbar\omega \qquad (2\text{-}41)$$

$$n = 0, 1, 2. \ . \ .$$

They are sometimes called Landau levels. ω is the "cyclotron resonance frequency", which is twice the Larmor frequency of an electron.
The susceptibility of this system is negative. The Landau magnetism leads to a diamagnetic contribution to the susceptibility which
is equal to $-\ 1/3$ of the Pauli paramagnetism due to the dipole moment of the spinning electron.

E. Magnetism of Nearly Free Electrons

The susceptibility of semimetals and semiconductors is not so
well understood as the susceptibility of most other elements. In
semiconductors and semimetals, the electrons in the valency band are
only loosely bound to the atom. Their diamagnetism cannot be described in the same way as the core diamagnetism of typical metals
or of insulators. They are half way in between the two extremes of
a completely free and a tightly bound state. Peierls tried to take
their magnetic properties into account in a modified form as the
Landau diamagnetism [88 and 89]. This is then referred to as a
Peierls-Landau diamagnetism. Calculations for such systems are, in
principle, straightforward. One has to use a second order perturbation theory. However, the resulting equations are very lengthy and
can be written down in different but equivalent forms. Wannier and

Upadnyaya [90] took into account in their susceptibility calculation the following effect of a magnetic field on the band structure.
Firstly, the magnetic fields lead to a gradual change in wave functions and energy values. Secondly, one obtains a breaking up of
bands into discrete states. This leads to the Landau diamagnetism
in metals. Unfortunately, Peierl's equation (and e.g. improvements
by Wannier and Upadnyaya which lead even to an additional paramagnetic term) are rather complicated. A simple interpretation of experimental data is not possible in view of the fact that the susceptibility of semiconductors and semimetals is strongly temperature
and orientation dependent.

   The magnetic properties of the Bi-Sb alloy system are described
in Section K. Its band structure changes with composition from the
semimetal state to the semiconductor state. It gives therefore a
good example for the effect of the energy band structure on the
Peierls-Landau diamagnetism. The strong structure dependence of
the magnetic properties of these elements is also apparent from the
fact that melting changes the susceptibility drastically. $|\chi|$ de-
creases to less than 1/10 when bismuth changes from the solid to
the liquid state, and $|\chi|$ increases by a factor of 3 during liquifi-
cation of germanium. Such changes are only of the order of a few
percent in silver or gold.

F. Susceptibility of Elements

   Figure 2-8 schematically shows the different paramagnetic and
diamagnetic contributions to the susceptibility of elements. Only
the Langevin paramagnetism is markedly temperature dependent. The
other terms are all of the same order of magnitude. The separation
of these contributions is therefore difficult. Further, susceptibility data given by different authors vary noticeably, indicating
that small amounts of impurities can affect the measurements markedly. Table 2-1 gives a survey of the susceptibility of elements.
Summarizing, one can distinguish among the following groups of metals.

TABLE 2-1

ATOMIC SUSCEPTIBILITY AT $\sim 290^\circ K$ [91]

$(\chi_{mole} \times 10^4)$

A Metals

IA	IIA	IIIA
Li	Be	B
+24	-10	-6.7
Na	Mg	Al
+14	+10	+17

T Metals / **B Elements**

	IV	V	VI	VII	VIII	IX	X	IB	IIB	IIIB	IVB	VB	VIB
K +18	Ti +160	V +290	Cr +165	Mn +530	Fe	Co Ferromagnetic	Ni	Cu -5.5	Zn -10	Ga -22	Ge -7.6	As -5.5	Se -22
Ca +50													
Sc +260													
Rb +18	Zr +120	Nb +210	Mo +85	Tc +270	Ru +43	Rh +105	Pd +560	Ag -20	Cd -20	In -10	Sn α, -37 β, +3.1	Sb -72	Te -37
Sr +85													
Y +190													
Cs +29	Hf +72	Ta +154	W +53	Re +65	Os +10	Ir +26	Pt +190	Au -28	Hg -33.5	Tl -51	Pb -23	Bi -280	Po
Ba +23													
La +112													
	Th +130		U +410		Pu +600								

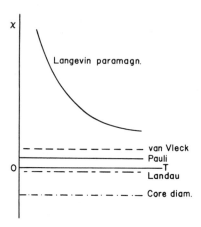

FIG. 2-8. Major contributions to the susceptibility in metals.

Elements of the alkali and alkaline earth groups are mostly
paramagnetic. Beryllium is an exception. The susceptibility of
these elements is nearly temperature independent, except for strontium and barium. The large $\partial\chi/\partial T$ values of these two elements have
been attributed to temperature dependent variations in the s- and
p-states. Figure 2-9 gives examples of $|\chi|$ of alkali earth metals.

Copper, gold, zinc and cadmium are typical examples of elements
in the B group. These elements have completely filled cores. The
diamagnetic contribution from these core states usually dominates
since the spin paramagnetism of a completely filled shell is zero.
In this group of elements, only β-tin is paramagnetic. Especially
large diamagnetism is found for semimetals like bismuth. Figure
2-10 gives experimental results.

Transition metals have partially filled d-shells. Electrons in
these shells are responsible for the strong paramagnetism of these
elements, or for spontaneous magnetic ordering. The χ(room temp)
values of paramagnetic elements follow similar lines in the periodic
system in the first, second and third long row (2-1a). This indicates
that a rigid band model may describe basic features in all three
rows.

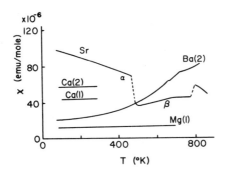

FIG. 2-9. $\chi(T)$ of alkali earth metals. After Ref. 91.

Let us now discuss the susceptibility of some elements in more detail. Table 2-2 gives data on various alkali metals in which one evaluated the paramagnetic contribution from the conduction electrons, their Landau diamagnetism, the diamagnetism of the ion core and then compare these values with experimentally observed susceptibilities. The Pauli and Landau contributions (χ_P and χ_L) have been taken into account by using the equation:

TABLE 2-2

Mass Susceptibility $\chi \cdot 10^6$ of the Alkalis [29]

	Li	Na	K	Rb	Cs
Spin susceptibility, (cal.)	1.5	0.68	0.60	0.32	0.24
Diamagnetism of conduction electrons, (cal.)		-0.5	-0.23	-0.20	-0.11	-0.08
Diamagnetism of ions, (obs.)	-0.1	-0.26	-0.34	-0.33	-0.29
Total susceptibility, (cal.)	0.9	0.2	0.06	-0.12	-0.15

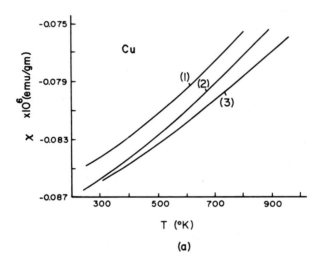

(a)

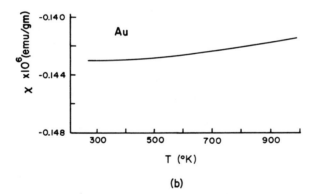

(b)

FIG. 2-10a + b. Susceptibility of copper (a). (3) gives data
by   Ref. 92, (1) and (2) give previous results (see Ref. 92 for
details). Susceptibility of gold (b). After Ref. 92, with permission.

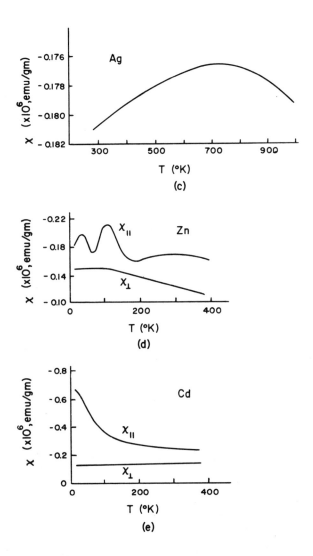

FIG. 2-10c to e. Susceptibility of silver (c). After Ref. 92, with permission. Susceptibility of zinc (d) and cadmium (e). The field is parallel ($||$) or perpendicular ($\perp$) to the hexagonal axis. After Ref. 93.

$$\chi_p + \chi_L = 1.26 \; n_o^{1/3} \; \rho^{-2/3} A^{+2/3} \; 10^{-6} \qquad \text{(cgs)} \qquad (2\text{-}42)$$

where n_o is the number of conduction electrons per atom, ρ the density and A the atomic weight. Variations between experimental
values (Table 2-1) and calculated values are difficult to explain.
It may be that electron-electron interaction effects are the most
important correction terms which have not been taken into account.

The magnetic susceptibility of copper, silver and gold has been
measured on very pure samples. Figure 2-10 a to c gives the experimental results as reported by Garber et al. [92]. The reproducibility of these measurements is estimated to be ± 0.15%. The
data show that the temperature variation of χ of silver shows a
maximum at 700°K. Ferromagnetic impurities could have a marked
effect on the measurements. Iron impurities were not detected in
silver, whereas the copper sample had less than 70 ppm (parts per
million) iron impurities and the gold sample less than 100 ppm iron
impurities. Sufficiently high fields give saturation for ferromagnetic impurities. This makes it possible to estimate their contribution to the susceptibility in noble metals.

These measurements were executed on polycrystalline samples.
This is adequate for cubic metals, where the susceptibility is isotropic. The susceptibility of zinc, which is hexagonally closed
packed, is a complex function of temperature and orientation, as
measurements by Markus [93] showed (2-10d). He obtained:

$$\chi = (\chi_{||} \sin^2\phi + \chi_{\perp} \cos^2\phi)\cos^2\theta + \chi_{\perp}\sin^2\theta \qquad (2\text{-}43)$$

where $\chi_{||}$ is the susceptibility parallel, and $\chi_{\perp}$ the suscepti-
bility perpendicular to the hexagonal c-axis. ϕ is the angle between the field and the projection of the hexagonal axis into the x_1-
x_2 plane. The anisotropy of χ for other non-cubic elements from
group II to IV in the periodic system is given in Table 2-3.

TABLE 2-3

Anisotropy of Susceptibility per Gram [91]

Element	$T(^{\circ}K)$	$\chi_{\parallel} \times 10^6$	$\chi_{\perp} \times 10^6$	$\chi_{\parallel}/\chi_{\perp}$
Zn	293	−0.169	−0.124	1.36
Cd	293	−0.243	−0.142	1.71
	14	−0.679	−0.130	5.2
Hg	80	−0.112	−0.121	0.93
In	293	−0.121	−0.054	2.22
Tl	293	−0.420	−0.164	2.56
Graphite	293	−22.8	−0.4	57
β−Sn	293	+0.0241	+0.0270	0.89
Sb	293	−1.42	−0.50	2.84
	90	−1.73	−0.50	3.46
Bi	294	−1.05	−1.48	0.71
	14	−1.20	−1.77	0.68
Ti	298	+3.35	+3.07	1.09
Ta	303	−0.329	−0.296	1.11
	493−633	−0.296	−0.296	1.00

	$T(^{\circ}K)$	$\chi_a \times 10^6$	$\chi_b \times 10^6$	$\chi_c \times 10^6$
Ga	290	−0.150	−0.506	−0.278
	80	−0.240	−0.467	−0.343
	298	−0.12	−0.42	−0.23

Subscripts $\parallel$ and $\perp$ indicate values parallel and perpendicular to the principal crystallographic axis.

The susceptibility of semimetals and semiconductors can be
rather large. Its theory is complicated. The experiments show
pronounced variations of χ with orientation and field strength
for bismuth. Shoenberg [94] showed that one should explain this
effect with electrons of different masses in different sections of

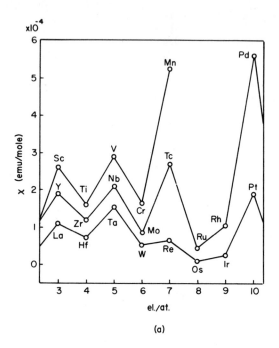

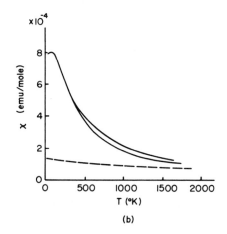

FIG. 2-11a + b. Susceptibility of transition elements at room temperature (a). After Ref. 91. Susceptibility of palladium (b). Broken line is calculated, full lines give experimental results. See Ref. 96 for more details.

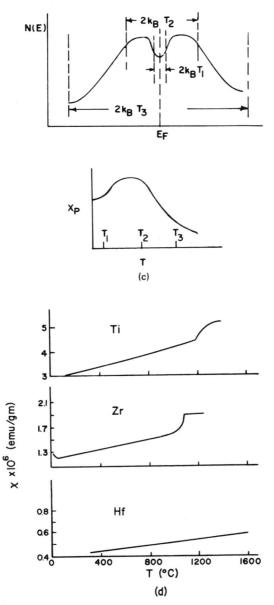

FIG. 2-11c + d. Pauli susceptibility of a metal with a N(E) curve which has a small minimum in a larger maximum near the Fermi energy level (c, top figure). This leads to a maximum in the susceptibility (c, bottom figure). See text for details. Susceptibility of titanium, zirconium and hafnium (d). After Ref. 95.

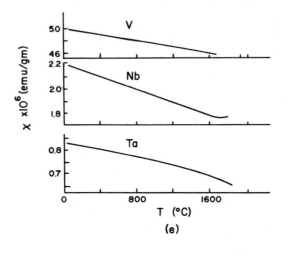

(e)

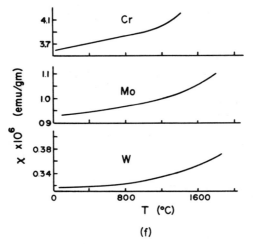

(f)

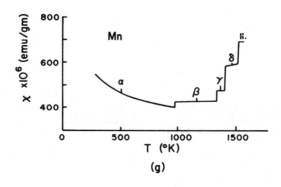

(g)

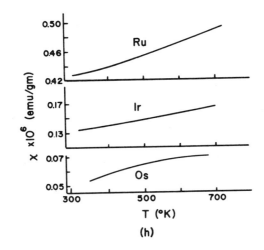

(h)

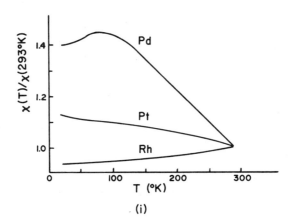

(i)

FIG. 2-11e to i. Susceptibility of vanadium, niobium and tanta-
lum (e, opposite page); chromium, molybdenum and tungsten (f, opposite page). After Ref. 95. Susceptibility of manganese (g, opposite
page). "li." represents the liquid state. After Ref. 12. Susceptibility of ruthenium, iridium and osmium (h); palladium, platinum
and rhodium (i). After Ref. 95.

the Fermi surface. It is presently believed that the strong anisotropy of χ has the same origin as the oscillation in the de Haas-
van Alphen effect. This agrees with the observation that the strong
anisotropy of χ is found in bismuth only at low temperatures.

Transition elements are paramagnetic and χ depends in a systematic way on their position in the periodic table. Alternating
rows of transition elements have either low or high susceptibility
values.(2-11a). Those with high χ values usually have low $\partial\chi/\partial T$
values and vice versa (naturally elements with spontaneous magnetic
ordering are excluded in this discussion). A qualitative understanding of this behavior has been given by Kriessmann and Callen
[95]. These authors used the equation for the Pauli susceptibility
and a weighted average of the density of states curve near the Fermi
potential for the calculation of χ . Near a maximum of $N(E)$, increasing temperatures should lead to a decrease of $N(E)_{aver.}$, near
a minimum of $N(E)$, increasing temperatures lead to an increase of
$N(E)_{aver.}$. Therefore, $\partial\chi/\partial T$ should be negative in the first and
positive for the second case.

Palladium is a transition metal where the absolute value of χ
as a function of temperature has been determined from $N(E)$ curves
as calculated from electronic specific heat measurements (see Chapter III). In this metal, the dominant contribution to χ comes
from the Pauli susceptibility. This is evident from the relatively
large temperature dependence of χ . $\chi(T)$ has a peak at 80°K.
Such a peak cannot be explained with a simple parabolic energy band.
Therefore, Shimizu and Katsuki [96] used as density of states curve
a curve obtained from low temperature specific heat measurements.
These measurements give information on the electronic specific heat,
which is proportional to the density of states at the Fermi energy
level. These authors then used the assumption of a rigid band, so
that they could calculate the density of states curve as a function
of energy. They calculated the $\chi(T)/\chi(T = 0^{\circ}K)$ for several arbitrary band shapes with a computer program to investigate in detail
the origin of the maximum in χ of palladium and showed the possibility of both maxima or minima in the $\chi = \chi(T)$ function. The

result of their calculation for palladium is given in Fig. 2-11b.
The calculation gives much lower values for χ than obtained from
experiments. The larger experimental values may be explained with
the concept of molecular fields acting on the spin of electrons.
Agreement between calculations and experiments could be obtained if
one assumed molecular fields with a molecular field constant W_m =
6.2 to $6.3 \cdot 10^4$ in cgs units. The calculation is not sensitive
enough to locate the maximum of χ at 80°K. However, one can show
with the rigid band model and the N(E) curve obtained from low
temperature specific heat measurements on Ag-Pd and Rh-Pd alloys
[97], that the N(E) curve should have a maximum at 4 at.% rhodium
in palladium. This is about 0.02 eV below the Fermi energy level of
palladium. A smaller maximum is found about 0.01 eV above E_F of
palladium. The minimum between these two maxima occurs close to
palladium. As mentioned above, a minimum in N(E) should be responsible for an increase in χ with increasing temperature. At
higher temperatures, one will average the density of states curve
over such a large energy interval that the small minimum in N(E)
at palladium becomes unimportant. The electrons 'see' only one
large average maximum in N(E) near the Fermi level, and the susceptibility decreases with increasing temperature. This is shown
schematically in Fig. 2-11c.

A similar calculation as that for palladium has been given by
Shimizu et al. [98] for chromium. The density of states curve has
a minimum at chromium as determined from low temperature specific
heat measurements of Cr-V and Cr-Fe alloys. The CrMn alloy system
yields lower electronic specific heats for the same electron concentrations used in the calculation by Shimizu et al. However, this
should not affect the general results of this calculation. The
calculations give the right temperature dependence, but the calculated χ values are much smaller than experimentally determined
values. Again, one may suspect that a molecular field may influence
the magnetization, or that enhancement effects may have to be
considered.

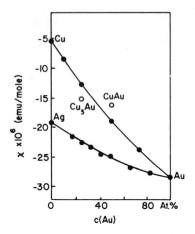

FIG. 2-12. χ_A of copper, silver, gold, ordered Cu_3Au and CuAu, disordered ^{A} Cu-Au and Ag-Au alloys. After Ref. 9, with permission.

G. Susceptibility of Non-transition Element Alloys

Simple models can be used for the analysis of the susceptibility of metallic alloys if the elements in the alloys have completely filled ion cores, and the outer electrons combine to a free electron gas of conduction electrons. For such a system, the diamagnetism of the cores of all atoms are added, and the susceptibility of the electron gas would be given as the sum of the Pauli paramagnetism and the Landau diamagnetism. If the Fermi surface of the conduction electrons of the two elements is very similar, and changes smoothly with composition, the χ(conduction electrons) and χ(ion cores) are also smooth functions of composition.

Copper and gold have the same crystal structure and form a complete series of solid solutions. They have also a very similar electronic structure, as found from Fermi surface determinations. One should therefore expect that the susceptibilities of these two alloys are linear functions of composition. Figure 2-12 shows

experimental results. Marked deviations from a smooth line in the
susceptibility versus composition plot are found only for ordered
copper-gold alloys. This deviation can be both positive and negative, but it is not very large. These changes in χ are not unexpected, since the Fermi surface of Cu_3Au differs markedly in the
ordered state from the Fermi surface of the disordered alloy. It
is unlikely that the diamagnetic contributions from the cores are
noticeably affected by atomic ordering. Minor changes are possible,
since atomic ordering may lead to changes in atomic radii.

Non-transition element impurities of the first long period dissolve as substitutional impurities in copper. Their ion cores have
the same number of completely filled shells as copper. Therefore,
the susceptibility in these alloys due to ion cores should be essentially the same as that of the copper core, since the radii of
the shells are similar. One of the outer electrons of the impurities will join the conduction electron gas, and the other will
form a screening charge around the impurity. This screening charge
should lead to an increase of the susceptibility, which should be
proportional to both the impurity concentration, provided
$c_{imp.}$ $\ll 1$, and to the difference in valency between copper and
the impurity atom.

The effect of non-transition element impurities of the first
long period on the susceptibility of copper base and silver base
alloys is given in Table 2-4. It can be seen that $d\chi/dc$ of the
copper base alloys systematically depends on the conduction electron per atom ratio, even if one includes CuAs alloys, where the
difference in valency between copper and arsenic is four. This
table shows that silver alloys with small amounts of impurity atoms
from the first long period also show a relationship in which $d\chi/dc$
is approximately proportional to the valency difference. The proportionality constant is much larger than in the case of the copper
alloys. This is not surprising since the diamagnetism of the silver
ion core differs markedly from the diamagnetism of the impurity core
of elements in the first long period.

TABLE 2-4

Change of χ of Cu and Ag by Small Additions
of Various Solute B Elements [91]

Solute	$(d\chi/dc)_{c \to 0} \times 10^6 \ (cm^3 gram\text{-}atom^{-1})$	
	Cu	Ag
Al	-6.9	
Cu		+15.7
Zn	-7.2	+3.5
Ga	-18.8	-21.7
Ge	-26.7	-50.0
As	-35.8	-81.8
Ag	-14.2	
Cd	-20.3	-6.2
In	-30.6	-30.8
Sn	-42.4	-54.9
Sb	-55.0	-80.9
Tl		∿-7

H. Susceptibility of a Noble Metal Element Matrix with Transition Element Impurities

The interaction of transition elements with noble metals is complex. The transition element, in its pure metallic state, has d-states in the neighborhood of the Fermi energy level. If a transition element is dissolved in a noble metal matrix, the d-electrons will be localized near the impurity core but their spacial distribution and their energy will be modified. The energy change will depend on the interaction of the impurity d-electrons with the

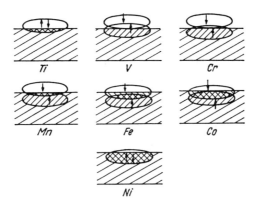

FIG. 2-13.   Localized d-states of 3-d impurities in copper and gold.   After Ref. 99.

electron gas of the noble metal matrix.   One obtains, therefore, as Friedel [55], Anderson [56] and Daniel and Friedel [57] pointed out, "virtual bound states".   This concept has been already discussed in Chapter I.

The virtual bound states are characterized by their "resonance energy", which differs from the energy in its pure metallic state, and the band width Γ of this virtual bound state.   The resonance energy may sometimes be independent of the electron spin orientation.   Daniel and Friedel expect for the case of transition element impurities in a noble metal matrix, where the Fermi energy level is much lower than in aluminum and closer to that of the transition element in its metallic state, that the interaction of electrons will lead to a splitting of the virtual energy state.   This is shown schematically in Fig. 2-13.   The characteristic parameter for the splitting seems to be the relation of the virtual energy band width Γ to the intra-atomic interaction energy of electrons at the impurity and the resonance energy.   Daniel and Friedel suggested tentatively that decreasing E_F values of the matrix should lead to an increase of the splitting.

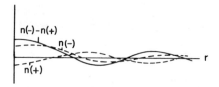

FIG. 2-14. Oscillation of electron densities with spin up or spin down around a virtual bound state. After Ref. 57 with permission.

The localized magnetic moments of the virtual energy states give, according to these authors, a susceptibility similar to that of localized electrons. The susceptibility should follow a Curie-Weiss type temperature dependence above a critical temperature. If the virtual energy state intersects the Fermi energy level, it will be only partly filled, and the number of Bohr magnetons per impurity atom will not be an integer. Localized states may give rise to long range (spacial) charge and spin oscillations, which can lead to coupling between impurity electrons and could be responsible for anti-ferromagnetic ordering below a critical temperature (2-14).

It was frequently found that the susceptibility of alloys follows the Curie-Weiss law only over limited temperature intervals. Extrapolating this short segment by a straight line to $1/\chi = 0$ gives an "apparent" critical temperature of the alloy for this temperature interval, and the slope would be proportional to the square of the "effective" number of Bohr magnetons per atom. This, and the "apparent" critical temperature which characterizes the "ferromagnetic" ($T_c > 0$, $H_m > 0$) or the "antiferromagnetic" ($T_c < 0$, $H_m < 0$) exchange interaction between electrons, are temperature dependent.

A discussion of a few selected alloy systems may illustrate the concepts discussed above. Measurements on gold samples with iron impurities [91], which should be typical for a system with localized impurity spins in a noble metal matrix (2-15), show that the susceptibility follows only approximately the predicted Curie-

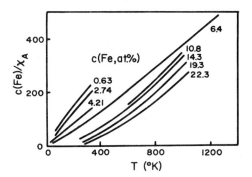

FIG. 2-15. $c_{Fe}/\chi_A(T)$ of Au-Fe alloys (see Refs. 23 and 35 for details).

Weiss relationship. $1/\chi$ is a nearly linear function of temperature. However, the slight change in slope in these curves with temperature (one finds both positive and negative curvatures) indicates that the effective number of Bohr magnetons per magnetic unit
is a function of temperature. A decrease in the number of Bohr magnetons per magnetic unit with increasing temperature can be explained in the following way. Since the alloys are not infinitely dilute,
not only individual iron impurities will be found, but also iron
pairs or larger groups of interacting iron atoms. Nearest neighbor
interaction, and even next nearest neighbor actions, may be important. With increasing temperature, loosely bound iron groups may
break up. This would lead to a decrease in the average number of
Bohr magnetons per atomic unit, which will more than compensate for
the increase in the number of magnetic units for the total magnetization. This leads to a decrease in $d\chi/dT$ with increasing temperature since χ is proportional to the square of p_{eff}, but only
a linear function of the number of magnetic units in a volume element.

However, there must also be other mechanisms which affect the
temperature dependence of χ, since experiments also give examples
of $d\chi/dT$ values which increase with increasing temperatures.

Figure 2-15 shows both types. It is possible that a temperature dependent coupling between impurity atoms with localized moments is
responsible for this effect. This coupling can be both ferromagnetic, which leads to a positive critical temperature in the Curie-
Weiss law T' > 0, or antiferromagnetic with T' < 0. If ferromagnetic coupling decreases with increasing temperature, it will shift
the $1/\chi$ curve to the left in a $1/\chi$ versus T plot. The opposite
effect would be found if antiferromagnetic coupling between magnetic
impurity atoms exists, and that this antiferromagnetic coupling decreases with increasing temperature. Shifts of $1/\chi$ curves with
temperature lead therefore to changes in $\partial\chi/\partial T$.

Alloys with iron impurities are not the only examples where the
number of Bohr magnetons per impurity atom depends on the matrix
and temperature. Measurements on alloys with manganese impurities
show that the effective number of Bohr magnetons is not only a function of temperature, but depends also on the solvent. One found
for solutions of manganese in copper [91]: 4.8 μ_B/Mn atom, in zinc:
4.0 μ_B/Mn atom, and in α phase Cu-Zn alloys: 4.83 μ_B/Mn atom
for the copper rich, and 4.5 μ_B/Mn atom at the zinc rich side of the
α-phase. The virtual energy state concept makes it possible to explain these fractional numbers of Bohr magnetons or electrons per
impurity atom. This does not mean that a fraction of an electron
can be found at the impurity site, but that electrons occupy this
site only for a fraction of the time. Then they will move to
another site. In other words, these electrons are not truly localized, but oscillate from place to place.

I. Susceptibility of Noble Metal and
Group VIII Element Alloys

Figure 2-16 [91] shows the susceptibility of noble metal-palla-
dium binary alloys. The χ values are negative and nearly composition independent for alloys with less than 55 at.% palladium.

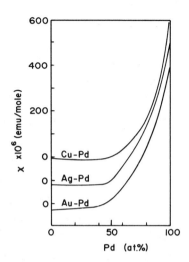

FIG. 2-16. $\chi(T)$ of disordered Cu-Pd, Ag-Pd and Au-Pd alloys.
Lines are shifted in respect to each other. After Ref. 9.

Larger palladium concentration leads to a rapid increase of the
susceptibility χ.

The high susceptibility of the palladium rich alloys is due to
incompletely filled d-states, leading to high density of states
values. Increasing the electron concentration in these alloys
leads to a filling of these bands and a corresponding decrease in
the Pauli susceptibility χ. The d-shell should be completely
filled at the el./at. ratio of 10.6 or larger since χ becomes
negative for these alloys. This can be interpreted with the model
that each noble metal atom in the palladium alloys adds 0.6 electrons to the d-band. The susceptibility of alloys with higher
electron concentration is mostly due to the diamagnetic contribution from the filled cores. Interestingly, small additions of
palladium to copper or silver increase the diamagnetism, indicating that palladium in its diamagnetic state (state with filled d-
band) is higher than the diamagnetism of copper or silver.

A quantitative correlation of the susceptibility and the density of states for alloy systems requires, however, computer calculations. The typical calculation of χ_P starts with the correlation $\chi_P = N(E_F)\mu_B^2$ taking into account that the Fermi potential ζ is temperature dependent, and that $N(E_F)_{aver.}$ has to be formed in a temperature interval of $k_B T$. The average value of $N(E)$ as a function of energy is obtained from low temperature specific heat measurements. One then uses the rigid band model to calculate the correlation of energy shift with concentration changes.

Several ternary alloy systems have been studied with similar methods to check if the rigid band model can be used to analyze more complex alloys. It was concluded in a study by Hahn and Treutmann [100] that the scattering of free electrons by the lattice had an effect on χ. The observed parallel composition dependence of both χ and the electronic specific heat coefficient γ (γ is equal to "electronic specific heat/temperature" which is proportional to $N(E)$ [see Chapter 3]) could be due to a composition dependent electron phonon interaction effect which would affect γ and χ_P in a similar way, since:

$$\chi_P = \mu_B^2 \, N(E)/[1-I \cdot N(E)]$$

$$\simeq \mu_B^2 \, N(E)[1+I \cdot N(E)] \quad \text{for} \quad I \cdot N(E) \ll 1 \qquad (2-44)$$

and

$$\gamma = \gamma_0 \, [1+J_{ph.-el.} N(E)] \qquad (2-45)$$

where I is due to an effective exchange integral (see page 150), and $J_{ph.-el.}$ is an electron-phonon interaction energy per energy state. If $J_{ph.-el.}$ and I depend in a similar way on composition, then χ and γ will also follow a parallel path. However, it is much more difficult to obtain a consistent model for the analysis of χ and the electronic specific heat with this concept than with the assumption that the composition dependence of χ and γ is due to the composition dependence of $N(E)$. Hahn and Treutmann

studied the binary Pd-Ag alloy system and "pseudo binary" alloy system $Pd_{0.94-2c}Rh_{0.06+c}Ag_c$ in which all alloys have the same electron concentration. They measured χ of these alloys over an extended temperature range and first corrected for ferromagnetic
impurities by using the corrected susceptibility $\chi(T)_{cor.}$:

$$\chi(T)_{cor.} = \chi_{exp.} - C/T \quad . \tag{2-46}$$

The Curie constant, C, was selected in such a way that $\chi(T)$ followed the relation:

$$\chi(T)_{cor.} = \chi(T=0)_{cor.} + AT^2 \quad . \tag{2-47}$$

Figure 2-17a and b give $\chi_{exp.}$ and $\chi(T)_{cor.}$ obtained in this way
for two samples, and Figs. 2-17b and c give $\chi(T)_{cor.}$. These curves
show that the maxima in $\chi(T)$ disappears if 1 at.% Rh or 2 at.%
Ag is added to palladium. In the first case, χ increases with
silver concentration. The addition of 0.8 at.% Rh - 0.8 at.% Ag
also eliminates the maximum.

Hahn and Treutmann qualitatively discuss the expected susceptibility from calculated N(E) curves by Freeman [101]. This
N(E) curve is shown in Fig. 2-17d. It shows a sharp peak just below the Fermi energy level. This peak can be affected by impurity
scattering of electrons. This was taken into account by changing
the area of the peak Γ with impurity concentration:

$$\Gamma = p_1 c_{Ag} + p_2 c_{Rh} \tag{2-48a}$$

where the values of p_1 and p_2 were obtained from electrical resistivity data. The peak light decreases with increasing Γ values,
as shown in Fig. 2-17e. This approach leads to χ values, which
agree very well with $\chi(T=0)_{exp.}$ obtained from experiments (2-18a
and b). These results indicate that if one takes scattering of
electrons by impurities and its effect on the effective density of

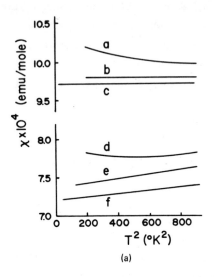

(a)

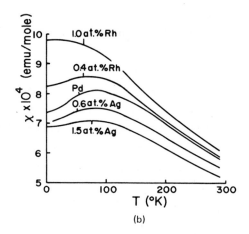

(b)

FIG. 2-17a + b. Susceptibilities of palladium and PdRh alloys (a). The curve d gives the measured, the curves e and f the corrected values for palladium. The curve a gives the measured, the curve b the corrected values for a PdRH alloy with 1 at.% Rh. The curve c gives the corrected values for a PdRh alloy with 0.0104 at.% Rh. The corrected susceptibilities of Pd-Ag and Pd-Rh alloys are given in (b). After Ref. 100.

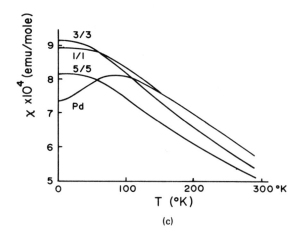

(c)

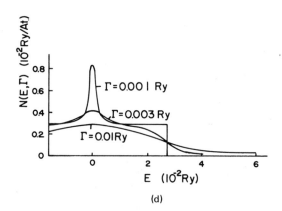

(d)

FIG. 2-17c + d. Temperature dependence of the corrected susceptibility of selected ternary Pd-Ag-Rh alloys (c). i/i gives the Ag/Rh
concentration in %. Density of states curve used for the calculation
of the susceptibility (d). The sharp peak at E = 0 corresponds to
the N(E) approximation for pure palladium. The width of this peak
Γ increases with increasing impurity concentration. The height of
the peak was selected in such a way that the area under the curve
is constant.

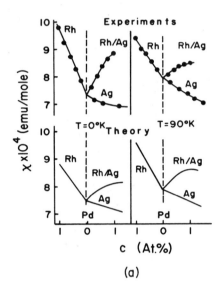

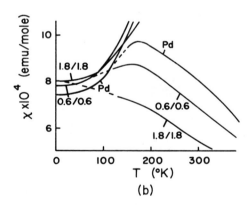

FIG. 2-18. Concentration dependence of the susceptibility at 0°K and 90°K of binary and ternary palladium base alloys (a). The top section gives experimental results, the bottom section calculated results. Temperature dependence of the susceptibility without scattering ($\alpha = 0$) on the left hand side and with scattering ($\alpha = 0.0000105$ Ry/deg) on the right hand side (b). i/i gives the Ag/Rh ratio in %. After Ref. 100.

states curve into account, the rigid band model gives even quantitatively a good description of susceptibility measurements below $100^{\circ}K$
(2-18a). $\chi(T)$ cannot be calculated accurately above $100^{\circ}K$ with
this model. Results of calculations and experimental results are
given in Fig. 2-18b, where thermal scattering of electrons is taken
into account to calculate Γ.

$$\Gamma = P_1 c_{Ag} + P_2 c_{Rh} + \alpha T \tag{2-48b}$$

This correlation leads to the expected increase of χ with T for
$T < 100^{\circ}K$, but it gives no maximum for $T \simeq 170^{\circ}K$. Several explana-
tions for the discrepancy between $\chi_{exp.}$ and $\chi_{theo.}$ are available.
For instance, $N(E)_{calc.}$ needs improvement, the temperature dependence
of I has to be taken into account, or the electron-phonon interaction is important. Even with these reservations, the generally
good agreement of $\chi(T=0^{\circ}K)$ over an extended composition range of
binary and ternary alloys is a strong argument for the application
of the rigid band model to alloys.

One should keep in mind that both χ and γ should follow a
similar concentration dependence, even if one uses the concept of
virtual energy states. Small amounts of palladium in a B-element
matrix would create virtual energy states both below and above the
Fermi level. Neither type of state would lead to a noticeable
contribution to the susceptibility or the electronic specific heat.
The resonance energy level of impurities can be concentration dependent. With 40% Pd impurities one may reach the critical concentration where the virtual energy states will be found at the Fermi
energy level. A corresponding result maybbe obtained, for example,
with the "coherent potential" approximation. It can be easily seen
that a suitable combination of two bands will give, up to a critical
composition, no change in the density of states curve at E_F, but
that a further increase in the composition leads then to a rapid
increase in $N(E)$. Presently, it is not possible to go beyond
such general statements in the correlation of energy band models
with experimental data from susceptibility measurements.

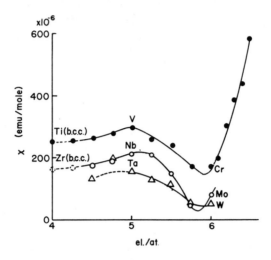

FIG. 2-19. Room temperature susceptibility of b.c.c. alloys, showing the minimum χ which occurs for all three periods. After Ref. 30.

J.   Susceptibility of Transition Element Alloys

The room temperature susceptibilities of several binary transition element alloys in the three long periods, with outer electron per atom ratios from 4 to 6.5, are given in Fig. 2-19. The composition dependence for all alloy systems is very similar. This is a strong argument for the concept of the "rigid band model", which assumes that the density of states of an alloy depends only on the outer electron concentration, not on the type of elements which are components in the alloy. For alloys with $4 \le$ el./at. ≤ 6.5, one would expect from these susceptibility curves that the density of states curve has a minimum for el./at. = 6, and a maximum for el./at. = 5. N(E) would increase rapidly with el./at. above 6. These interpretations agree qualitatively with electronic specific heat measurements which are a more accurate way to determine N(E) as a susceptibility measurement.

Calculations of χ with $N(E)$ determined from low temperature specific heat data give frequently the right composition dependence of χ of alloy systems like silver-palladium or chromium-
vanadium. However, the theoretical $\chi_{calc.}$ values are usually
much smaller than measured $\chi_{exp.}$ values. Therefore one has to
use additional effects like the molecular field (Weiss field) to
explain the data. Other contributions to χ may come from the orbital movement of electrons, $\chi_{orb.}$, and the diamagnetic contributions from the core electrons. The separation of χ into these
various contributions is only possible if one takes a number of
physical phenomena such as the Knight shift into account. Only a
few calculations of χ will be mentioned here.

Extensive computer calculations were conducted by Katsuki on
binary alloy systems [103]. He used the equation $\chi = \chi_{spin}$ +
$\chi_{orb.}$ + $\chi_{diam.}$ with $\chi_{spin}^{-1} = \chi_o^{-1} - \alpha$ where α is the molecular field coefficient, $\chi_{orb.}$ the orbital paramagnetic and $\chi_{diam.}$
the diamagnetic susceptibilities.

Katsuki calculated the susceptibility for several alloy systems. Some results are given in Fig. 2-20. For V-Cr alloys,

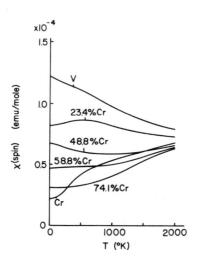

FIG. 2-20. Calculated susceptibility of V-Cr alloys. After
Ref. 103.

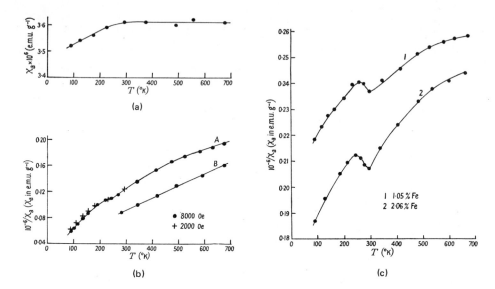

FIG. 2-21. Temperature variation of the susceptibility of chromium (a).   Temperature variation of the reciprocal susceptibility of
a sample with 5.76% Fe (b).   The specimen tested in curve A was annealed at 1200°K and quenched, in B the specimen was aged by cold-
working and annealed at 700°K.   Temperature variation of the reciprocal susceptibility of two CrRe alloys (c).   After Ref. 104.

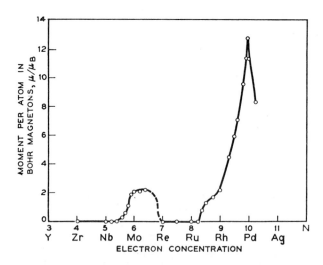

FIG. 2-22. Effective moment of Fe-impurities in alloys of 4d-
elements adjacent in the periodic system.   After Ref. 105.

Katsuki suggests that the orbital paramagnetic contribution is important to explain $\chi_{exp.}$. Table 2-3 shows values he selected. He concluded that this correction was sufficient to obtain good agreement between experiments and his calculation. The molecular field coefficient was set equal to zero. This seems reasonable, since vanadium is paramagnetic, and chromium shows antiferromagnetic ordering due to the conduction electrons only below 310°K. Calculations for the V-Ti system require corrections due to both molecular fields and the orbital paramagnetism. It seems difficult to see why an α correction is necessary in these alloys, and not for the V-Cr. However, even if the detailed features of the temperature and composition dependence of χ may be subject to further investigation, it is obvious that there is a strong correlation between density of states curves determined from low temperature specific heat measurements and susceptibility measurements.

The model of localized impurity spins in a non-magnetic matrix has been also used to evaluate χ of alloys with iron impurities in a chromium matrix (2-21). The susceptibility due to chromium atoms was assumed to be the same as that of pure chromium [104] (this can be justified only for temperatures above the Néel temperature of chromium). The remainder of the susceptibility was associated with localized electrons of the iron impurities. These yielded 2.85 to 2.90 as the number of Bohr magnetons per iron atom, in reasonable agreement with values for pure iron.

Electrons of ferromagnetic impurity atoms may affect not only conduction electrons in the matrix, but may also interact with adjacent d-electrons in a transition element matrix. This has been observed in ternary alloys [105], which were prepared from binary alloys of elements adjacent to each other in the second long row of the periodic system by adding a small amount of iron. The number of Bohr magnetons per iron atom in these alloys is plotted in Fig. 2-22 as a function of the binary alloy composition. This curve shows a peak at palladium, with more than 12 μ_B/iron atom. This cannot be attributed to an exceptionally large number of localized electrons for each iron atom. Therefore, this effect has been

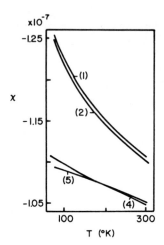

FIG. 2-23. Magnetic susceptibility of germanium as a function
of temperature. Curve 4 is for a p-type sample with hole concentrations of 2.1 x 10^{17}/cm^3, curves 1, 2 and 5 are for n-type samples with electron concentrations of 7 x10^{17}, 1 x 10^{18}, and 1.3 x
10^{14}/cm^3 respectively. After Refs. 109 and 169.

explained with the concept that some of the electron spins of the

palladium atoms line up preferrably parallel to the spin of the

magnetic iron electrons. The giant moment per iron atom is, in

other words, due to "polarization".

K.   Susceptibility of Semiconductor and Semimetal Alloys

Since it is difficult to analyze the susceptibility of pure

semiconductors, it is not surprising that the analysis of the ef-

fect of impurities on χ is not easy. This is obvious from the re-

sults shown in Figs. 2-23 and 2-24 which give the susceptibilities of

germanium and of InSb with various amounts of dopants. The sus-

ceptibility of a number of III - V compounds and of some group IV

elements is given in Fig. 2-25. The data in Fig. 2-23 indicate

that germanium samples change their susceptibility more if doped

for electron conduction (n-type samples), then doped for hole

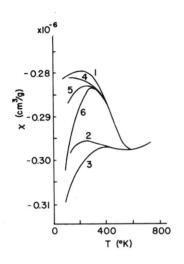

FIG. 2-24. Temperature dependence of the magnetic susceptibility of InSb samples with various
doping:

1, $n = 6.2 \cdot 10^{15} cm^{-3}$

2, $n = 5.5 \cdot 10^{17} cm^{-3}$

3, $n = 8.0 \cdot 10^{18} cm^{-3}$

4, $p = 1.4 \cdot 10^{16} cm^{-3}$

5, $p = 3.6 \cdot 10^{16} cm^{-3}$

6, $p = 2.2 \cdot 10^{17} cm^{-3}$.

After Refs. 107, 227.

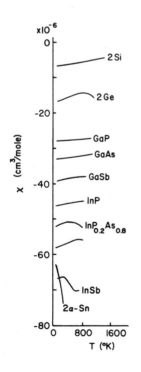

FIG. 2-25. Molar susceptibility as a function of temperature
for a number of III-V compounds
and group IV elements. After Refs.
107, 227.

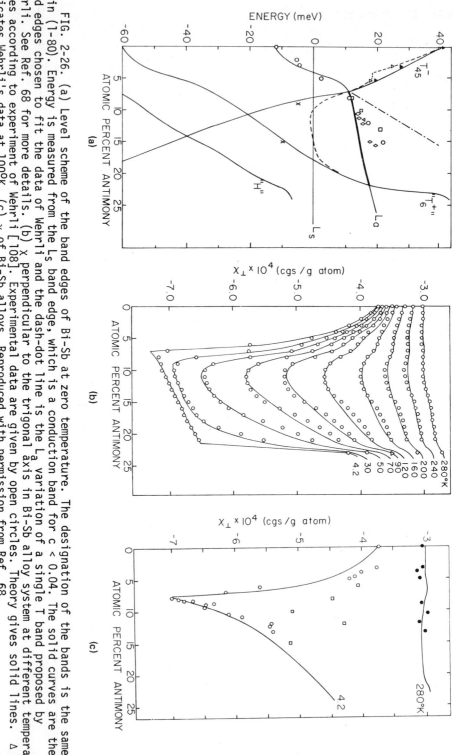

FIG. 2-26. (a) Level scheme of the band edges of Bi-Sb at zero temperature. The designation of the bands is the same as in (1-80). Energy is measured from the Ls band edge, which is a conduction band for c < 0.04. The solid curves are the band edges chosen to fit the data of Wehrli and the dash-dot line is the La variation of a single T band proposed by Wehrli. See Ref. 68 for more details. (b) χ perpendicular to the trigonal axis in Bi-Sb alloy system at different temperatures according to experiment of Wehrli [108]. Experimental data are given by open circles. Theory gives solid lines. Δ indicates Wehrli's data at 100°K, (c) χ of Bi-Sb alloys. Reproduced with permission from Ref. 68.

conduction (p-type sample). An increase in the number of electrons
from $6.2 \cdot 10^{15}$ to $5.5 \cdot 10^{17}$, cm^{-3} for doped InSb leads at room temperature to a change of the susceptibility by $0.015 \cdot 10^{-6}$ in cgs units
per gram. In other words, the change of the susceptibility due to
the addition of 10^{-5} electrons per atom in this semiconductor has
the same effect as if one replaces 10% copper atoms with zinc atoms in
a copper matrix. Clearly, the simple models used to describe the
effect of alloying on the susceptibility of metals cannot be used
to describe the effect of doping in semiconductors.

A study of the effect of the energy band structure on the susceptibility can be illustrated in alloy systems in which the gradual
change of the energy band structure with composition can be followed,
as in the case of bismuth-antimony alloys. Experiments on the magnetic susceptibility of bismuth-antimony alloys show that the magnetic susceptibility depends strongly on sample composition, orientation and temperature. Figures 26b and c give experimental results
obtained by Wehrli [108] as individual points for both susceptibility values measured with H parallel and perpendicular to the trigonal axis, below room temperature. The composition range covers
the section where the alloys become semiconducting. This leads to
especially low susceptibility values at very low temperatures. At
$4.2^{\circ}K$, the susceptibility, measured in a field perpendicular to the
trigonal axis, $\chi_{\perp}$, is a nearly linear function of composition from
about 7 to 22 at.% antimony. This is the composition range for
which the samples are semiconductors. There is a sharp minimum in
$\chi_{\perp}$ at 7 to 8 at.% As. This is an exceptionally high value if one
calculates it per individual hole in the valency band or per electron
in the conduction band. The full lines are results of calculations
by Buot [68]. At room temperature, the $|\chi|$ values are nearly a
linear function of composition over the complete composition range.
The thermal energy $k_{B}T$ is at room temperature much larger than the
energy gap of the semiconductor. It is also much larger than the
band overlap of the conduction and valency bands of the semimetal
bismuth as shown in Fig. 2-26a. The distinction between semiconductor and semimetal is therefore lost at higher temperatures.

The susceptibility of the system was calculated by Buot with a
three term function:

$$\chi = \chi_A + \chi_G + \chi_C \qquad (2\text{-}49a)$$

where χ_A is the ionic susceptibility (the susceptibility of Bi^{5+}
in this case), χ_G is the background diamagnetic term, which depends on the energy gap, and χ_C is the paramagnetic carrier contribution. This separation of terms is not the only possible way
to analyze the data. For instance, sometimes one uses the equation:

$$\chi = \chi_A + \chi_{LP} + \chi_{CP} + C_{ID} \qquad (2\text{-}49b)$$

where χ_{LP} is the usual Landau Peierls term, χ_{CP} is the "crystalline paramagnetic" term associated with the effective g-factor (it
approaches the Pauli susceptibility for the case of free electrons),
and χ_{ID} is an induced diamagnetism term. χ_A is again the ionic
susceptibility. Wehrli [108] analyzed his data with the three term
equation. χ_A was no more than 10% of the total susceptibility.
The diamagnetic contribution of the full bands was obtained by a
linear extrapolation of experimental data.

III.   FERROMAGNETIC, ANTIFERROMAGNETIC, AND FERRIMAGNETIC
ORDERING IN ELEMENTS AND COMPOUNDS

A.   The Molecular Field Theory

The elements iron, cobalt, nickel, gadolinium, and dysprosium
are ferromagnetically ordered below a critical temperature, the
Curie temperature T_c. This means that some of their electrons are
coupled in such a way that the average magnetic dipole moment of
the crystal is lined up in a given crystallographic direction. At
absolute zero temperature, 2.22 μ_B/atom are lined up in the [1 0 0]

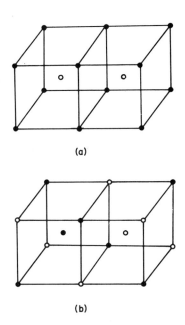

(a)

(b)

FIG. 2-27. Antiferromagnetic arrangement of atomic moments in a body-centered cubic lattice. The first kind of order is given in (a), the second kind of order is given in (b). After Ref. 114.

direction in iron, 0.606 μ_B/atom are lined up in the $[1\ 1\ 1]$ direction in nickel, and 1.27 μ_B/atom are lined up in the direction of the hexagonal axis of cobalt. A large number of alloys containing one or more ferromagnetic elements show ferromagnetism over an extended composition range. Even some alloys with only non-ferromagnetic elements are ferromagnetic. Ferromagnetism is also found in compounds.

In some crystalline systems magnetic spins line up parallel to each other at absolute zero temperature in sublattices. The sublattices may be lined up antiparallel to each other, so that the total magnetization of the sample is zero without an applied magnetic field. These materials are defined as antiferromagnets. Figures 2-27a and b show spin arrangements in an antiferromagnetic material, where the atomic order is bcc. The example given in Fig.

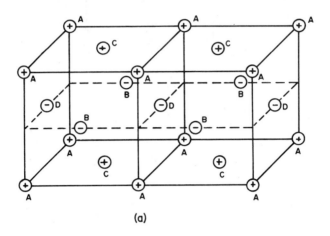

(a)

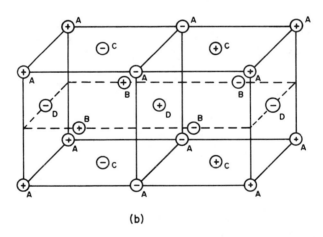

(b)

FIG. 2-28. Antiferromagnetic arrangement of atomic moments in an fcc structure. First kind order is given in (a), second kind order is given in (b).

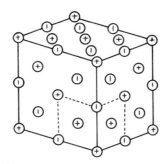

FIG. 2-29. Magnetic moment in MnO below the Néel temperature. The lattice parameter of the magnetic unit cell is twice as large as that of the chemical unit cell given by a dashed line. After Ref. 114.

2-27a shows an antiferromagnetic system with two sublattices and in Fig. 2-27b a system with four sublattices, in which the two A-sublattices and the two B-sublattices are coupled antiferromagnetically to each other. The A-sublattices can be coupled in several ways to the B-sublattices. Figure 2-28 gives possible spin orientations in an fcc crystal. These figures do not represent magnetic unit cells. The unit cell of the antiferromagnetically ordered MnO crystal as shown in Fig. 2-29 has twice the lattice parameter as the "atomic" unit cell.

Ferrimagnetic crystals consist of two or more different sublattices which are ferromagnetically ordered, but where the total moment is not zero nor equal to the sum of the absolute values of the individual magnetic moments. Typical examples of ferrimagnetic materials are compounds like spinels, PQ_2X_4, where Q represents a trivalent element, and P a divalent element (2-30 and 2-31). Another example is a garnet, $P_3Q_2R_3O_{12}$, a cubic crystal with 180 atoms/unit cell, or a hexagonal compound of the chemical formula $PO \cdot 6 \cdot Fe_2O_3$. P could be barium, strontium, or lead. Ferric ions may be partly replaced by gallium or aluminum ions.

These compounds are usually electrical insulators. Insulators have only localized electrons. Therefore, they are good examples for most theoretical models which are based on the Heisenberg model which associates ferromagnetism with localized electrons.

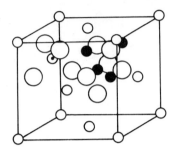

FIG. 2-30. Atom positions in the unit cell of the spinel lattice. The large sphere inside the unit cell represents oxygen ions, the small light sphere with center dot inside the unit cell represents ions in tetrehedral sites and the small dark spheres ions in octahedral sites. The positions of all ions are shown only in two octants. The other octants have one or the other of these two structures and are arranged so that no two adjacent octants have the same configuration.

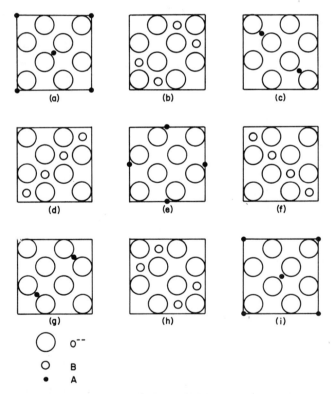

FIG. 2-31. Sequences of planes parallel to the (100) plane in the spinel lattice, 1/8 of the lattice parameter apart. After Ref. 114.

The first phenomenological theory to describe ferromagnetism was proposed by Weiss, who suggested that the interaction of magnetic dipoles could be given by a molecular field H_m, which is proportional to the magnetization of the sample:

$$H_m = W_m M. \tag{2-50}$$

Since the spontaneous magnetization of a ferromagnet decreases with increasing temperature and is zero at the critical temperature T_c, one would expect that the magnetic energy per electron $\mu_B H_m$ should be of the order of the thermal energy $k_B T_c$:

$$\mu_B H_m \simeq k_B T_c. \tag{2-51}$$

One obtains for iron with $T_c = 1043^{\circ}K$: $H_m \simeq 10^7$ Oe, and $W_m \simeq 10^4$.

The magnetization of ferromagnetic materials would again be determined from the Brillouin function as for paramagnetic materials, except that the field H is replaced by $H_{appl.} + H_m$. One obtains (see Eq. 2-24):

$$M = N\mu_B j B_j(x) \tag{2-52}$$

where j is the total angular momentum quantum number. x is given by:

$$x = j\mu_B g H/k_B T \tag{2-53}$$

$$= j\mu_B g(H_{appl.} + W_m M)/k_B T. \tag{2-54}$$

For zero applied field, one obtains for the temperature dependence of $M = M(T)$:

$$M(T)/M(T=0) = B_j(x) \tag{2-55}$$

and:

$$M(T)/M(T=0) = k_B Tx/NW_m g^2 \mu_B^2 j^2 \qquad (2-56)$$

where $M(T=0) = Ng\mu_B j$.

A solution for these two simultaneous equations can be obtained graphically, as shown in Fig. 2-32. The figure shows that there exist two solutions for $T < T_c$. However, the solution $M(T) = 0$ is unstable for $T \neq 0$, since fluctuations of M from its zero value will lead to a non-zero H_m value, which will lead to an increase of M up to the moment when $Ax = B_j(x)$. The intersection of $B_j(x)$ and $k_B Tx/W_m Ng^2 \mu_B^2 j^2$ moves to lower values with increasing temperature and is zero at a critical temperature T_c. At this temperature, the slope of $B_j(x)$ is equal to $k_B T_c/W_m g^2 \mu_B^2 j^2$, since the slope of $B_j(x)$ for $x \to 0$ is equal to $(j + 1)/3$. This gives for T_c:

$$T_c = W_m g^2 \mu_B^2 j(j + 1)N/3k_B. \qquad (2-57)$$

This should be compared with the value of the Curie temperature T_c given in Eq. 2-51.

Figure 2-33 gives the $M(T)/M(T=0)$ curves as calculated for several j values, and experimental data for iron, cobalt and nickel. Agreement between experiments and calculations is good if one uses values of $1/2$ or 1 for j. By a series expansion of the Brillouin function, one can show that:

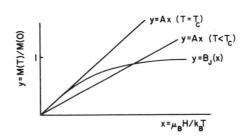

FIG. 2-32. Illustration of a graphical method for the determination of the spontaneous magnetization at temperature T.

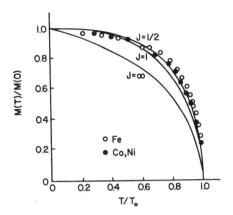

FIG. 2-33. The spontaneous magnetization as a function of temperature. The curves are obtained from theory, the points represent experimental data. After Ref. 110.

$$M(T)/M(T=0) = 1 - \exp(-x/j)/j - \cdots$$

$$= 1 - \exp(-3/\{j + 1\})(T_c/T) - \cdots \quad (2-58)$$

for low temperatures.

One obtains near the Curie temperature:

$$\{M(T)/M(T=0)\}^2 = (10/3)(j + 1)^2(1 - T/T_c)/(j^2 + (j+1)^2). \quad (2-59)$$

As mentioned before, the Weiss theory predicts that the susceptibility in the paramagnetic region is given by $1/\chi \propto (T - T_c')$. Figure 2-34 shows $1/\chi$ for iron, cobalt and nickel. Well above T_c, the Curie-Weiss law describes the data adequately. Deviations from this law close to T_c may be due to magnetic short range ordering. Such ordering has been observed experimentally in iron by neutron diffraction experiments. Interestingly, $1/\chi$ of γ-iron extrapolates for $1/\chi \to 0$ to negative temperatures, indicating antiferromagnetic ordering in fcc iron. Antiferromagnetic ordering has been observed in fcc iron at very low temperatures, where this crystal structure was stabilized by the formation of small, coherent iron

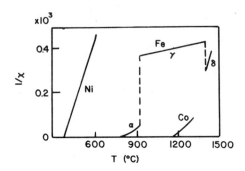

FIG. 2-34. Reciprocal susceptibility of iron, cobalt, and nickel above the Curie temperature. After Ref. 111.

precipitates in an fcc copper matrix, which prevented the iron precipitates from transforming into the bcc structure.

The Weiss theory describes within a few percent the magnetization of ferromagnetic elements as a function of temperature. It has been applied successfully to the analysis of the magnetization of antiferromagnetic and ferrimagnetic materials. As mentioned before, ferrimagnetic materials are crystals with two or more sublattices with different magnetization. Each sublattice can have a different saturation magnetization and critical temperature than the other sublattices. The sum of the magnetic moments in a ferrimagnetic material is finite. The bulk features of a ferrimagnet seem therefore to be similar to those of a ferromagnet. However, $M = M(T)$ will be much more complicated and one may find $M(T) = 0$ for $0 < T < T_c$. This is shown in Fig. 2-35.

Let us discuss briefly as an example the Weiss molecular field theory for an antiferromagnetic system with a bcc lattice, where the center atom is part of the A-sublattice, and the corner atoms are part of the B-sublattice. The molecular field acting on A- or B-atoms is:

$$H_{mA} = -N_{AA}M_A - N_{AB}M_B \qquad (2-60)$$

$$H_{mB} = -N_{BA}M_A - N_{BB}M_B, \qquad (2-61)$$

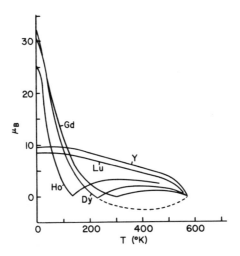

FIG. 2-35. Magnetization of some rare-earth iron garnets,
$5 \cdot Fe_2O_3 \cdot M_2O_3$. M is the rare earth metal. After Ref. 112.

where N_{ij} are constants, and M_i the magnetization of sublattice
"i". Symmetry requires that $N_{AB} = N_{BA}$, and $N_{AA} = N_{BB}$. The total
field acting on an atom in sublattice "i" would be:

$$H_i = H + H_{mi}. \qquad (2\text{-}62)$$

The magnetization in each sublattice is a function of temperature.
It is found in the same way as the magnetization of a ferromagnetic
crystal. One again uses the Brillouin function and replaces the
field by H_i. This gives as before:

$$M_i = \tfrac{1}{2} Ng\mu_B jB_j(x_i) \qquad (2\text{-}63)$$

with:

$$x_i = jg\mu_B H_i/k_B T \qquad (2\text{-}64)$$

where j is the total angular quantum number of the magnetic electron.

For small magnetization values at $T > T_c$, the Brillouin function is again replaced by the first term of a series expansion [82]:

$$M_i = Ng^2\mu_B^2 j(j + 1)H_i/6k_BT. \tag{2-65}$$

One obtains, since H_i and M_i are parallel in the paramagnetic region:

$$H_i = H - N_{ij}M_j - N_{ii}M_i. \tag{2-66}$$

i or j correspond to A or B. Inserting this result into the previous equation yields:

$$\begin{aligned}M &= M_A + M_B \\ &= Ng^2\mu_B^2 j(j + 1)(2H - \{N_{AA} + N_{AB}\}M)/6k_BT\end{aligned} \tag{2-67}$$

and one obtains for the susceptibility:

$$\chi = M/H = C/(T + T') \tag{2-68}$$

with:

$$\begin{aligned}C &= Ng^2\mu_B^2 j(j + 1)/3k_B \\ T' &= (N_{AA} + N_{AB})/2, \end{aligned} \tag{2-69}$$

One would expect that the interaction of one atom with nearest neighbors is larger than with next nearest neighbors. Therefore, one would expect that $N_{AB} > N_{AA}$. This gives a positive "Néel temperature" T_N, since:

$$T_N = C(N_{AB} - N_{AA})/2. \tag{2-70}$$

The Curie-Weiss temperature for this system is:

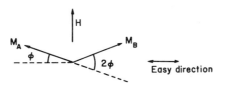

FIG. 2-36. An illustration for the calculation of the susceptibility of an antiferromagnetic material, if H lies perpendicular
to the easy axis.

$$T' = T_N(N_{AB} + N_{AA})/(N_{AB} - N_{AA}). \tag{2-71}$$

Weak next nearest neighbor interaction would be characterized by
$N_{AA} \ll N_{AB}$. This would lead to $T' \to T_N$.

Below T_N, for antiferromagnetic single crystals, χ is orientation dependent. Let us consider the case where H is nearly perpendicular to $\vec{M}_A$ or $\vec{M}_B$ (2-36). It was stated before that:

$$\vec{H}_A = \vec{H} - N_{AA}\vec{M}_A - N_{AB}\vec{M}_B. \tag{2-72}$$

At equilibrium, the torque on each sublattice is zero:

$$|\vec{M}_A \times (\vec{H} + \vec{H}_{mA})| = 0. \tag{2-73}$$

The last two equations yield:

$$M_A H\cos\phi - N_{AB}M_A M_B \sin2\phi = 0. \tag{2-74}$$

or:

$$2M_B \sin\phi = H/N_{AB}. \tag{2-75}$$

The magnetization in the H-direction is:

$$M = (M_A + M_B)\sin\phi = 2M_A \sin\phi. \tag{2-76}$$

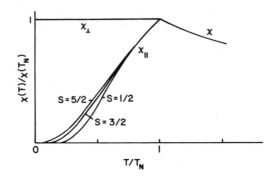

FIG. 2-37. The susceptibility of an antiferromagnetic material as a function of temperature in reduced units. $N_{ii} = 0$. After Ref. 113.

This gives finally:

$$\chi_\perp = M/H = 1/N_{AB}. \qquad (2\text{-}77)$$

$\chi_\perp$ is temperature independent.

A more complicated calculation leads to the susceptibility of a single crystal with H parallel to the magnetization direction of the sublattices [82]:

$$\chi_\parallel = N\mu_B^2 g^2 j^2 B_j'(x_o)/\{k_B T + (N_{AA} + N_{AB})\mu_B^2 g^2 j^2 N B_j'(x_o)/2\} \qquad (2\text{-}78)$$

where $B_j'(x) = \partial B_j(x)/\partial x$. x_o is:

$$x_o = g\mu_B j\{(N_{AB} - N_{AA})M_A\}/k_B T. \qquad (2\text{-}79)$$

Figure 2-37 gives results of calculations for different spin values.

The derivation given above assumes that the applied fields are small. One finds for larger fields that the spin orientation can switch from one direction to another at a critical field H_f. This process is called "spin flopping" or "spin flipping". Let us look at a system where the crystal anisotropy keeps M_A in a preferred orientation, and where one applies a field parallel to M_A. M_A and

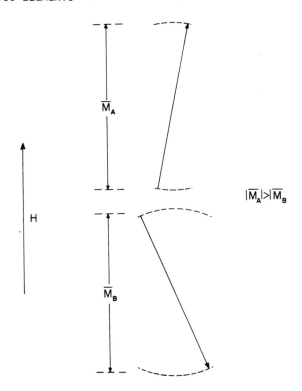

FIG. 2-38. Spins of an antiferromagnet and their oscillations
in an applied field.

M_B will fluctuate around their equilibrium position. The fluctuation
for M_A will decrease with increasing field, whereas the fluctuation
for M_B will increase (2-38). This leads to the net magnetization
of the antiferromagnet in an applied field.

Since the susceptibility of antiferromagnets is orientation dependent, the magnetic energy will be different for H parallel or
perpendicular to the magnetization of the sublattices. One obtains:

$$E_{\parallel} = - (1/2)\chi_{\parallel} H^2$$

$$E_{\perp} = - (1/2)\chi_{\perp} H^2.$$

(2-80)

At the critical field for spin flopping, $E_\perp = E_\parallel + K$, where K is
the energy due to the crystal anisotropy, assumed to be a constant.
This gives:

$$-(1/2)\chi_\parallel H_f^2 = K - (1/2)\chi_\perp H_f^2 \qquad\qquad (2-81)$$

or:

$$H_f = [2/(\chi_\perp - \chi_\parallel)]^{1/2} \cdot K^{1/2} . \qquad\qquad (2-82)$$

The susceptibility of ferrimagnetic materials can be calculated
in a similar way with the molecular field theory. The calculation
shows that $1/\chi$ is given by [114]:

$$1/\chi = T/C_1 + 1/\chi_o + C_2/(T - C_3). \qquad T > T_c. \qquad (2-83)$$

$1/\chi$ is only well above the critical temperature T_c, a linear function of T. The magnetization below the critical temperature should
be given by a superposition of several Brillouin functions, where
the effective field in a given sublattice would be a linear function
of the magnetizations of all sublattices.

B.   The Heisenberg and Ising Models

The Weiss molecular field theory gives a good description of
the magnetization and its temperature dependence. However, Weiss
was not able to show the origin of the spontaneous spin alignment.
The explanation for this phenomena could be obtained only from quantum mechanical considerations, as shown by Heisenberg. He used the
following approach to calculate the energy difference between two
electrons with parallel or antiparallel spin orientations (see
Ref. 82 for more details). Heisenberg investigated a two electron
system like the hydrogen molecule. Let us try to follow the general

line of his argument. Let us assume that one knows the solution of
the Schrödinger equation for the system consisting of electron "I"
and ion "a", and for the electron "II" and ion "b". The Schrödinger
equation would have for the first system the form:

$$[-(\hbar^2/2m)\nabla_I^2 + V(\vec{r})]\psi_a(I) = E_a\psi_a(I). \tag{2-84}$$

A similar equation would be given for the system of electron
"II" and ion "b". The solutions for the wave functions are $\psi_a(I)$
and $\psi_b(II)$, and the electron energies E_a and E_b. The Schrödinger
equation for the molecule is:

$$[-(\hbar^2/2m)(\nabla_I^2 + \nabla_{II}^2) + V(\vec{r}^I) + V(\vec{r}^{II})]\psi = E\psi \tag{2-85}$$

if one neglects electron-electron and ion-ion interaction.
$E = E_a + E_b$, and the wave functions $\psi_a(I)\psi_b(II)$ and $\psi_a(II)\psi_b(I)$
are solutions to this equation. One can construct a linear combina-
tion of these wave functions, which is either symmetric or antisymmetric if the two electrons "I" and "II" are exchanged:

$$\begin{aligned}\psi_{sym.} &= \{\psi_a(I)\psi_b(II) + \psi_a(II)\psi_b(I)\}/\sqrt{2} \\ \psi_{antis.} &= \{\psi_a(I)\psi_b(II) - \psi_a(II)\psi_b(I)\}/\sqrt{2}\end{aligned} \tag{2-86}$$

These solutions contain no information on the spin orientation
of the two electrons. Solutions to the Schrödinger equation which
contain this information could be written as:

$$\psi = \phi(\vec{r})\chi(\alpha) \tag{2-87}$$

if the orbital movement of the electron is "quenched", which means
it is so small that it can be neglected. $\phi(\vec{r})$ contains the information on the spacial distribution of electrons and $\chi(\alpha)$ states
whether a specific spin direction is occupied ($\chi(\alpha) = 1$), or if the
spin is aligned in the opposite direction ($\chi(\alpha) = 0$). $\chi(\alpha)$ is an

antisymmetric wave function. Since ψ has to be antisymmetric, one
has to find symmetric wave functions $\phi(\vec{r})$ for the spacial distribution of the electrons. These wave functions are $\phi_a(I)$, $\phi_b(I)$,
$\phi_a(II)$, and $\phi_b(II)$.

The investigation up to now neglects completely the interaction
between the two electrons, the two ions, and the interaction of electron "I" with ion "b" and electron "II" with ion "a". This interaction is responsible for either parallel or antiparallel alignment of
the electron spins. The Hamiltonian for the interaction is:

$$H_{12} = e^2[1/r_{ab} + 1/r_{12} - (1/r_{1b} + 1/r_{2a})] \qquad (2\text{-}88)$$

where r_{ij} is the distance between particles "i" and "j". This
gives in first approximation an additional energy term for the elec-
tron-ion system:

$$E_{12} = \int \psi^* H_{12} \psi dx_1 dx_2 dx_3 \qquad (2\text{-}89)$$

as shown from variational computational methods. This integral leads
to energy terms of the form:

$$K_{12} = \int \phi_a^*(I)\phi_b^*(II)H_{12}\phi_b(II)\phi_a(I)dx_1 dx_2 dx_3 dx_1' dx_2' dx_3' \qquad (2\text{-}90a)$$

and:

$$J_{12} = \int \phi_a^*(I)\phi_b^*(II)H_{12}\phi_a(II)\phi_b(I)dx_1 dx_2 dx_3 dx_1' dx_2' dx_3'. \qquad (2\text{-}90b)$$

K_{12} is the average Coulomb energy, and J_{12} is the exchange integral.
The difference in energy for the electron system with both electron
spins parallel compared with the system where the two spins are
aligned antiparallel is:

$$E_{exch.} = J_{exch.} = J_{12} \qquad (2\text{-}91)$$

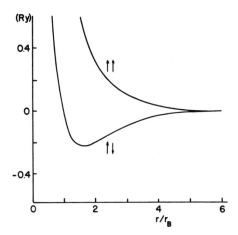

FIG. 2-39. Energy of molecular hydrogen for parallel and anti-parallel line up of spins.The antiferromagnetic line-up leads to the lowest energy state. After Ref. 114.

For hydrogen, $J_{exch.}$ is negative. Figure 2-39 shows the energy for the hydrogen molecule for the two possible spin orientations. Since the antiparallel arrangement of spins leads to a lower energy state, the system in its ground state is antiferromagnetic.

The possibility for ferromagnetic ordering in metals will, as the integral for J_{12} indicates, depend on the electron distribution and the Coulomb energy. Large J_{12} values are expected in the section between the two ions, since $1/r$ can become very large. The wave functions of two adjacent transition elements may overlap extensively in this region, since these states are incompletely filled. Therefore, $\phi_j(i)$ can have large values where $1/r_{12}$ is large. If ϕ_i has no nodes in this region, then $\phi_a^*(I)\phi_b^*(II)\phi_a(II)\phi_b(I)$ will always be positive. Therefore, J_{12} will be large and positive. Separating the ions should decrease J_{12}. The expected dependency of J_{12} on d/r is shown in Fig. 2-40. There d is the distance between nearest neighbors, and r is the radius of the open d-shell.

The exchange energy of a system of electrons depends on the relative orientation of the electron spins. Since the spin of an electron is usually characterized by the spin quantum number s, which

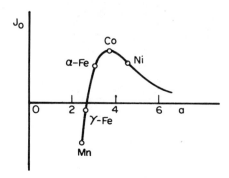

FIG. 2-40. Exchange integral as a function of $a = d/r$. d is the distance of nearest neigbor nucleii, and r the radius of the open 3-d shell. After Ref. 114.

is either $+1/2$ or $-1/2$, it is frequently convenient to write for the energy associated with the orientation of spins in a two electron system:

$$E_{exch.} = -2J_{12} \vec{s}_1 \cdot \vec{s}_2 \tag{2-93}$$

since this gives:

$$E(\uparrow\uparrow) - E(\uparrow\downarrow) = -2J_{12}[(1/4) + (1/4)] = -J_{12}. \tag{2-94}$$

For a multiple electron system, one writes:

$$H = -2J_{ij} \Sigma' \vec{s}_i \cdot \vec{s}_j = -2J_{ij} \vec{S}_i \cdot \vec{S}_j, \quad i \neq j \tag{2-95}$$

with:

$$\Sigma \vec{s}_i = \vec{S}_i, \qquad \Sigma \vec{s}_j = \vec{S}_j. \tag{2-96}$$

To make this expression more manageable, one usually considers only nearest neighbor interaction. This gives for the atom "i" with its neighbor "j":

$$H = -2J_{exch.} \sum_j \vec{s}_i \cdot \vec{s}_j \qquad\qquad (2\text{-}97)$$

assuming that the exchange energy for all nearest neighbor atoms
with the center atom is the same. A physically interesting solution
for this equation can be obtained rigorously only for a one or two
dimensional case, in which the spin of the electrons can be lined
up only parallel or antiparallel to a given direction. This highly
anisotropic model is the Ising model, which has been studied also
for the 3-dimensional case, where only approximate solutions are presently available. Its Hamiltonian is given by:

$$H = -2J_{exch.} \sum \vec{s}_{x_3 i} \cdot \vec{s}_{x_3 j} \qquad\qquad (2\text{-}98)$$

if the x_3 direction is the direction for the line up of electron
spins. The two-dimensional system was analyzed by Onsager [114].

It is sometimes assumed that $\vec{s}_i$ is a classical vector. Then
the spin orientation is not lined up into a few selected directions,
but can take any arbitrary direction. The term $\sum \vec{s}_i \cdot \vec{s}_j$ can be evaluated for small angles between $\vec{s}_i$ and $\vec{s}_2$. One obtains:

$$E = -2J_{exch.} \, s^2 \sum \cos\alpha_{ij} \qquad\qquad (2\text{-}99)$$

where α_{ij} is the angle betwen $\vec{s}_i$ and $\vec{s}_j$. For small angles one
can set: $\cos\alpha_{ij} = 1 - \alpha_{ij}^2/2$. Therefore, the orientation dependence of the energy is given by:

$$E(\alpha_{ij}) = J_{exch.} \, s^2 \sum \alpha_{ij}^2 . \qquad\qquad (2\text{-}100)$$

One uses this equation to study magnetic domain boundaries, where the
spin orientation changes gradually over a few hundred lattice sites.

The previous equations seem to imply that the energy of a system
depends only on the relative orientation of spins. However, one can
show with a classical model that spins should line up in specific
orientations. One can show for a planar arrangement of dipoles that

its energy depends on its orientation in respect to a square lattice.
One obtains, if all spins are lined up parallel [115]:

$$E(\cos\beta) = g_0 + g_1(r)[\cos^2\beta - 1/3]$$
$$+ g_2(r)[\cos^4\beta - (6/7)\cos^2\beta + 3/35] + .$$

(2-101)

$g_i(r)$ are the Legendre polynomials, β the angle between spin and
[10] direction. A minimum of the energy for a given direction would
correspond to a direction of "easy magnetization". The dipole energy
calculated with this equation is several orders of magnitude smaller
than found in ferromagnetic materials.

The energy associated with the deflection of the magnetization
vector with respect to the direction of easy magnetization is the
"energy of anisotropy" or "crystal energy". It is not possible to
calculate it for a three dimensional crystal. However, one can show
from symmetry considerations that the energy of crystal anisotropy,
$E_{anis.}$, should have the following form for a cubic lattice, if one
neglects higher terms [114]:

$$E_{anis.} = K_1 (\underset{i \neq j}{\Sigma} \cos^2\alpha_i \cdot \cos^2\alpha_j) + K_2\cos^2\alpha_1\cos^2\alpha_2\cos^2\alpha_3.$$

(2-102)

α_i is the angle between the direction of magnetization and the
x_i-direction. For hexagonal metals like cobalt, symmetry arguments
lead to the form:

$$E_{anis.} = K_3\sin^2\theta + K_4\sin^4\theta$$

(2-103)

where θ is the angle between the direction of magnetization and the
hexagonal axis perpendicular to the basal plane. Again, higher terms
are neglected. The crystal anisotropy energy depends also on the
angle between the directon of densest packed lines in the basal plane
and the projection of the magnetization direction on the basal plane.
However, this orientation dependence can be frequently neglected.

C. Series Expansion Calculations and Spin Waves

It is not possible to obtain the temperature dependence of the magnetic energy for a 3-dimensional lattice for the Heisenberg or Ising model in a closed form. One has to use perturbation calculations. Only for very low temperatures is it possible to obtain an exact solution for spin excitations. There one can show that thermal fluctuations of the spin orientation will travel through the crystal. It leads to a disturbance, which is periodic in space and time. This is a "spin wave".

A linear ferromagnetic lattice in its ground state consists of a row of spins, lined up in one direction. The total exchange energy for such a system is:

$$E_{exch.} = -2J_{exch.} \sum_{i \neq 1}^{N} \vec{s}_i \cdot \vec{s}_{i+1} \qquad (2\text{-}104)$$

if N is the number of atoms with spin s. One replaces the exchange interaction by an effective field H_i, which acts on atom "i". This gives:

$$E_{exch.} = -(1/2) \sum_i \mu_i H_i \qquad (2\text{-}105)$$

with:

$$\vec{H}_i = -2J_{exch.} (\vec{s}_{i-1} + \vec{s}_{i+1}) g^{-1} \mu_B^{-1}. \qquad (2\text{-}106)$$

Thermal excitations will change the spin orientation of some electrons. The potential energy increases. If the spin of one electron is tilted from its lowest energy position, it will interact with its nearest neighbors and they will change their orientation. This disturbance, as shown in Fig. 2-41, will propagate as a spin wave through the lattice. The torque acting on the spin is proportional to the rate of change of the angular momentum, $\hbar \partial s_i / \partial t = \vec{\mu}_i \times \vec{H}_i$, as shown in

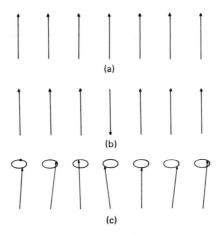

FIG. 2-41. Classical picture of the ground state of a simple
ferromagnet; all spins parallel (a). A possible excitation, one spin
is reversed (b). The ends of spin vectors precess on the surface of
cones in spin waves (c).

classical mechanics. One obtains therefore:

$$\partial \vec{s}_i / \partial t = -(g\mu_B/\hbar)\ \vec{s}_i \times \vec{H}_i.$$ (2-107)

The solution of this differential equation shows that the spin of the
electron will move by precession around the spin orientation of the
ground state. This angular frequency should be the Larmor precession
in the "effective field". The electron spins go through the same
precession. However, there is a phase shift in this precession from
lattice site to lattice site. For the long wavelength limit (ka << 1
with a the lattice parameter and k the wave number of the spin
wave), one obtains:

$$\hbar\omega = 2J_{exch.}\ sa^2 k^2$$ (2-108)

as the energy of the spin wave. A system of N atoms with spin s
each can have spin numbers −Ns, −Ns + 1, ... Ns − 1, Ns. The energy levels of the spin waves are quantized and can have values of:

$$E_k = 4J_{exch.} s(1 - cos[ka])n_k = n_k \hbar\omega_k .$$ (2-109)

Since each spin wave has a quantized energy level and a quantized momentum, a spin wave is also called a "magnon".

Several spin waves with the same quantum number can exist in a crystal. Therefore, the Pauli principle does not apply. One has to use the Bose-Einstein statistic to determine the the the average occupancy of energy states for magnons (see Chapter 3). The Bose-Einstein statistic is obtained by assuming that the number of particles with energy E_k is proportional to the Boltzmann factor, $exp(-E_k/k_B T)$. This gives for the number of particles at the energy level E_k: $<n_k> = 1/[exp(E_k/k_B T) - 1]$. The total number of magnons is then obtained with the equation:

$$\Sigma n_k = \int d\omega g (\omega) <n (\omega_k)>$$ (2-110)

where $g(\omega)$ is the number of spin wave states or magnon states per frequency interval. Magnons have only a single direction of polarization. Therefore, the number of states below k is $(2\pi)^{-3} 4\pi k^3/3$. The number of magnon states between k and $k + dk$ is $(2\pi)^{-3} 4\pi k^2 dk$. Counting the number of states in a given angular frequency interval gives:

$$g(\omega) = (2\pi)^{-2} (\hbar/2J_{exch.} sa^2)^{3/2} \omega^{1/2}.$$ (2-111)

The evaluation of this equation gives [3]:

$$\Sigma n_k = (2\pi)^{-2} (k_B T/2J_{exch.} sa^2)^{3/2} \int_o^\infty dx \cdot x^{1/2} (exp\{x\} - 1)^{-1}.$$ (2-112)

The last integral is equal to $0.0587 \cdot 4\pi^2$. The relative magnetization changes due to spin waves is $\Sigma n_k/Ns = \Delta M/M$. One obtains finally:

$$\Delta M/M = (0.0587/sQ) (k_B T/2J_{exch.} s)^{3/2}.$$ (2-113)

Q is the number of atoms in the cubic unit cell. This $T^{3/2}$ temperature dependence of the magnetization was first derived by Bloch.

Perturbation calculations of magnetic properties for temperatures above the range where the Bloch law is valid are very complex.
They have a special name: "series expansion calculations" [117 to
119], and are basically of three different types: (A) a high temperature "moment" type calculation for $T \gg T_c$, (B) a low density
"cluster" type near T_c , and (C) a low temperature "excitation" type.
At low temperatures, only small thermal excitations have to be considered. Near the critical temperature, long range order breaks
down, but short range magnetic order leads to a parallel line-up of
spins in small volume elements. At high temperatures only the interaction of individual moments of nearest neighbor spins is important.

The series expansion method has been applied to several models.
Calculations with this method predict typically the following temper-
ature dependence for the magnetic specific heat $c_{mag.}$ (see also
Chapter 3), the magnetization M and the susceptibility χ:

$$c_{mag.} \propto \left|1 - T_c/T\right|^{-\alpha^{\pm}} \tag{2-114}$$

$$M \propto (1 - T/T_c)^{\beta} \tag{2-115}$$

$$\chi \propto (1 - T_c/T)^{-\gamma} \tag{2-116}$$

+ means that $T > T_c$, − means that $T < T_c$. α, β, and γ are criti-
cal exponents. Equations of the form $c_{mag.}/k_B = A\ln\left|1 - T/T_c\right|$ are
also used. A detailed discussion of these and other calculations
can be found in Domb [117, 118] and Fisher [119].

A combination of the Heisenberg or Ising model and the Weiss
molecular field model has been used to determine the critical temperature. In this model, one consideres a cluster of atoms and assumes
that the center atom of this cluster "i" interacts with its nearest
neighbors "j". The interaction of these neighboring atoms with the
rest of the crystal is described by an effective field $H_{eff.}$, just

as in the Weiss molecular field model. The Hamiltonian of the cluster is given by [82]:

$$H = -2J_{exch.}\vec{S}_i \cdot \sum_j \vec{S}_j - g\mu_B \vec{H}_{eff.} \cdot \sum_j \vec{S}_j - g\mu_B \vec{H} \cdot \vec{S}_i \qquad (2\text{-}117)$$

where $\vec{S}_i$ is the spin of the central atom, and $\vec{S}_j$ the spin of
a nearest neighbor. Table 2-5 gives $J_{exch.}/k_B T_c$ values as have been
found with this Bethe-Peierls-Weiss method for simple cubic and
bcc crystals. Table 2-6 gives a survey of the characteristic parameters of ferromagnetic elements and compounds.

D. Collective Electron Model

The Heisenberg model assumes that localized electrons are responsible for ferromagnetism. However, the magnetization of ferromagnetic iron, cobalt and nickel , and of antiferromagnetic chromium
cannot be given by integer multiples of μ_B/atom. This may be explained in the case of iron with a model in which the conduction electrons are partly polarized by the magnetically aligned d-electrons.
Since the saturation magnetization of iron is 2.22 μ_B/atom, and
2.04 μ_B/atom are due to localized d-electrons, the polarization of
conduction electrons leads to a moment of 0.18 μ_B/atom. This description cannot be used for nickel, which has a saturation moment
of 0.606 μ_B/atom, since one needs at least one μ_B/atom as a localized
moment. One could assume in nickel, which has ten electrons in the
conduction and d-band per atom, that 0.6 el./at. are in the conduction band. That leaves 0.6 holes/atom in the d-band. These 0.6
holes per atom are responsible for ferromagnetic order. Naturally,
this is only an average number of holes. In reality, only an integer
number of holes can be found at one lattice site. This means that
60% of the lattice sites will have only nine electrons in d-states,
and 40% of the lattice sites will have ten electrons in d-states, if
one excludes states with less than nine d-states.

TABLE 2-5

$J_{exch.}/k_B T_c$ according to the Bethe-Peierls-Weiss Calculation

Spin quantum number		1/2	1	3/2
Simple cubic lattice	$J_{exch.}/k_B T_c$ =	0.540	0.169	0.086
Body centered cubic	$J_{exch.}/k_B T_c$ =	0.343	0.148	0.060

TABLE 2-6

Ferromagnetic Elements and Compounds [3,114]

	Saturation moment (Gauss)		μ_B/unit	Curie temperature T_c
	Room temperature	$0^\circ K$		
Fe	1707	1740	2.22	$1043^\circ K$
Co	1400	1430	1.72	1403
Ni	485	510	0.606	631
Gd		2010	7.12	290
Dy		2920	10.0	85
Cu_2MnAl	500			710
MnAs	670	870	3.4	318
MnBi	620	720	3.84	
Mn_4N	183		1.0	743
MnB	152	163	1.92	578
CrTe	247		2.5	339
CrO_2	515		2.03	393
$MnO \cdot Fe_2O_3$	410		5.0	573
$NiO \cdot Fe_2O_3$	270		2.4	858
$CoO \cdot Fe_2O_3$	400		3.7	793
$MgO \cdot Fe_2O_3$	110		1.1	713
$GdMn_2$		215	2.8	303
$Gd_3Fe_5O_{12}$	0	605	16	564

A completely different approach has been used therefore by
Stoner [84 to 86]. He used a concept in which the electrons occupy a
partly filled energy band. Their contribution to the magnetization
would be described similarly as in the case of the Pauli susceptibility, except that not only an outside field shifts some electrons
from one band to another with opposite spin alignment, but that ad-
ditionally a Weiss molecular field has to be considered.

Stoner proposed for the energy difference of electrons in bands
with spin down and spin up in respect to energy levels in the non-
magnetic metal [82]:

$$E = \pm W'_m M \mu_B. \tag{2-118}$$

The spontaneous magnetization of the sample would be given as a function of temperature by:

$$M = \mu_B \int \{f^{\circ}(E - W'_m n \mu_B^2) - f^{\circ}(E + W'_m n \mu_B^2)\} N(E) dE. \tag{2-119}$$

Stoner obtained for the case of a parabolic band with $E = \hbar^2 k^2 / 2m*$:

$$M = (\mu_B/6\pi^2)(2m*/\hbar^2)^{3/2} \{(E_F + W'_m n \mu_B^2)^{3/2} - (E_F - W'_m n \mu_B^2)^{3/2}\} \tag{2-120}$$

The critical parameter, which indicates para- or ferromagnetic ordering, is the ratio of the magnetic energy $W'_m n \mu_B^2$ to the Fermi energy E_F. Stoner showed that the relative magnetization M/M(sat.)
could have values between zero and one for 2/3 < $W'_m n \mu_B^2$ < $2^{-1/3}$.
The sample would be ferromagnetic for $W'_m n \mu_B^2 / E_F$ larger than $2^{-1/3}$.
Results of computer calculations are given in Fig. 2-42, which shows
both the relative magnetization M(T)/M(0) as a function of temperature below T_c, and the inverse susceptibility as a function of
temperature above T_c for different values of the parameter $W'_m n \mu_B^2 / E_F$.
One interesting result of this calculation is the slight curvature
of the reciprocal susceptibility curve, which is also found experimentally for nickel.

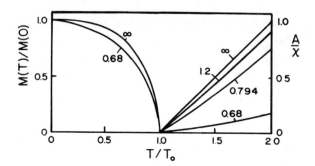

FIG. 2-42. The magnetization below and the reciprocal susceptibility above the Curie temperature according to the collective
electron model of ferromagnetism. $A = \mu_B M(0)/k_B T_0$. After Ref. 84.

The polarization of conduction electrons by localized electrons
is only a minor effect for the magnetization of iron. However, for
rare earth elements, where a direct exchange between localized electrons responsible for ferromagnetism is not possible, the interaction
between conduction electrons and localized magnetic electrons in
partly filled electron states is essential for ferromagnetic ordering.
A theory in which the polarization of conduction electrons is important was first proposed by Vonsovsky and Zener [120 to 122] to explain ferromagnetism of 3d-electrons in transition elements, not
for rare earth elements.   The spin energy for such a system was given
as:

$$E = (1/2)\alpha S_d^2 - \kappa S_d \cdot S_c + (1/2)\xi S_c^2 \qquad (2\text{-}121)$$

where S_d is the average value of the spin components of d-electrons,
and S_c is the average spin of the conduction electrons.   The first
term corresponds to d-d, the second term to d-c, and the third term
to c-c interaction.   It is now believed that this theory may be
applicable to explain ferromagnetism in rare earth elements; it is
not applicable to transition elements.
   Another example of a metal, where the spontaneous magnetization
is due to collective electrons, is chromium.   Antiferromagnetism in
this element is partly due to an anomaly of the Fermi surface.   The

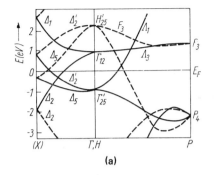
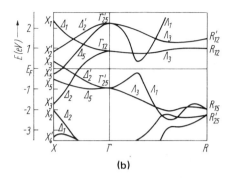

FIG. 2-43. E(k) curves for paramagnetic (a) and antiferromagnetic chromium (b). After Ref. 123.

surface consists of several segments (1-66a). Some contain electrons, others contain holes. One hole-segment has nearly the same size and shape as an electron segment. These two segments combine and annihilate each other, which leads to antiferromagnetic ordering. Figure 2-43 shows parts of a recent energy band calculation by Asano and Yamashita. They determined the E = E(k) relationship both for the paramagnetic and the perfect antiferromagnetic state using the Green function method. The paramagnetic state was calculated on the basis of a self consistent procedure. They then used the exchange energies to determine the E(k) curves for the antiferromagnetic state. The exchange term contains an adjustable parameter, which is selected in such a way that the calculations give the experimentally observed magnetization. The effect of magnetic ordering can be seen near the Fermi energy level between points Γ and P in the paramagnetic state, and Γ and R in the antiferromagnetic state. The Λ_1 curve in the paramagnetic state goes smoothly through the Fermi energy level. Antiferromagnetic order induces an energy gap in this segment.

The antiferromagnetic structure of chromium has been determined from neutron diffraction experiments. These measurements show that chromium does have a perfect antiferromagnetic structure. It does not consist of two simple sublattices of equal but opposite

moment, but it has a helical structure. This means that the magnetization vectors of two adjacent planes are not opposite to each
other, but are rotated by $180°$ $- \delta$ (or $\pi - \delta*$). δ is about $6°$ for
pure chromium at zero temperature. Chromium exists in two modifi-
cations. From $0°K$ to $153°K$, the magnetic moments are lined up
parallel to the modulating axis (AF_2 structure), from $153°K$ to the
Néel temperature of $311°K$, the magnetization is perpendicular to
the modulating axis. $153°K$ is the "spin flop" temperature of chromium. The periodicity of $180°$ $- \delta$ does not mean that the electrons
responsible for magnetic ordering occupy a lattice with this periodicity. It only states that the spins have a periodic structure with
this periodicity. Figure 2-44 gives a schematic diagram for such a
system, in which the spin orientation is lined up perpendicular to
the modulating axis, which is the x_3-direction. The spins are lined
up in the plane containing the x_1- and x_2- direction, and the vector
of magnetization rotates in this plane. This would correspond to
the AF_1 structure.

Since planes in real space are conveniently described by points
in reciprocal space, the antiferromagnetic structure of chromium
can be characterized by a vector 'Q' in reciprocal space. The direction of the vector would be parallel to the normal of the plane and
its length would be inversely proportional to the distance between
adjacent planes. Figure 2-44 shows the points in reciprocal space,
which characterize the planar waves. The magnetization in the
sample would be given by sine or cosine   functions, which would
be of the form (for the AF_1 structure) [124]:

$$P(\vec{r}) = P_o(\vec{\epsilon}_1 \cos Q \cdot x_1 + \vec{\epsilon}_2 \cos Q \cdot x_2 + \vec{\epsilon}_3 \cos Q \cdot x_3) \qquad (2\text{-}122)$$

where $\vec{\epsilon}_i$ is a unit vector in reciprocal space. For the AF_2
structure, one obtains:

$$P(\vec{r}) = P_o{}'(x_{1.0} \cos Q \cdot x_1 + x_{2.0} \cos Q \cdot x_2 + x_{3.0} \cos Q \cdot x_3)$$

$$(2\text{-}123)$$

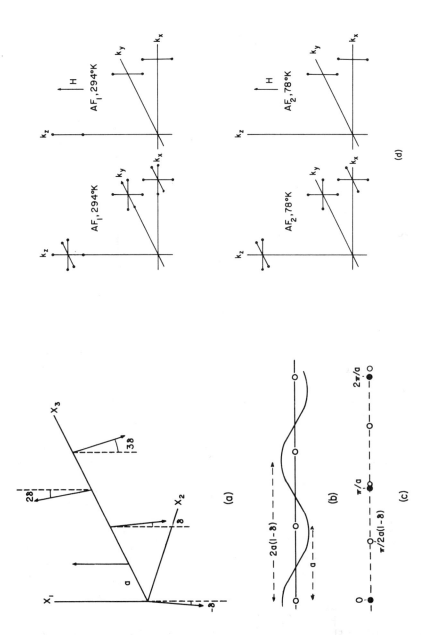

FIG. 2-44. Spin orientation of helical antiferromagnetism (a). Longitudinal modulation (b). Open circles represent lattice points. The magnetic wave length is $2a(1-\delta)$, and a is the lattice parameter. Reciprocal lattice points for the crystal lattice (full circles) and for the magnetic lattice (open circles) (c). Reciprocal space plots of neutron-diffraction reflections from normal chromium in the AF_1 and AF_2 phases (d, left hand side) and similar plots from field cooled chromium (d, right hand side) at 78°K and 294°K. After Ref. 124.

where $x_{1.0}$, $x_{2.0}$, and $x_{3.0}$ are unit vectors in the x_1, x_2, and x_3 direction respectively. These waves are "spin density waves" (SDW). The SWD are stable because they reduce the energy of the system [125]. Neutron diffraction measurements above the Néel temperature confirm this model since they show no localized spins [126].

IV. MAGNETICALLY ORDERED ALLOYS AND PSEUDO BINARY SOLID SOLUTIONS

A. Critical Temperatures and Magnetization of Ferromagnetic Alloys

The saturation magnetization of alloys of elements in the first long period of the periodic system is shown in Fig. 2-45. This is the Slater-Pauling curve, which shows that the number of Bohr magnetons per atom, μ_B/atom, is usually a smooth function of the el./at. ratio, provided the components of the alloy are close to each other in the periodic system. The μ_B/atom curve in first approximation is frequently a nearly linear function of the el./at. ratio. However, μ_B/atom has a maximum for the Fe-Co systems. Figure 2-46 shows the magnetization of binary fcc iron-nickel alloys. It seems that the increase of iron concentrations leads to an increase in the quantum number, if one tries to analyze the data with the Weiss theory. However, the relative magnetization of the alloy with 67.7% Fe is lower than even $J = \infty$ would give. Figure 2-47 shows the saturation magnetization of ternary Fe-Co-Ni alloys. In spite of the fact that this is a multi-phase system, M_s is a rather smooth function of the average composition. If elements differ noticeably in their el./at. ratio, the μ_B/atom curve will usually decrease below the Slater-Pauling curve . Similarly, plots of the Curie temperature of alloys as a function of the el/at. ratio do not seem to depend on the specific element, but only on the el./at. ratio.

This simple correlation between magnetization or Curie temperature and the el./at. ratio seems to show that a "rigid band model" may be appropriate to describe magnetic properties of alloys. For instance, the onset of magnetism in copper-nickel and zinc-nickel alloys occurs at nearly the same el./at. ratio if one extends the curves linearly (2-45), in spite of the fact that the nickel concentration in both alloys is quite different. On the other hand, the $N(E)$ curves are rather complex functions of the el./at. ratio and should depend markedly on the crystal structure; whereas the magnetization as a function of electron concentration curve is smooth, and the change in crystal structure leads only to small changes in the saturation magnetization.

The discussion of the Heisenberg model and the exchange integral shows that not the number of electrons at the Fermi energy level is important for ferromagnetic ordering, but the spacial distribution of electrons in overlapping electron orbits and their Coulomb energies. Such overlapping is only possible if one has partly filled states as in transition elements, which leads also to high density of states values. Therefore, a certain amount of correlation between the $N(E)$ curve and μ_B/atom is not surprising. However, one should not expect a similar composition dependence for these two parameters.

One knows [30] that several parameters enter into the calculation of the energies associated with magnetic ordering. One is the correlation energy U, which represents the energy of electrostatic repulsion between two electrons with antiparallel spin on the same atom, the other is the intra-atomic exchange energy $J_{exch.}$, which gives the difference in energy of two electrons in states on the same atom; when their spins are either parallel or antiparallel. Several authors concluded that U is of the order of 10 eV, and $J_{exch.}$ is of the order of 1 eV. Then ferromagnetism may be due to the strong repulsion between electrons with antiparallel spin. However, others assume that U is not large enough to prevent two electrons being at the same atom. Therefore, the origin of

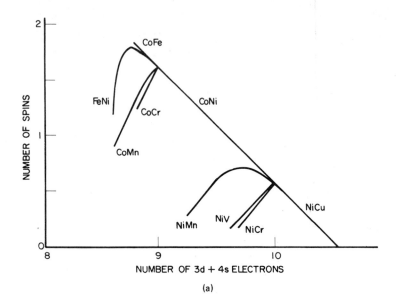

(a)

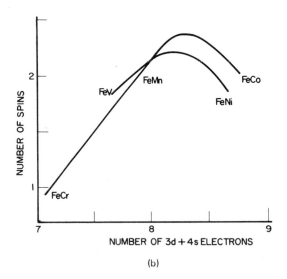

(b)

FIG. 2-45a + b. Average moment of binary fcc alloys (a) and
bcc alloys (b).   After Ref. 301.

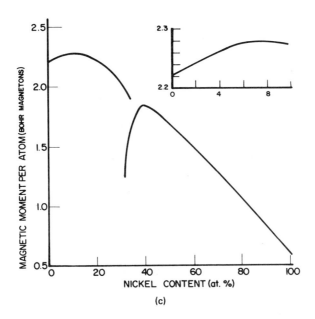

FIG. 2-45c. Average moment of Fe-Ni alloys. The insert gives data in iron rich alloys. Iron rich alloys are bcc. The line extending to nickel represents fcc alloys. After Ref. 301.

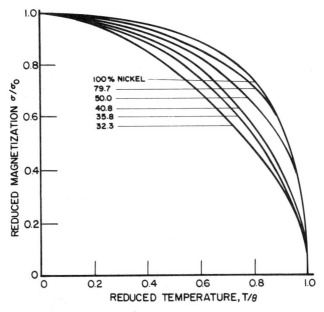

FIG. 2-46. Reduced magnetization of Fe-Ni alloys. After Ref. 301.

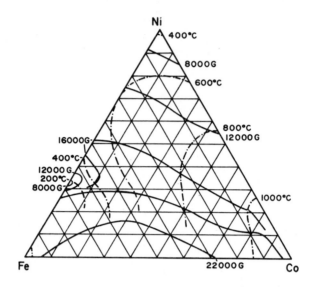

FIG. 2-47. Saturation magnetization $4\pi M_S$ and Curie temperatures of ternary Fe-Co-Ni alloys. Full line gives constant magnetization. Broken line gives constant Curie temperatures. After Ref. 114.

ferromagnetism may lie in intra-atomic exchange coupling. Since even a pure element like iron, a good example of a metal with localized electrons, shows both ferromagnetic and antiferromagnetic features, it is not surprising that the interpretation of the origin of the ferromagnetism in alloys is complex.

The Weiss molecular field theory can be easily adapted to describe properties of some alloys. It is only necessary to know the compositional dependence of the molecular field. In its simplest form, one would expect that $H_m \propto M$ or T_c would be a linear function of composition. This would be a good first approximation for nickel-palladium alloys, since T_c decreases continuously with increasing palladium concentration and reaches $T_c = 0$ close to palladium. This curve shows a positive curvature. As Fig. 2-45 shows, M decreases linearly with increasing non-magnetic impurities for a large number of iron, cobalt and nickel alloys. M_s will be

zero even with large amounts of ferromagnetic elements in the alloy
as in the case of Co-Cr or Ni-Cu. This is not surprising, if one
keeps in mind that ferromagnetic ordering is essentially due to
nearest neigbor interaction. Continuous lines of ferromagnetic elements have to exist in a crystal to give ferromagnetic ordering.
One should therefore expect that one needs a minimum concentration
of ferromagnetic impurities to obtain ferromagnetic ordering in an
alloy. For instance, in a bcc alloy like Cr-Fe, one would need at
least 12.5 % iron atoms to have on the average one iron nearest neighbor for each iron atom in a disordered alloy, since there are eight
nearest neighbors to each atom. It is therefore unlikely to have
ferromagnetism in Cr-Fe alloys with less than an iron concentration
of $c_{Fe} = 0.125$.

Oguchi and Obokata [127] calculated the Curie and Néel temperatures of disordered solid solutions with the Heisenberg and Ising
model. They assumed that the exchange interaction between two nearest neighbor magnetic atoms at R and $R + \Delta$ could be given by:

$$-2J_{exch.} (u)(S_R \cdot S_{R+\Delta}) = -2J_{exch.} [u(S_R^{x_1} \cdot S_{R+\Delta}^{x_1} + S^{x_2} \cdot S_{R+\Delta}^{x_2})$$
$$+ S_R^{x_3} \cdot S_{R+\Delta}^{x_3}] \qquad (2-124)$$

where $J_{exch.} > 0$, and where "u" represents the anisotropy of the exchange interaction. $u = 0$ for the Ising model, and $u = 1$ for the
Heisenberg model. The effective Hamiltonian for the two magnetic
atoms at R and $R + \Delta$ is given in their model by:

$$H_{II} = -2J_{exch.} (u) S_R \cdot S_{R+\Delta} - (z - 1)p'\lambda(S_R^{x_3} + S_{R+\Delta}^{x_3}) \qquad (2-125)$$

where z is the number of nearest neighbors, and p' is the concentration of magnetic atoms which are not isolated. Isolated magnetic atoms have only non-magnetic neighbors. The effective Hamiltonian on site R would be:

$$H_I = -[1 + (z - 1)p']\lambda S_R^{x_3} \qquad (2-126)$$

where the first term on the right hand site is the contribution from
$R + \Delta$ sites with magnetic atoms, and the $(z - 1)p'$ term comes
from the non-magnetic atoms.

The investigation by Oguchi and Obokata showed that there is
a simple relation between p', z, $J_{exch.}$ and T_c, which is given by:

$$J_{exch.}/k_B T_c = \log\{[p'(z - 1) + 1]/[p(z - 1) - 1]\} \qquad (2\text{-}127)$$

for the Ising model. The Heisenberg model gives:

$$J_{exch.}/k_B T_c = \tfrac{1}{2}\log\{[p'(z - 1) + 1]/[p'(z - 1) - 3]\}. \qquad (2\text{-}128)$$

If the total concentration of magnetic ions is c, then the total
number of magnetic atoms is Nc. The number of isolated magnetic
atoms for a disordered alloy is $Nc(1 - c)^z$, so that $c(1-[1-c]^z)=p'$.
The calculation then gives for the onset of magnetic ordering, c':

$$c'[1 - (1 - c')^z] = \begin{cases} 1/(z - 1), & \text{Ising model} \\ \\ 3/(z - 1), & \text{Heisenberg model.} \end{cases} \qquad (2\text{-}129)$$

Results of a calculation for $k_B T_c/J_{exch.}$ as a function of composition are given in Fig. 2-48 for both ferromagnetic and antiferromagnetic alloys. It shows a nearly linear decrease of $k_B T_c/J_{exch.}$
near the pure magnetic element with increasing concentration of the
non-magnetic element, and a rapid decrease near c'. $J_{exch.}$ may depend on the number of magnetic nearest neighbors, but one may assume
as first approximation that $J_{exch.}$ is constant. Therefore, Fig.
2-48 gives in first approximation the composition dependence of T_c.
It gives a reasonable description of experimental results for the
Cr-Fe alloy system, if one assumes the Heisenberg model.

Several examples of alloy systems can be given which cannot be
explained on the basis of the Oguchi-Obokata model. For instance,
it would be difficult to explain why the onset of ferromagnetism

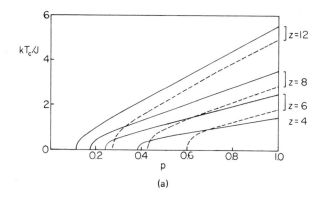

(a)

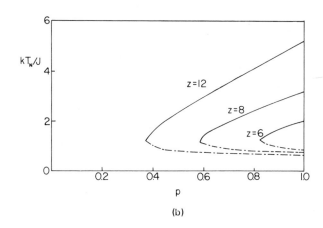

(b)

FIG. 2-48. The Curie temperature T_c for the Heisenberg model (dashed line) and Ising model (full line) for various coordination numbers z (a). The variation of the Néel temperature T_N with concentration p in Heisenberg antiferromagnet for various coordination numbers (b). Chain lines are anti-Néel temperature. After Ref. 127, with permission.

occurs at different iron concentrations for Fe-Cr and Fe-V alloys. This occurs in chromium alloys at about 84 at.% iron, and in Fe-V at about 72 at.% iron. It is possible that iron in vanadium-rich alloys has a different structure than in chromium-rich alloys. d-wave functions in vanadium are not localized. d-electrons of iron impurities may fill up these states. Localized states can exist

only after the conduction type electron states are filled up. This
implies that localized electron states can be found only for V-Fe
alloys with more than 28 at.% Fe, the composition where the density
of states curve begins to increase rapidly. Again, the onset of
ferromagnetism should occur at slightly higher iron concentration
than 28 at.%, since one has to fill at least enough localized electron states with electrons that chains of spins can form through the
lattice. Oguchi and Obokata assumed that the alloy was disordered.
Real alloys will show some degree of short range ordering. This could
affect the magnetization markedly.

Effects of atomic ordering on the magnetization of alloys has
been studied in Ni-Cu alloys. One would expect that a detailed analysis of short range order parameters could lead to a statistical analysis of the average magnetization of a nickel-rich cluster as a
function of order parameters. The short range order parameter gives
information on the concentration of the different components of the
alloy in the neighborhood of a given atom. Let us take a nickel atom.
One would look at its nearest neighbor atoms, which form the first
shell around this atom, and define as short range order parameter
for this first shell:

$$\alpha_1 = (c_1 - c_o)/(1 - c_o), \tag{2-130}$$

where c_o is the average concentration of nickel atoms in the crystal, and c_1 is the average concentration of nickel atoms around
all nickel atoms (this is a slightly different approach than used
in Chapter III for the definition of short range order in the Bethe
model). For the n-th shell, one defines accordingly as short range
order parameter α_n:

$$\alpha_n = (c_n - c_o)/(1 - c_o). \tag{2-131}$$

For a perfectly disordered alloy, α_n is zero since $c_o = c_n$. If
there are more nickel atoms around another nickel atom than in a

completely disordered alloy, then α_n is larger than zero (this
corresponds to "clustering"), whereas negative α_n values indicate
deficiencies of nickel atoms around a given nickel atom. Short range
order parameters have been measured in neutron diffraction studies on
furnace cooled alloys near the equiatomic composition. It was found
that this heat treatment gave $\alpha_1 = 0.121$, and $\alpha_n = 0$ for $1 < n \leq 9$.
It is possible to calculate with these parameters the possibility
P_{nm} of finding a nickel atom with n nearest and m next nearest
nickel atoms. A calculation by Robins et al. [128] gives:

$$P_{nm} = 12!6!c_1^n(1-c_1)^{12-n}c_o^{m+1}(1-c_o)^{6-m}/(12-n)!(6-m)!n!m!. \quad (2-132)$$

The average Bohr magneton number in the alloy is:

$$\bar{\mu}_B = \sum_{n=0}^{12} \sum_{m=0}^{6} P_{nm}\mu_{nm}. \quad (2-133)$$

μ_{nm} is the moment of a nickel atom with n nearest and m next
nearest nickel atoms. This equation gives with reasonable accuracy
the experimentally found average ferromagnetic moment as shown in
Fig. 2-49a, if μ_{nm} values for various local atomic configurations
were chosen according to Fig. 2-49b. The fit could be improved by
slight changes in the μ_{nm} values.

Figure 2-49 shows that it is possible to interpret the average
atomic moment in Ni-Cu alloys in terms of nearest and next nearest
neighbor interaction. This model indicates that it is reasonable to
have nickel-rich clusters near the composition $Ni_{0.5}Cu_{0.5}$. This has
been confirmed in neutron scattering experiments in weakly ferromagnetic alloys. The study showed that even in the ferromagnetic state
the alloys under investigation contained magnetic clusters which
overlap only slightly. The ferromagnetic bulk moment is about the
sum of the moments of the clusters, with about $8\mu_B$/cluster. Neutron
experiments suggest that magnetic clusters of this size are present
in paramagnetic alloys with 40 at.% nickel. Even $Ni_{.3}Cu_{.7}$ shows pronounced superparamagnetism, which is due to clustering.

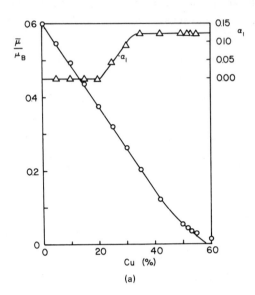

(a)

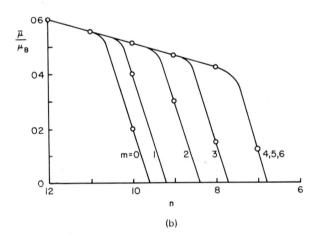

(b)

FIG. 2-49. Short range order parameter α_1 and average number of Bohr magnetons per atom $\bar{\mu}/\mu_B$ for copper-nickel alloys (a). Model average number of Bohr magnetons per nickel atom as a function of the number of nearest neighbors n and next nearest neighbors m (b). After Ref. 128.

This strong evidence of magnetic clusters in Cu-Ni alloys near
the onset of ferromagnetism is further strengthened through experimental studies, in which several different techniques are used to
determine cluster concentrations. The cluster concentration was
varied by different sample preparations. Plastic deformation in-
variably increased cluster concentrations. This is to be expected,
since the slip of dislocations in slip planes will section clusters
into smaller fragments, and each cluster fragment is a new cluster.
This shows that the magnetic properties of alloys should depend
markedly on short range order. Single phase materials which show
ferromagnetic and antiferromagnetic properties due to variations
in short range order are called mictomagnetic. These materials
usually exhibit at high temperature super paramagnetic properties.
At low temperatures, the magnetic structure freezes into configurations which depend on applied magnetic fields during cooling. Hysteresis loops may be displaced with respect to those found without
a magnetic field during cooling.

The complexity of magnetic interaction in alloys has been demonstrated in the ternary copper base alloy copper-nickel with iron
impurities [129]. These experiments will be discussed as an example in more detail. The susceptibility of the samples with no
iron impurities and with 10, 100, and 1000 ppm Fe impurities is
given in Fig. 2-50. The susceptibility of the purified $\underline{Cu}Ni$ alloy
is nearly temperature independent above $160^{o}K$, as typical for the
band susceptibility. An additional temperature dependent contribution is noticeable at low temperatures. It follows the Curie-Weiss
law, with a Curie temperature of $-3^{o}K$, and a Curie-Weiss constant
$C = 0.38(10^{-6} emu^{o}K/g)$. If one assumes that the Curie-Weiss component of the susceptibility is due to moments of individual nickel
atoms, and that the contribution from large clusters can be neglected, one then finds the experimentally observed Curie-Weiss susceptibility if the short range order parameter $\alpha_1 = 0.1$, and $\alpha_2 = 0$.

The effect of iron impurities on the magnetization was obtained
from the experimental data by assuming that the magnetization due

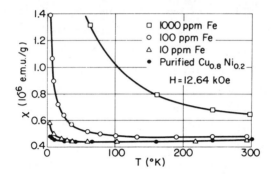

FIG. 2-50. Susceptibility vs temperature for Cu-Ni alloys with various amounts of Fe impurities, quenched from 1050°C. After Ref. 129, with permission.

to the iron impurities again followed a Curie-Weiss law (this required in the calculation that the Curie-Weiss contribution of the iron impurities follow the Brillouin function, and that there are 97 ppm iron atoms with a 3.7 μ_B iron atom). Agreement between the calculation and the corrected data points is very good. Similar measurements and calculations were made for alloys with 10 and 1000 ppm Fe in the Cu-Ni alloy.

B. Magnetization and Critical Temperatures of Antiferromagnetic and Ferrimagnetic Alloys

Extensive investigations on the magnetic structure of chromium alloys have been made with neutron diffraction measurements. One would expect that magnetic order in these alloys is strongly composition dependent since it is associated with conduction electrons. Adding or subtracting electrons will change the shape of the Fermi surface. The peculiarity, that one hole segment and one electron segment of the Fermi surfaces have nearly the same shape, will be lost by changing the electron concentration. One would therefore expect that small amounts of impurities can change the antiferromagnetic structure of chromium markedly.

The addition of manganese to chromium will increase the el./at. ratio. The alloy is not compensated. The number of electrons is larger than the number of holes. Using the rigid band model, Koehler et al. [126] show that the wave vector of the electrons should increase by one percent, if one percent manganese is added to the alloy. Vanadium alloys with the el./at. ratio below 6 should show a corresponding decrease of the wave vector by 1% if 1% vanadium is added to the sample. Experiments give approximately twice this change in the wave vector for both vanadium-chromium and chromium-manganese alloys. Figure 2-51c gives experimental results. It is interesting to note that the wave vector changes discontinuously at a critical impurity concentration from the non-commensurate spin wave system (the rigid band model calculations are applicable only in this range) to a spin wave commensurate with the lattice, without a discontinuous change in the number of Bohr magnetons per atom. Figure 2-51a + b shows experimental results where the Néel temperature and the number of Bohr magnetons/atom is given as a function of impurity concentration. Naturally, the spin-flop temperature T_{SF} depends also on impurity concentration, since both the crystal energy and the susceptibility are functions of alloy concentration. The spin-flop temperature seems to decrease for all alloy systems with increasing impurity concentration. The neutron-diffraction measurements seem to show that the number of μ_B/atom decreases for all alloy systems where the el./at. ratio of the impurity is $\leq$ 6, and it increases for el./at. values larger than 6. Electrical resistivity measurements show, however, that T_N decreases if iron and nickel are added to chromium, whereas cobalt changes T_N only slightly (see Chapter 4). The μ_B/atom value for chromium-manganese alloys increases rapidly with manganese concentration. 7 at.% Mn are sufficient to increase the Bohr magneton/atom ratio to 0.8. This cannot be explained with the addition of "magnetic atoms" of manganese to the alloy, since only 7 at.% Mn are added to the alloy. Therefore, manganese changes the electronic structure of the system, maybe by some type of polarization.

Usually ferrimagnetic compounds have magnetic moments due to
localized electrons, since they are typically insulators. It therefore requires that the number of metallic elements is a fixed ratio
of the total number of atoms, as indicated by the low integer ratio
of elements in a chemical formula. However, it is possible to replace one metallic element in a compound gradually by another metallic element, which keeps the ratio of metallic elements to non-
metallic elements constant. These materials would be "pseudo-binary
alloys". An example is the "ferrite" $Co_{1-c}Zn_cO\ Fe_2O_3$, where c
ranges from zero to one. Typical ferrimagnetic materials of this
type are cubic ferrites, which are found in the spinel structure
$Mg \cdot O \cdot Al_2O_3$, and the corresponding ferrimagnetic compound $Me \cdot O \cdot Fe_2O_3$,
where Me could be any of the twice ionized elements Mg, Mn, Fe,
Co, Ni, Cu, Zn, or Cd. Excluded are Ca, Sr, and Ba. The basic
structure for these ferrites is given in Fig. 2-30. The crystal
consists of a cubic closely packed lattice with 32 oxygen ions and
24 metallic ions. Eight of these metallic ions occupy positions in
the center of oxygen tetrahedrons, and 16 are found in the center of
oxygen octahedrons. If all twice-ionized metal ions occupy tetrahedral positions, and the three times ionized metal ions the octahedral positions, then the ferrite is "normal". However, it is
possible that all tetrahedral voids are occupied by Me^{+++} ions, half
the octahedral voids are occupied by Me^{++} ions, and the other half
again occupied by Me^{+++} ions. This is the "inverse" structure,
which is found for $Cu \cdot O \cdot Fe_2O_3$ and $Mg \cdot O \cdot Fe_2O_3$, whereas $Zn \cdot O \cdot Fe_2O_3$
and $Cd \cdot O \cdot Fe_2O_3$ are "normal" ferrites. In an inverted spinel, the
sum of the magnetization of all spins is equal to the spins of the
Me^{++} ions, since the number of Me^{+++} ions in the octahedral voids
is equal to the number of Me^{+++} ions in tetrahedral voids, and these
should cancel since they are lined up antiparallel.

The magnetic properties of ferrites are determined by strong
antiparallel coupling of the two sublattices of either octahedral or
tetrahedral metal ion positions. This leads to a ferrimagnetic system with parallel alignment of spin within each sublattice, and

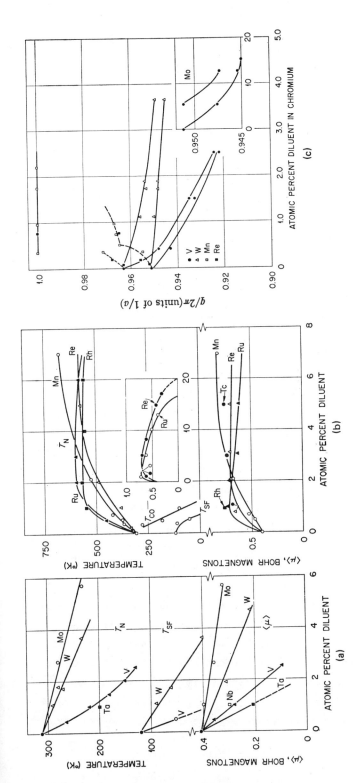

FIG. 2-51. Transition temperatures and average-moment values for chromium base alloys (a + b). Transition temperatures are designated T_N = Neel temperature and T_{SP} spin flip temperature. The symbol $\langle\mu\rangle$ stands for the root-mean-square moment per atom in the non-commensurate alloy. The dependence of the wave vector q on composition for chromium alloys (c). The value of $q = 2\pi/a$ corresponds to the commensurable structure. The upper two curves refer to measurements made near the Neel temperature, the lower to very low temperature data. Reproduced with permission from Ref. 126.

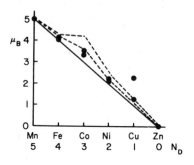

FIG. 2-52. Molecular magnetic moment of some inverted ferrites.
After Ref. 130.

antiparallel alignment of the two sublattices with respect to each
other, with the magnetization of the "octahedral" sublattice $M_{s,o}$
usually different from the magnetization of the "tetrahedral" sublattice $M_{s,t}$. This model was first proposed by Néel in 1948.

Figure 2-52 gives the number of Bohr magnetons as a function
of uncompensated spins of the Me^{++} ions. The experimental data are
slightly higher than expected from the spin of the electrons. A
calculation of possible orbital contributions to the moment of the
ions is given by the dashed lines in this figure. Since the experimental points are close to or within this theoretical prediction,
except for one point of the copper compound, one can conclude that
the magnetic moment of the spinel is theoretically well described.

Figure 2-52 shows for the copper compound two magnetization
values. The lower value was obtained on a sample which was annealed
for a long time at $300^{\circ}C$. The higher value of 2.56 μ_B was measured
on a sample which was quenched from $1000^{\circ}C$. This point differs
markedly from the linear magnetization vs composition relationship
observed in the other compounds. Néel explained the increase in
the magnetization due to the high temperature anneal in the following way. The energy of an Me-ion depends on its lattice position.
The difference in energy in the crystal, if a Me^{++} ion and a Me^{+++}
ion are exchanged, is responsible for changes in the ion distribution in the crystal with temperature. Let us assume the energy

change due to this exchange of metal ions is E^o. If c is the
concentration of Me^{++} ions in tetrahedral voids, and 1-c its concentration in octahedral voids, then thermal equilibrium leads to
using the Boltzmann distribution factor [114]:

$$c(1 + c)/(1 - c)^2 = \exp(- E^o/k_B T) \quad . \tag{2-134}$$

One obtains the "normal" ferrite for T = 0 and E^o < 0, and the
"inverse" ferrite for T = 0 and E^o > 0. The molecular moment for
the two sublattices as a function of composition c is, if one
characterizes the magnetic moment of a twice-ionized ion by m, and
of the three times ionized metal ion as m':

$$M_{s,t} = (1-c)m' + cm \qquad \text{sublattice in tetrahedral voids}$$
$$\tag{2-135}$$
$$M_{s,o} = (1+c)m' + (1-c)m \qquad \text{sublattice in octahedral voids}$$

of the oxygen lattice. m' is 5 μ_B, the moment of the Fe^{+++} ion.
The saturation magnetization $M_s = M_{s,o} - M_{s,t}$ is

$$M_s = m + 2c(m' - m) \tag{2-136}$$

This value is temperature dependent, since c is temperature dependent. Inserting Eqs. (2-134) and (2-135) into Eq. (2-136) gives this
temperature dependence. A comparison of calculations with experimental data is given in Fig. 2-53 for two E^o values. Agreement
between theory and experiments is good.

The composition dependence of the saturation magnetization of
pseudo binary solutions of ferrites depends on their magnetic structure. If both ferrites are inverse, then

$$M_s = c\, m_i + (1 - c)\, m_k \tag{2-137}$$

if Me_i and Me_k ions occupy positions in octahedral voids. A
typical example is $Ni_c Mn_{1-c} OFe_2 O_3$ (2-54). The Curie temperature

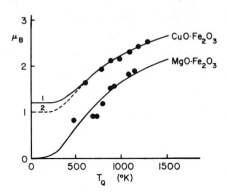

FIG. 2-53. Magnetic moment of CuO·Fe$_2$O$_3$ and MgO·Fe$_2$O$_3$ as a function of quench temperature. Full lines are calculated. After Ref. 131.

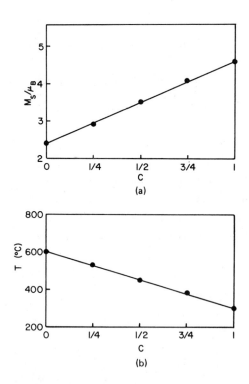

FIG. 2-54. Molecular magnetic moment (a) and Curie temperature (b) of Ni$_{1-c}$Mn$_c$O·Fe$_2$O$_3$. After Ref. 132.

is also a linear function of c. If the Me_i-ferrite is "normal",
and the Me_k-ferrite is "inverse", the saturation magnetization M_s
can show a maximum if M_s is plotted as a function of composition
c. Let us assume that Me_i is a diamagnetic ion ($m_i = 0$), and m_k
is equal to m. Me_k should be "inverse". Therefore, all Me_k ions
are in octahedral positions. Replacing Me_k by Me_i, the Me_i ion
occupies a tetrahedral position, and an Me^{+++} (Fe^{+++}) ion changes
from a tetrahedral position to an octahedral position. Since $m_i = 0$,
the substitution of Me_k by Me_i leads to a magnetization change
$2m' - m$. This gives, if c is the concentration of Me_i ions:

$$M_{t,s} = m' (1 - c)$$

$$M_{o,s} = m' (1 + c) + m (1 - c)$$

(2-138)

and

$$M_s = M_{o,s} - M_{t,s} = (1 - c)m + 2 c m'$$

(2-139)

Since $m < 2 m'$, M_s will increase with the addition of a non-ferromagnetic impurity. The equations (2-138) and (2-139) are applicable
only for $c \ll 1$. For $c \rightarrow 1$, the magnetic structure changes from
the inverse to the normal spin alignment of the spinel. This ferrite can be non-magnetic as shown by Néel with the molecular field
theory. In our example, M_s should be zero for $c = 1$. Therefore,
M_s should reach a maximum for $0 < c < 1$. Figure 2-55 gives experimental results for $Me_{1-c}Zn_c O Fe_2 O_3$ ferrites with Me = Mn, Co, or
Ni.

The molecular field theory for ferrites can be written in the
same way as before for ferromagnetic or antiferromagnetic systems.
However, we will use now a slightly different notation. This notation, proposed by Néel, makes it possible to give a systematic clas-
sification for different types of ferrimagnets (2-56 and 57). One
writes for a two sublattice system with sublattices A and B for
the molecular field:

$$H_{Am} = n(\alpha\lambda M_{As.} + \varepsilon\mu M_{Bs.})$$

(2-140a)

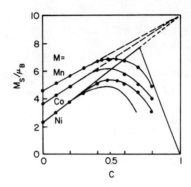

FIG. 2-55. Magnetic moment of $M_{1-c}Zn_cO \cdot Fe_2O_3$ (M = Mn, Co, Ni).
After Ref. 132.

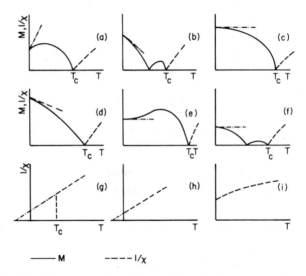

FIG. 2-56. Magnetization M and $1/\chi$ of ferrimagnetic crystals
according to the molecular field model. $-\cdot-$ line gives slope at T =
0^0K. Types of curves "j" given in Fig. 2-57. After Ref. 133.

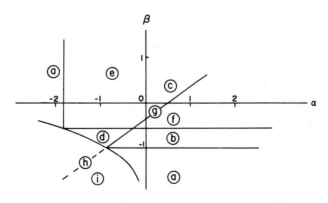

FIG. 2-57. α-β diagram for $\lambda/\mu = 2/3$. After Ref. 133.

$$H_{Bm} = n(\beta\mu \ M_{Bs} + \epsilon\lambda \ M_{As}) \tag{2-140b}$$

$\epsilon = -1$ gives a negative and $\epsilon = +1$a a positive coupling between sublattices A and B. n is always positive, and α and β are both positive for ferromagnetic, and both negative for antiferromagnetic coupling within each sublattice. n, nα and nβ are the exchange parameters, which characterize the magnetic behavior of a simple "two-sublattice" ferrimagnet, where the total magnetization would be given by [114]:

$$M_s = \lambda \ M_{As} + \mu \ M_{Bs} \tag{2-141}$$

The magnetization above T_c for the two sublattices would be given by:

$$M_A = (C/T) \ (H + H_{Am})$$
$$M_B = (C/T) \ (H + H_{Bm}) \tag{2-142}$$

with C the Curie constant. Elimination of M_A, M_B, H_{Am}, H_{Bm} in the above set of equations leads to:

$$\chi = M/H = [T^2 + n\{\alpha\lambda + \mu\beta\}T +$$
$$n^2C^2 \ \lambda\mu\{\alpha\beta - 1\}]/[T - nC \ \lambda\mu\{2 + \alpha + \beta\}] \tag{2-143}$$

which gives for $1/\chi$:

$$1/\chi = T/C + 1/\chi_o - \sigma/(T - \theta) \tag{2-144}$$

with:

$$1/\chi_o = n(2 \ \lambda\mu - \lambda^2\alpha - \mu^2\beta) \tag{2-145}$$

$$\sigma = n^2(\lambda\mu\{\lambda \ [1 + \alpha] - \mu[1 + \beta]\}^2 \cdot C \tag{2-146}$$

$$\theta = nC\lambda\mu \ [2 + \alpha + \beta] \tag{2-147}$$

Figure 2-56 gives typical plots of M and $1/\chi$ as a function of T. There exist combinations of α, β, λ, μ, n for which $T_c \leq 0$. Then the magnetization in zero field is always zero. For $T_c > 0$, the saturation magnetization for each sublattice as a function of temperature is given by the sum of two Brillouin functions:

$$M_s = M_{As,\infty} \, B_j\{c'[\alpha\lambda \, M_{As} + \mu \, M_{Bs}]\}$$
$$+ \, M_{Bs,\infty} \, B_j\{c'[\beta\lambda \, M_{Bs} + \lambda \, M_{As}]\} \qquad (2\text{-}148)$$

with $c' = jg \, \mu_B \, n/k_B T$. Néel showed that the different types of M or $1/\chi$ vs T functions which are possible can be systematically correlated with α and β values. α and β characterize the exchange interaction. Figures 2-56 and 2-57 show the correlation between α, β and different types of M and $1/\chi$ curves for $\epsilon = 1$ and $\lambda/\mu = 2/3$. For α, β combinations in section 'i', $T_c < 0$. These materials show no macroscopically noticeable spontaneous magnetization. M is zero and $1/\chi$ is a linear function of T along the line going through 'h' and 'g' in (2-57).

C. Magnetic Anisotropy in Alloys

The Curie and Néel temperatures are one of the few examples of magnetic properties which are independent of crystal orientation. However, most magnetic properties such as the magnetization in zero field depend on the crystal orientation. This was predicted from the classical model of a periodic array of classic magnetic dipoles given in Eq. (2-101).

The orientation dependence of the magnetic energy is associated with several types of anisotropies. Already discussed is the crystal anisotropy of pure metals. If the different types of atoms in an alloy are randomly distributed, then the crystal anisotropy of alloys can be given with the same equations as given for pure elements. However, if the atoms in the crystal show short range or

long range ordering which influences their energy, then the magnetization for a specific sample may be more easily achieved in a
[1 0 0] direction than in a [0 1 0] direction [134]. Since the non-
random distribution of different types of atoms is frequently obtained through a diffusion process at elevated temperatures in a
magnetic field, the phenomena is called "diffusion anisotropy", and
the annealing process is "magnetic annealing". The annealing temperature has to be below the Curie temperature to be effective.

Changes in the magnetization can also lead to changes in the
shape of the sample. If one heats a single crystal sample above
the critical temperature for magnetic ordering, machines it to a
sphere, and lets it cool to below the critical temperature, it will
usually have the shape of an ellipsoid if it is a "magnetic single
crystal" or one magnetic domain. Relative length changes are typically of the order of 10^{-5} to 10^{-6}. Magnetic free poles on the
surface of a sample lead also to magnetic energies, which depend on
the shape of the sample and can produce changes in shape. Needles,
for instance, prefer a magnetization parallel to their long axis,
since this reduces the magnetic energy of the system. Length changes
of the sample due to magnetic fields associated with free poles of
this sample are due to "shape anisotropy".

Length or volume changes due to magnetic effects are named
"magnetostriction". One frequently subdivides it into three effects:
"volume magnetostriction", $(\Delta V/V)_M$, which gives the volume change of
the sample with changes in the magnetization; "crystal magnetostriction", $(\Delta V/V)_C$, which describes the volume dependence on the direction
of the magnetization with respect to the crystallographic directions
without changes in the magnetization; and "shape magnetostriction",
$(\Delta V/V)_S$, due to demagnetizing field. Changes in volume or length
with changes in magnetization are technically very interesting,
because the magnetization, and with it the magnetostriction, is
temperature dependent. This means that one can control the thermal
expansion coefficient of a material by selecting the proper alloy
composition. Figures 2-58 and 2-59 give the thermal expansion

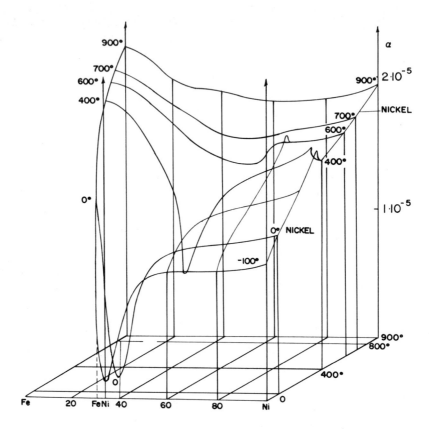

FIG. 2-58. Composition and temperature dependence of the thermal expansion coefficient α of Fe-Ni alloys. After Ref. 135.

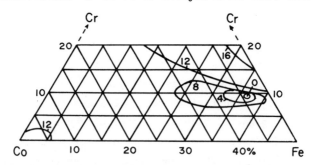

FIG. 2-59. Concentration dependence of the thermal expansion coefficient α of ternary Co-Fe-Cr alloys. α is given in units of 10^{-6}. The thermal expansion coefficient is close to zero for $Fe_{48}Co_{43}Cr_9$. After Ref. 136.

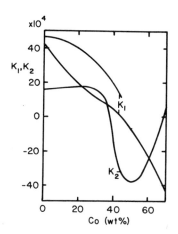

FIG. 2-60. Composition dependence of the constants of crystal energy K_1 and K_2 in Fe-Co alloys. (See Ref. 114 for details).

coefficient of some binary and ternary ferromagnetic alloys. The thermal expansion coefficient can be reduced to less than 10^{-2} of the value found in corresponding non-magnetic systems. With proper mechanical deformation, it is even possible to obtain negative thermal expansion coefficients. Therefore, one can build components which show practically no length change due to room temperature fluctuations, and it is possible to match the thermal expansion of some glasses over an extended temperature range with the thermal expansion of magnetic alloys for vacuum feedthroughs.

Since the crystal anisotropy depends on the chemical composition of the alloy (2-60), it is not surprising that the crystal anisotropy will depend also on long and short range ordering. The most noticeable effect of the interaction of atomic and magnetic ordering is found in systems which show unsymmetric magnetization curves. Figure 2-61a gives an example of a magnetization vs field strength curve of oxidized cobalt particles, where the M_s - H curve is shifted from the symmetry line H = 0. The explanation for such behavior can be deduced from the schematic diagram in Fig. 2-61b. In this figure, the material consists of two phases, a ferromagnetic

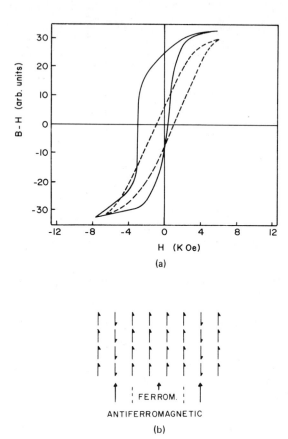

FIG. 2-61. Hysteresis loop of small oxidized cobalt particles
at 77°K. Full line: cooled in field. Dashed line: cooled without
field (a). After Ref. 114. Hypothetical magnetic moment orientation of a small ferromagnetic particle in an antiferromagnetic
matrix, which leads to an unsymmetric hysteresis loop (b).

metal and an antiferromagnetic oxide. The row of spins in the interface of these two phases (dashed line) matches the spin order of
both the ferromagnetic and antiferromagnetic phases.   One could
reverse the orientation of the spins in the ferromagnetic phase
by applying a field of a few hundred Oersted; however, such a field
would not affect the spin orientation of the antiferromagnetic

phase. Therefore, the reversal of the spins in the ferromagnetic phase would require a very high energy near the interface of the two phases, since there the spins are lined up in the "wrong" direction.

The exchange of two different atoms A and B in an alloy system could change the magnetic energy. Therefore, the magnetization can affect short (or long) range order, since some atomic configurations will have a lower energy than others in a magnetized sample. An applied field, which affects the orientation of the magnetization and its value, can therefore lead to changes in ordering through a diffusion process. Typically, one anneals the sample at a temperature T_1, high enough for diffusion (T_1 has to be below the Curie temperature T_c), and then cools it rapidly to a temperature T_0 where diffusion effects can be neglected. Calculations of this type of "diffusion" anisotropy have been given by Néel [137, 138] and Taniguchi and Yamamoto [139].

Figure 2-62 shows the ternary Fe-Co-Ni composition diagram. Shaded areas indicate which alloys show magnetic annealing. Heavily shaded areas give strong magnetic annealing effects. This diagram indicates clearly that magnetic annealing effects should be associated with alloys, and probably with their ordering, since the pure elements and samples with small amounts of second elements show no magnetic annealing effects. Further, if the sample is fully atomically ordered, then it will not respond to magnetic annealing. Figure 2-63 shows an example of changes in the magnetization curve due to a magnetic annealing treatment. The sample tested without this treatment shows a narrow waist in the M_s - H curve. Materials with such a magnetic characteristic frequently respond to a magnetic annealing treatment, which gives them a very narrow, nearly rectangular hysteresis loop.

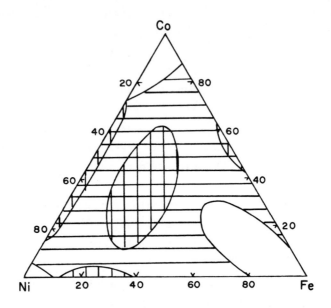

FIG. 2-62. Magnetic annealing in Fe-Co-Ni alloys. Shaded areas represent compositions which respond to magnetic annealing. Dark shadings are compositions which show the effect very strongly. After Ref. 140.

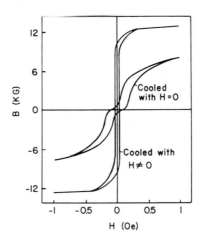

FIG. 2-63. Effect of magnetic annealing on $Ni_{65}Fe_{35}$. After Ref. 140.

Chapter 3

THERMAL AND ORDERING ENERGIES

I. INTRODUCTION

The interpretation of the internal energy values, or of the specific heat of solids was very important for the development of atomic models of metals and alloys. The Dulong-Petit rule, which states that the specific heat of metals is 6.2 cal/mole°K, could be best explained if one assumed that solids are built up of individual atoms, and that each atom contributes the same amount to the specific heat. Discrepancies between measured specific heats at low temperatures and this rule indicated by the end of the 19th century that classical physics could not give an adequate description of nature. Drastic changes of concepts in physics were needed to describe the properties of materials.

Planck showed that the model of "quantized energy states" would give an adequate theoretical background to solve the riddle of the "black body radiation", and Einstein used this model in 1907 to explain why the specific heats of solids decrease below a critical temperature rapidly with decreasing temperatures. Another twenty years passed before Sommerfeld applied the Fermi-Dirac statistic to an analysis of energy states of conduction electrons in metals. This explained why the very mobile conduction electrons in metals give only a small contribution to the specific heat in most temperature ranges. The classical model predicts that the conduction electrons contribute about 3 cal/mole°K to the specific heat, whereas the

measured values near room temperature could be best explained with
thermal energy contributions from ions alone.

The heat capacity of a sample is defined by:

$$C = \Delta Q / \Delta T \qquad\qquad (3\text{-}1)$$

where ΔQ is the amount of heat added to the sample, and ΔT is
the temperature increase of the sample due to the heat input. This
definition is not meaningful if the sample undergoes a first order
phase transformation in the temperature interval ΔT. Investigations
of the stability of phases, in which the heat of fusion associated
with phase changes is important, are outside the scope of this book.
We are concerned here with the question of how the specific heat or
internal energy of metals and alloys due to thermal excitations of
electrons, ions and atoms, and due to changes in the atomic configuration depends systematically on elements, chemical composition, and
the state of atomic or magnetic ordering.

The heat capacity of a sample can be determined per unit volume,
per unit mass, or per mole. It can be measured at constant volume
or at constant pressure. Unless otherwise stated, we will use the
specific heat per mole, c, and indicate by subscript "p" if the
specific heat is measured at constant pressure, or by subscript "v",
if the specific heat is measured at constant volume. The difference
between both terms is given by:

$$c_p - c_v = 9\alpha^2 V_A T / \beta \qquad\qquad (3\text{-}2)$$

with V_A the volume per mole, $\alpha = + (d\ell/\ell dT)_p$ the linear expansion
coefficient, and $\beta = - (dV/Vdp)_T$ the compressibility. Here the subscript p or T indicates which parameter is kept constant. Frequently, α and β are not known over the temperature range under
investigation in specific heat measurements. Then one often
uses the correlation:

$$c_p - c_v = A^* c(T)_p^2 T. \qquad\qquad (3\text{-}3)$$

A* is a constant, evaluated at room temperature. A* is equal to $9(\alpha^2 V_A/\beta c_p^2)_{\text{room temp.}}$. c_v is preferred in theoretical considerations; c_p is much more easily measured.

c_p measurements are frequently determined with adiabatic techniques in which the heat loss of the sample to the surroundings is so small that it can be neglected. Therefore, heat added to the sample, typically in the form of Ohmic heating, is equal to $\Delta Q = I^2 R\Delta t$ (I is the heater current, R the heater resistance) in Eq. 3-1. For temperatures in the liquid He temperature region, the temperature of the vessel surrounding the sample is usually kept constant during a heating pulse Δt. Both the heat pulse method and continuous heating techniques are used at liquid He temperatures. Liquid He-4 can be used down to 1.3°K, and with He-3 one is able to reach 0.4°K. The essential features of an adiabatic calorimeter used from 100°K to 600°K are shown schematically in Fig. 3-1.

II. LATTICE SPECIFIC HEAT

A. The Perfect Lattice

The Dulong-Petit rule states that c_p of metals is approximately equal to 6.2 cal/mole$^\circ$K. It is probably just coincidence that this rule works so well for a large number of metals at room temperature, since the classical model of a crystal can explain only why the lattice specific heat measured at constant volume should be constant. One assumes in this model that each atom in a crystal is a harmonic oscillator. Forces which are not proportional to the displacements of atoms lead to "anharmonic specific heat" contributions.

Each atom has three degrees of freedom because it can oscillate independently in three directions perpendicular to each other. Each atom can have both kinetic and potential energies. The thermal energy of a harmonic oscillator per degree of freedom, and per mode of oscillation, is in the classical model equal to:

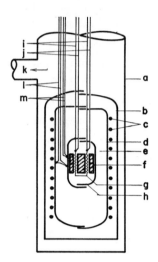

FIG. 3-1. Schematic diagram of a vacuum calorimeter. Stainless
steel tube (a), radiation shield (b), heater for radiation shield (c),
radiation shield (d), specimen assembly (e), specimen (f), specimen
heater (g), copper foil (h), specimen heater current leads (i), potential leads (j), to pump (k), differential thermocouple (l). Temperature range: 100 to 600°K.

$$E^*_{osc.} = k_B T/2. \qquad\qquad\qquad\qquad (3-4)$$

One assumes in the classical model of a solid that each oscillator

operates independently from all others. One assumes also that all

electrons of one atom are tightly bound to this atom. They cannot

absorb any thermal energy. Therefore, the total thermal energy of

a crystal lattice containing one mole of atoms with three degrees of

freedom and two modes of energy should be from absolute zero tempera-

ture to the melting point:

$$E_{1a.} = 3 \cdot 2 A_o k_B T/2 = 3RT, \qquad\qquad (3-5)$$

where A_o is the Avogadro number, and R the gas constant. The specific heat per mole of this lattice, $c_{la.}$, should therefore be:

$$dE_{la.}/dT = c_{la.} = 3R = 5.96 \text{ cal/mole}^o K. \tag{3-6}$$

This value is very close to the value given by the Dulong-Petit rule. This derivation would indicate that c_v should be constant, not c_p. Experimental data indicate, however, that the specific heat at constant pressure, and not the specific heat at constant volume at room temperature is nearly the same for numerous metals.

The decrease of the lattice specific heat with decreasing temperatures cannot be explained with a classical model. Einstein [16] showed that one would expect such a decrease if one assumes that the energy states of an atom in a crystal are quantized. He proposed that the energy of an oscillator should be a multiple of $h\nu$:

$$E^o_{osc.} = nh\nu \tag{3-7}$$

where n is an integer and ν the frequency of oscillation of the atom. The Boltzmann factor:

$$n_i/n_j = \exp\{-(E_i - E_j)/k_B T\} \tag{3-8}$$

determines the ratio of the number of atoms in states with energy levels E_i and E_j. This gives for the total energy of one oscillator:

$$E_{osc.} = \{\Sigma_i E_i \exp(-E_i/k_B T)\}/\Sigma_i \exp(-E_i/k_B T) \tag{3-9}$$

where the denominator $\Sigma_i \exp(-E_i/k_B T)$ normalizes the equation. Replacing E_i by $ih\nu$ gives as the thermal energy per oscillator:

$$E_{osc.} = \{\Sigma_i ih\nu \cdot \exp(-ih\nu/k_B T)\}/\Sigma_i \exp(-ih\nu/k_B T). \tag{3-10}$$

The sum extends from one to infinity. $\Sigma_i \exp(-ih\nu/k_B T)$ is the sum

of an infinite geometric progression. One can write $\Sigma \exp(-ih\nu/k_BT)$
$= \Sigma x^i = 1/(1 - x)$, with $x = \exp(-h\nu/k_BT)$. This gives $\Sigma \exp(-ih\nu/k_BT)$
$= 1/(1 - \exp\{-h\nu/k_BT\})$. Further: $\Sigma i x^i = x \cdot d(\Sigma x^i)/dx = x/(1-x)^2$
$= \exp(-h\nu/k_BT)/\{1 - \exp(-h\nu/k_BT)\}^2$. Therefore the average thermal
energy of each oscillator is:

$$E_{osc.} = h\nu/\{\exp(h\nu/k_BT) - 1\} \tag{3-11}$$

(A more advanced quantum mechanical treatment gives an additional
term of $h\nu/2$.) One can rewrite Eq. 3-11 by introducing the "average
phonon occupancy number" $<n>$ which is the average number of phonons
per oscillator:

$$E_{osc.} = <n>h\nu. \tag{3-12}$$

This gives:

$$<n> = \{\exp(h\nu/k_BT) - 1\}^{-1}. \tag{3-13}$$

$<n>$ is the Bose–Einstein distribution function.
     The specific heat per mole in the Einstein model, c_E, is obtained
from this equation by differentiation of Eq. 3-11, and multiplying
it with $3A_o$. This gives:

$$c_E = 3R(h\nu/k_BT)^2 \exp(h\nu/k_BT)/\{\exp(h\nu/k_BT) - 1\}^2. \tag{3-14}$$

Instead of characterizing the harmonic oscillator by its Eigenfrequency ν, which is also called the "Einstein frequency" ν_E, one uses
frequently the "Einstein temperature" T_E as characteristic parameter.
It is defined by:

$$T_E = h\nu_E/k_B. \tag{3-15}$$

The specific heat c_E is a unique function of T/T_E (sometimes
one uses its reciprocal T_E/T as independent variable), and one

obtains from Eqs. 3-14 and 3-15:

$$c_E = 3R(T_E/T)^2 \exp(T_E/T)\{\exp(T_E/T) - 1\}^{-2} = 3R \cdot E(T/T_E) \qquad (3\text{-}16)$$

where $E(T/T_E)$ is the Einstein function. This function approaches
the value of one for high temperatures, so that the high temperature
limit of the Einstein model is equal to the classical Dulong-Petit
law. At $T/T_E = 1$, c_E is equal to 5.4 cal/mole^{o}K. Figure 3-2
shows the Einstein specific heat as a function of temperature. At
very low temperatures, the Einstein model predicts an exponential
temperature dependence of the specific heat. This could not be confirmed experimentally. One found that a function which is the sum
of two Einstein functions:

$$c = (3/2)R\{E(T/T_E(1)) + E(T/T_E(2))\} \qquad (3\text{-}17)$$

gives a good description of the specific heat down to liquid nitrogen
temperatures. Equation 3-17 is the Nernst-Lindemann equation [141].
These authors suggested that $T_E(1) = 2T_E(2)$.

In spite of its shortcomings, the Einstein model frequently
gives an adequate description of the lattice specific heat. It is
very useful if one correlates properties of a sample with the atomic
structure. The Einstein model predicts that the specific heat
of an alloy would be the sum of the contributions of all atoms, a
rule obeyed frequently in simple alloy systems without magnetic or
atomic ordering. The Einstein model makes it easy also to describe
different energy states of atoms on surfaces and interfaces. These
atoms may have different Eigenfrequencies as atoms below the surface,
since the spring constants can be different due to a smaller number
of nearest neighbors.

The Nernst-Lindemannn equation proved to be inadequate at very
low temperatures. So, just a few years after the Einstein model,
two different theories were proposed to describe thermal lattice
vibrations. Born and von Karman [19] used an atomic model, in which

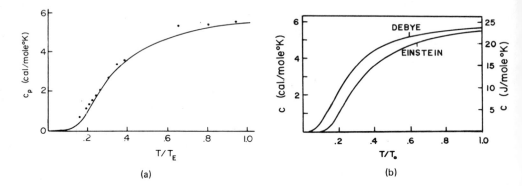

FIG. 3-2. Specific heat of diamond. Experimental results given by points, the Einstein specific heat is given by the full line (a, after Ref. 16). Comparison of the Einstein and Debye specific heats. $T_0 = T_E = T_D$ (b).

atoms with a given mass were connected by springs. They analyzed possible modes of oscillation in such a system. A second theory was proposed by Debye. He described the sample as an elastic continuum, in which the excitations of sound waves are investigated. The Born-von Karman theory would naturally give a more detailed picture of the atomic structure of solids. However, since the Debye theory is simple and frequently gives an adequate description of experimental results, most low temperature specific heat data were and are ana-lyzed with this model.

The thermal energy in the Debye model is obtained in a manner similar to the Einstein model. Again, one counts the number of pos-sible vibrations. However, in the Debye model one takes the sum over all possible elastic vibrational states instead of summing overall atoms. Further, one assumes that the sample is elastically isotropic and shows no dispersion. One obtains then by using $dn = g(\omega)d\omega$ (see Eq. 1-26):

$$E = \int dn <n> \hbar\omega = \int d\omega g(\omega) <n> \hbar\omega \qquad (3-18)$$

since the energy for elastic waves is equal to the average phonon occupancy number $<n>$ multiplied with the energy $\hbar\omega$. This has to be multiplied by three (neglecting differences in transverse and longitudinal wave velocity) since one finds one longitudinal and two transverse modes of acoustic waves in a crystal. One obtains therefore for this simplified model for each mode [3];

$$E = \int_{o}^{\omega_D} d\omega \, (\omega^2 L^3 / 2\pi^2 v^3) \, \hbar\omega / \{\exp(\hbar\omega/k_B T) - 1\}^{-1} \qquad (3-19)$$

where $\omega_D = 2\pi\nu_D$ is the angular Debye frequency. It is the maximum angular frequency (cut off frequency) in the Debye model. ω_D is proportional to the cut off wave vector q_D. L is the length of the cubic sample, and v the sound velocity. The Debye temperature is:

$$T_D = \nu_D h/k_B. \qquad (3-20)$$

Since this corresponds to the cut-off wave vector $q_D = \omega_D/v$, it is possible to determine the Debye temperature from the sound velocity in solids. One obtains, with $q_D = (6\pi^2)^{1/3} N^{1/3}/L$ [3]:

$$T_D = (\hbar v/k_B L)(6\pi^2)^{1/3} N^{1/3}. \qquad (3-21)$$

Here N is the number of atoms in a cube of length L.

A simple estimate of the low temperature lattice specific heat can be obtained in the following way. Only lattice modes with $h\nu < k_B T$ will be excited. Their energy is of the order of $k_B T$. The total energy should be of the order of the products of these terms. This is $(\nu/\nu_D)^3 Nk_B T = (T/T_D)^3 Nk_B T$. Here N is the number of atoms per mole. This correlation leads by differentiation with respect to T to the heat capacity $c_{la.} \simeq 4(T/T_D)^3 Nk_B$. This equation gives the right temperature dependence of the lattice specific heat at low temperatures. However, the constant factor is off by 50. The correct value is obtained by differentiating Eq. 3-19 with respect to T.

Numerical values of the Debye function can be found in Ref. 18. Figure 3-2 gives a plot of this function. The high temperature

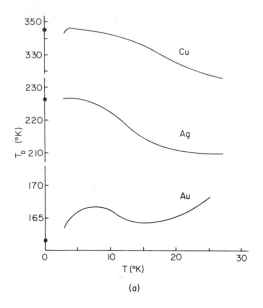
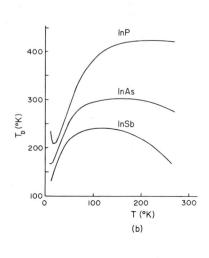

FIG. 3-3. The temperature dependence of the effective Debye temperature of copper, silver, and gold (a, after Ref. 143), and of several III-V compounds (b, after Ref. 107).

limit in the Debye theory is the same as the classical result. For low temperatures one obtains:

$$c_D = (12\pi^4 R/5)(T/T_D)^3. \tag{3-22}$$

This agrees well with data determined in liquid He-temperature experiments.

The lattice specific heat in the Debye theory is a unique function of T/T_D. Therefore, if $c_{la.}$ is known at a specific temperature, it is possible to determine the Debye temperature. Figure 3-3 gives typical plots of the Debye temperature as a function of temperature for several metals and semiconductors. A constant T_D value means that the frequency spectrum of the sample is the same as that in the Debye model which assumes $f(\nu) \propto \nu^2$ below the Debye frequency. If the Debye temperature decreases with increasing sample temperature, it indicates that the frequency distribution of the lattice vibrations increase more rapidly than ν^2. An increase of T_D with temperature would indicate the opposite character of the frequency spectrum.

It is in principle possible to obtain the frequency spectrum of a sample from the specific heat. However, c_v data are rather insensitive to the details of the frequency spectrum. Therefore, this so called "inversion problem" has not been attacked vigorously.

B. Impurities and Lattice Defects

Just as impurity atoms affect the vibrational frequency spectrum of alloys, one would expect that metals with some impurity atoms or extended solid solutions would have different values of the specific heat than the pure element. These effects should be noticeable only at temperatures below the Debye temperature, since c_{la} will approach the limiting case of independent harmonic oscillators for higher temperatures.

Effects of light impurities or of impurities with high spring constants lead to extra peaks in the frequency spectrum above the cut-off frequency ν_D of the host. At low temperatures, these frequencies are not excited. Heavy impurity atoms, or atoms with low spring constants, will lead to peaks in the frequency spectrum below the cut-off frequency. They will lead to additional lattice contributions, provided $\nu_{imp.}/\nu_D \lesssim T/T_D$. A calculation by Kagan et al. [144] showed:

$$\Delta c_v(T)/c_v^*(T) \rightarrow 3|\epsilon|c/2 \quad \text{for} \quad T \rightarrow 0^{\circ}K \qquad (3-23)$$

where $\Delta c_v(T) = c_v(T) - c_v^*(T)$ is the difference in the specific heat of the lattice with impurity atoms minus the specific heat of the perfect lattice. c is the impurity concentration, and $\epsilon + 1$ is the ratio of impurity mass to host mass. Kagan et al. [144] found the following results for the "specific heat enhancement" (the expression

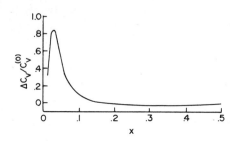

FIG. 3-4. Calculated specific heat enhancement due to heavy impurities. The impurity atomes are nine times heavier than the host atoms. Their concentration is 2%. $x = k_B T/h\nu_D$. After Ref. 145.

"enhancement" is frequently used to indicate an increase in a bulk property):

$$\Delta c_v(T)/c_v^*(T_o) \sim c(5/4\pi^4)(T_D/T)^3$$
$$\Delta c_v(T_o)/c_v^*(T_o) \sim c(5/4\pi^4)|3\varepsilon|^{3/2} \qquad T \gtrsim T_o \qquad (3\text{-}24)$$

where $T_o = h\nu_r/k_B$ is the characteristic temperature associated with the resonance frequency ν_r of the impurity atom. Figure 3-4 gives the results of a calculation in which the impurity atom is nine times heavier than the matrix atoms. It shows that a concentration of only 2% impurity atoms leads to a large specific heat enhancement in the liquid He-temperature range, where $T/T_D \sim 0.03$. One should note that the enhancement can be negative for higher temperatures, so that the total energy of the crystal below T_D will not change as drastically as a first look at the peak of this figure may indicate.

The frequency spectrum of an alloy can differ noticeable from that of pure elements (1-42). However, the effect of these changes in the frequency spectrum is usually not important for the lattice specific heat. Specific heats close to the melting point of solids may be affected by the formation of vacancies. The vacancy concentration of metals is temperature dependent and increases with

temperature. Close to the melting point concentrations of the order of 10^{-3} to 10^{-4} can be expected in typical metals.

Let us assume that the energy of formation of a vacancy is $E_{va.}$.
Then the energy associated with n vacancies is:

$$E = nE_{va.} .$$ (3-25)

The entropy associated with the formation of n vacancies in a
crystal with N lattice sites is:

$$S = k_B \ln W,$$ (3-26)

where W is the number of possible atomic arrangements. N atoms
can be placed in N! different arrangements. If n vacancies exist
in the lattice, they can be arranged in n! different ways, without
leading to any change in the atomic configuration. Similarly, the
N - n lattice sites occupied by atoms correspond to (N - n)!
identical configurations. Therefore, W for a system with n vacancies and N - n identical atoms is given by:

$$W = N!/n!(N - n)!$$
$$S = k_B \ln(N!/n!(N - n)!).$$ (3-27)

The free energy is therefore:

$$F = E - TS = nE_{va.} - Tk_B \ln(N!/n!(N - n)!).$$ (3-28)

Since the free energy of a system should be a minimum at equilibrium,
one sets dF/dn = 0. This yields, if one uses the Stirling approximation $\ln X! = X \ln X - X$ for $X \gg 1$:

$$n/(N - n) = \exp(-E_{va.}/k_B T).$$ (3-29)

This gives for $n \ll N$:

$$n/N = \exp(-E_{va.}/k_B T). \tag{3-30}$$

A similar calculation for the concentration of Frenkel defects, n_F, where one atom or ion is moved into a interstitial position gives:

$$n_F/(N \cdot N')^{1/2} = \exp(-E_F/2k_B T). \tag{3-31}$$

Here, N' is the number of interstitial positions.

Vacancy concentrations have been determined experimentally with several techniques. For instance, one may use electrical resistivity measurements, where the vacancy concentration found at high temperatures is frozen in by rapid cooling of the sample. One then measures the change of the resistivity at low temperature both for the sample with the frozen in vacancies, and after the vacancy concentration has reached equilibrium values. From calculated values of the resistance per vacancy one can calculate the vacancy concentration.

This approach is not very accurate. Much better measurements of vacancy concentrations have been obtained by Simmons and Balluffi [146], who measured both the thermal expansion and the lattice parameter of aluminum and gold as a function of temperature. The experiments showed that the relative length change of the sample increased more rapidly with temperature than the relative lattice parameter (3-5). This is essentially due to the formation of vacancies. The volume of the sample should be equal to the volume of the atoms plus the volume of the vacancies. One obtains, neglecting di-, tri- and higher groups of vacancies:

$$V = V_o(1 + \Delta L/L_o)^3 \sim V_o(1 + \Delta a/a_o)^3(1 + n/N) \tag{3-32}$$

with $\Delta L/L_o$ the relative length change, and $\Delta a/a_o$ the relative lattice parameter change of the sample. This gives, neglecting higher terms:

$$3\Delta L/L_o = 3\Delta a/a_o + n/N \tag{3-33a}$$

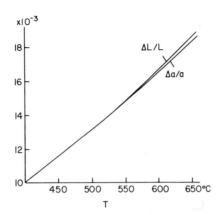

FIG. 3-5. Relative length changes ΔL/L and relative lattice
parameter changes Δa/a for aluminum. ΔL and Δa is set equal to zero
at 20°C. After Ref. 146, with permission.

Therefore:

$$n/N = 3(\Delta L/L_o - \Delta a/a_o).$$ (3-33b)

Experimental results of $\Delta L/L_o$ and $\Delta a/a_o$ measurements are shown in
Fig. 3-5. They give: $n/N = \exp(2.4 - 0.76/k_B T)$ if k_B is given
in eV/°K. It gives as vacancy concentration at the melting point
of aluminum the value $9.6 \cdot 10^{-4}$. The enthalpy of formation of one
vacancy is 0.76 eV/vacancy.

Density measurements, and measurements of the ionic conductivity,
show that Schottky defects are dominant in pure alkali halides.
Frenkel defects are most frequently found in silver halides.
Schottky defects should decrease the density ρ, with $\Delta\rho/\rho \sim n/N$. In
Frenkel defects, the volume change due to the formation of the vacancy
is mostly compensated by elastic compressions of the lattice around
the atom in the interstitial position.

The specific heat contribution due to the formation of vacancies
$c_{va.}$ in metals is very small. It is difficult to separate it from
the total specific heat, since other contributions to the specific

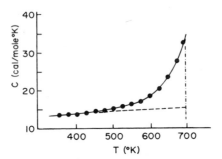

FIG. 3-6. Heat capacity of silver bromide at constant pressure. The specific heat follows from 350°K to 420°K a function of the form 12 + 0.005T, given as a dashed line. After Ref. 148.

heat like the anharmonic specific heat terms are not well known. $c_{va.}$ in the case of aluminum, for instance, is close to the melting point less than 1% of the anharmonic specific heat term $c_{anh.}$, which again is only about 10% of c_p [147]. Much larger $c_{va.}$ values have been found in silver bromide, where $c_{va.}$ near the melting point is of the same order as the lattice specific heat (3-6).

The total energy of a crystal depends also on the dislocation density in the sample. However, one does not use the concept of a specific heat contribution due to dislocations since they are usually not in thermal equilibrium with the lattice. Therefore, the dislocation density is not a unique function of temperature. A metal or alloy may be heavily deformed mechanically to increase the number of dislocations. This can lead to a dislocation density of the order of 10^{11} to 10^{12} dislocations/cm^2. Heating such a sample would lead to a change in the dislocation structure. Some would be annihilated by combination, or by moving to the surface. The energy released during this process would seem to lead to a small change in the specific heat, typically less than 1% of the total specific heat. Since this is of the same order as the error in most specific heat measurements, special calorimetric techniques have been developed for the measurement of the energy release or of the "stored energy", $Q_{st.}$. Typically, one compares the energy required to heat up two nearly

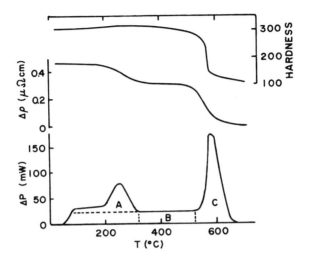

FIG. 3-7. Power difference ΔP, resistivity change Δρ, and hard-
ness V.h.n. for specimens of nickel deformed in torsion and heated
at 6°C/min. After Ref. 149, with permission.

identical samples of the same material, one with a high dislocation

density, and the other with a low dislocation density. Both samples

are heated up simultaneously in an adiabatic calorimeter, and the

temperature difference of these two samples is kept as close to zero

as possible. The run is repeated after a high temperature anneal.

This gives the difference of the heat capacity of both samples in

their low dislocation density state. Figure 3-7 shows the energy

release in a nickel sample due to the release of stored energy [149].

This energy release per time interval is equal to $\partial Q_{stored}/\partial t$. The

area under the curve in Fig. 3-7 is proportional to the total stored

energy from the deformation process. In conjunction with information

from resistivity and density measurements, and from theoretical con-

siderations, one explains these curves in the following way. The

first part of the curve from 100°C to 500°C corresponds mostly to a

rearrangement of dislocations. The peak in ΔP near 200°C has been

attributed to the elimination of vacancies. The peak in the power

curve at high temperatures is associated with a drastic change in
the number and shape of the small crystallites or "grains" which
make up a polycrystalline sample.

III. ELECTRONIC SPECIFIC HEAT

A. The Degenerate Electron Gas

Classical models predict that the electronic specific heat per
mole of conduction electrons should be equal to 3R/2. These electrons
have only kinetic energies, no potential energies. The calculation
of the energy of electrons on quantum mechanical principles shows
that the electronic specific heat of most metals is only a small
fraction of what the classical theory predicts. This explains why
the room temperature specific heat of copper is mostly due to lattice
vibrations, with c_p close to the value predicted by the Dulong-
Petit rule.

One can obtain the essential results of the electronic specific
heat calculations in the following way. Electrons occupy energy levels in metals much higher than the thermal energy $k_B T$. The Fermi
energy level is of the order of a few electron volts, whereas $k_B T$
at room temperature is 0.026 eV. Only electrons near the Fermi level
in an energy interval of the order of $k_B T$ can absorb thermal energies of the order of $k_B T$ per electron. This energy is insufficient
to move electrons at the energy level E_i far below E_F to an empty
state, since all energy states below $\sim(E - k_B T)$ are already occupied by two electrons. The absorbed thermal energy of the electrons
will be equal to the number of electrons which can be thermally
excited dn times their thermal energy $(3/2)k_B T$. One assumes then
that $dn/A_o \simeq k_B T/E_F$. This gives:

$$E \simeq dn(3/2)k_B T \simeq (3/2)A_o k_B^2 T^2 / E_F. \qquad (3\text{-}34)$$

This gives for the electronic specific heat:

$$c_{el.} = \partial E/\partial T \simeq 3A_o k_B^2 T/E_F \simeq 3R(k_B T/E_F). \tag{3-35}$$

This equation shows that $c_{el.}$ is a linear function of temperature. $c_{el.}/c_{la.}$ at room temperature is typically of the order of 0.01.

Equation 3-35 assumes that the density of states is essentially constant. For arbitrary density of states values one uses dn = N(E)dE and sets dE equal to $k_B T$. This gives in Eq. 3-34:

$$E \simeq N(E)dE(3/2)k_B T \tag{3-36}$$

and if N(E) is the density of states per mole:

$$c_{el.} \simeq 3N(E)k_B^2 T. \tag{3-37}$$

Accurate results for the electronic specific heat can be obtained only if one uses the Fermi-Dirac distribution:

$$f^o(E) = \{\exp(E - \zeta)/k_B T + 1\}^{-1} \tag{3-38}$$

which gives the probability that an electron state with energy E is occupied at the temperature T. ζ is the Fermi potential. It is equal to E_F at absolute zero temperature.

The total number of electrons is:

$$n = 2\int_o^\infty N(E)f^o(E)dE. \tag{3-39}$$

Equations 3-38 and 39 determine ζ. Its position with respect to E_F for a highly degenerated electron gas ($k_B T \ll E_F$) is given by:

$$\zeta \simeq E_F - (\pi^2 k_B^2 T^2/6)\partial \ln N(E_F)/\partial E. \tag{3-40}$$

The difference between ζ and E_F is so small that it can be frequently neglected.

The Fermi-Dirac statistic is an extension of the Pauli principle which states that each energy state available is occupied either by one electron or it is empty.   It can be derived in the following way. Let us assume that we have a system, in which energy states are closely spaced, with $N(E_i)$ energy levels with energies between E_i and $E_i + dE_i$. The number of electrons in these states is $n(E_i)$. The number of holes in this energy interval is therefore $N(E_i) - n(E_i)$. The number of possible arrangements for electrons and holes in this energy interval is:

$$W(E_i) = N(E_i)!/n(E_i)!\{N(E_i) - n(E_i)\}!. \qquad (3\text{-}41)$$

The total number of possible arrangements, W, is $W = \overset{i}{\Pi}W(E_i)$.   The most likely system is the system with the largest number of possible arrangements. This system has the maximum W value. For convenience, one determines not the maximum of W, but that of lnW. Lagrange's method of undetermined multipliers shows that an extremum of lnW with the conditions $n = \Sigma n(E_i)$ and $E = \Sigma E_i n(E_i)$ is obtained, if lnW $- \alpha n - \beta E$ is an extremum. α and β are the undetermined multipliers. One obtains [82]:

$$\delta\{lnW - \alpha n - \beta E\} = \delta lnW - \alpha\delta n - \beta\delta E \qquad (3\text{-}42)$$
$$= \Sigma\{\delta ln[N(E_i)!/n(E_i)!\{N(E_i) - n(E_i)\}!] - \alpha\delta n(E_i) - \beta\delta E_i\delta n(E_i)\}$$
$$= \Sigma\{ln[\{N(E_i) - n(E_i)\}/n(E_i)] - \alpha - \beta E_i\}\delta n(E_i)$$
$$= 0.$$

Since this equation is zero for all possible $\delta n(E_i)$ values, each bracket has to be zero. One obtains therefore, if one solves for $n(E_i)$:

$$n(E_i) = N(E_i)/\{\exp(\alpha + \beta E_i) + 1\} = N(E_i)f^o(E_i). \qquad (3\text{-}43)$$

$f^o(E_i)$ is the Fermi-Dirac statistic. Figure 3-8 shows $f^o(E)$ and $\partial f^o(E)/\partial E$ as a function of E. For small values of $n(E)/N(E)$,

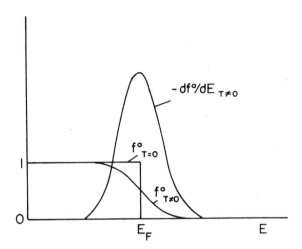

        FIG. 3-8. Fermi-Dirac statistic and its derivative as a function
of E.

the Fermi-Dirac statistic approaches the classical Boltzmann statistic. This shows that $\beta = 1/k_B T$. The Fermi-Dirac function should
lead at absolute zero temperature to the Pauli principle. This leads
to $\alpha = E_F/k_B T$.

        At room temperature, the Fermi-Dirac statistic is applicable to
systems where the carrier concentration is of the order of 10^{18} cm^{-3}
or larger, provided that the effective mass of the carrier is close
to the free electron mass. For lower carrier concentrations, the
Fermi-Dirac statistic is replaced by the Boltzmann distribution.   In
this case, the electron gas is non-degenerate. Such low carrier
concentrations are found in semiconductors.

        The Fermi energy level in an intrinsic semiconductor ($n_{el.}$ =
$n_{ho.}$) lies in the middle of the energy gap. Since the thermal energy
of the electrons or holes is small compared with the Fermi energy,
the Fermi-Dirac statistic reduces to:

        $f^{**} \simeq \exp(\zeta - E)/k_B T.$ (3-44)

This gives the probability that an electron state is occupied.   The energy of electrons in the conduction band is given by [3]:

$$E = E_g + \hbar^2 k^2 / 2m_{el.}$$ (3-45)

E_g is the gap width between valency and conduction band. The top of the valency band is set as zero energy reference. The density of states per unit volume is given by:

$$N(E) = (2^{1/2} m_{el.}^{3/2} / \pi^2 \hbar^3)(E - E_g)^{1/2}$$ (3-46)

if the electron band is parabolic. The total number of excited electrons is therefore:

$$n_{el.} = \int_{E_g}^{\infty} N(E) f^{**} dE = (\pi 2^{5/3} m_{el.} k_B T/h^2)^{3/2} \exp\{(E_F - E_g)/k_B T\}$$ (3-47)

A similar calculation for holes gives:

$$n_{ho.} = (\pi 2^{5/3} m_{ho.} k_B T/h^2)^{3/2} \exp(-E_F/k_B T)$$ (3-48)

This gives for the product:

$$n_{el.} \cdot n_{ho.} = (\pi 2^{5/3} k_B T/h^2)^3 (m_{el.} \cdot m_{ho.})^{3/2} \exp(-E_g/k_B T).$$ (3-49)

This derivation makes no assumption about the value of the Fermi energy, nor about whether the material is an intrinsic or extrinsic conductor. Therefore, this correlation is valid also for extrinsic semiconductors.

The energy of the degenerate electron gas in a metal is:

$$E^* = \int_0^{\infty} E N(E) f^o(E) dE.$$ (3-50)

The specific heat per mole of a degenerate electron gas is therefore:

$$c_{el.} = dE^*/dT = (1/3)\pi^2 k_B^2 N(E_F) T = \gamma T.$$ (3-51)

γ is the electronic specific heat coefficient. As mentioned before, $c_{el.}$ is usually much smaller than the lattice specific heat at room temperature. At liquid He temperatures, $c_{el.}$ and $c_{la.}$ are frequently of the same order of magnitude. Therefore, $c_{el.}$ is usually determined at these low temperatures. The separation of $c_{el.}$ and $c_{la.}$ is then quite easy if no other specific heat contributions have to be taken into account. One plots c/T as a function of T^2 and obtains a linear plot, with $\beta = 12\pi^4 R/5T_D^3$ (see Eq. 3-22) as slope and γ as intercept of the c/T axis:

$$c/T = \gamma + \beta T^2. \tag{3-52}$$

A linear temperature dependence of $c_{el.}$ is to be expected only for temperatures that are low with respect to the degeneracy temperature of the electron gas, $T_{deg.} = E_F/k_B$. The calculation of $c_{el.}$ over an extended temperature for arbitrary density of states curves requires a major computational effort. For the specific case of a parabolic band with $N(E) \propto E^{1/2}$ one obtains for T below $T_{deg.}$ [150]:

$$c_{el.} = (1/2)\pi^2 n_o R[T/T_{deg.} - 2.96(T/T_{deg.})^3 + ..] \tag{3-53}$$

where n_o is the number of free electrons per atom. One obtains for a parabolic band at high temperatures $(T \gg T_{deg.})$:

$$c_{el.} = (3/2)n_o R[1 - (T_{deg.}/2\pi T)^{3/2}/6 + ...]. \tag{3-54}$$

These equations are also correct for a nearly filled band, with $N(E) = C(E_o - E)^{1/2}$. n_o would then be the number of holes per atom. Figure 3-9 gives a plot of $c_{el.}$ as a function of temperature for electrons in a parabolic band. It shows that $c_{el.}$ approaches the value of the classical electron gas with increasing temperature.

The correlation between γ and $N(E)$ in Eq. 3-51 may be modified through "phonon enhancement", which is due to the interaction of the electron gas with a thermally disordered lattice. Since

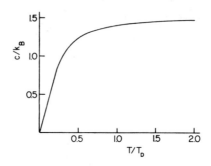

FIG. 3-9. Heat capacity per electron of a degenerate electron gas. $T_D = T_{deg}$. After Ref. 150.

thermal vibrations are slow compared with the relaxation time of electrons, the calculation of the electron enhancement determines the electronic specific heat of a thermally disordered lattice by assuming that the ions are frozen into irregular positions. This leads to an increase in the specific heat, as shown in Eq. 2-45.

B.   Elements

Table 3-1 shows γ and Debye temperatures of elements. γ depends systematically on the postion of the element in the periodic system.   Noble metals have low γ values.   However, the values are larger than expected from the free electron model.   This could mean that the gradient of   E   with respect to   k,   which is proportional to the Fermi velocity (velocity of electrons at the Fermi surface) is probably   lower   near the contact points with the first Brillouin zone than the Fermi velocity at the large belly section.   This influences N(E) as Eq. 1-59 shows [2].

γ of IIB metals is larger than expected.   The Fermi surface area of hcp crystals like zinc should be much smaller than predicted from the free electron model, since the two electrons per atom nearly fill the first Brillouin zone [2]. Therefore, the Fermi surface of zinc consists only of the surface of small electron and hole sections.

TABLE 3-1

ELECTRONIC SPECIFIC HEAT COEFFICIENT γ AND

DEBYE TEMPERATURE T_c [134]

Element	γ mJ/mole°K^2	T_c °K	Element	γ mJ/mole°K^2	T_c °K
Li	1.65	355(hcp)	Bi	0.008	120.4
		335(bcc)	Sc	10.7	360
Na	1.38	159(hcp)	Y	8.2	256
		153(bcc)	La	9.4	152(dhcp)
K	2.1	91		11.5	140(fcc)
Rb	2.6	56	Ti	3.34	429
Cs	4.0	40	Zr	2.80	291
Be	0.171	1481	Hf	2.15	252
Mg	1.23	405	V	9.82	400
Ca	2.9	230	Nb	7.81	277
Sr	3.6	147	Ta	6.02	258
Ba	2.7	110	Cr	1.42	600
Cu	0.695	344.5	Mo	1.84	470
Ag	0.647	226.0	W	1.00	380
Au	0.70	162.3	α-Mn	12.7	385
Zn	0.653	327	γ-Mn	9.2	370
Cd	0.687	209	Re	2.26	415
Hg	1.82	72	Fe	4.78	477
Al	1.36	430	Ru	3.00	555
Ga	0.598	322	Os	2.3	500
In	1.63	110	Co	4.5	460
Tl	1.47	78.5	Rh	4.7	500
Sn	1.77	198	Ir	3.2	420
Pb	2.99	105	Ni	7.05	473
As	0.192	282	Pd	9.40	274
Sb	0.110	210	Pt	6.55	235

γ of semiconductors is zero at absolute temperature. Even $c_{el.}$ of semimetals can usually be neglected, since $N(E)$ in these elements is orders of magnitude smaller than in noble metals. The highest γ values are found for transition elements. Only transition elements in the chromium group have γ values of less than 2 mJ/mole$^{\circ}$K^2, indicating that the energy band of the d-electrons has at least one sharp minimum.

The degeneracy temperature $T_{deg.}$ of non-transition elements is usually very high. Table 3-2 gives examples. $T_{deg.}$ of copper is for instance 84,000°K. Therefore, $c_{el.} = \gamma T$ should be a good approximation for these elements even at high temperatures. Only transition elements have such low $T_{deg.}$ values (typically 10^3 to 10^4 $^{\circ}$K) that the temperature dependence of $c_{el.}/T = \gamma(T)$ has to be taken into account. This temperature dependence is very noticeably if one plots $\gamma' = c_{el.}/\gamma_o T$ as a function of temperature. Here γ_o is the electronic specific heat coefficient at liquid He temperatures. Infinitely high $T_{deg.}$ values would give constant γ' values. γ' should decrease noticeably with increasing temperature above $T_{deg.}$. Figure 3-10 shows calculated γ' values for several transition elements. γ' for several elements is larger than one. This unexpected result (according to the free electron model for a parabolic band, γ' can only decrease with increasing temperature) is much more noticeable in a γ' plot than in the standard separation of c_p into its various components, as shown in Fig. 3-11 for chromium. Figure 3-10 shows that γ' in a given group of transition elements in the

TABLE 3-2

Degeneracy Temperatures of Metals

	Li	Na	K	Rb	Cs	Cu	Ag	Au
$T_{deg.}$	5.5	3.7	2.4	2.1	1.8	8.4	6.4	$6.4 \cdot 10^4$ K

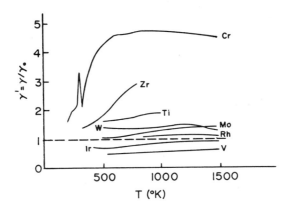

FIG. 3-10. Temperature variations of the electronic specific heats of transition elements γ/γ_0. After Ref. 98, with permission.

periodic system is either always larger or always smaller than one (rhodium excepted). Further, if γ' is larger than one in a given group, it will be smaller than one in adjacent groups. Figure 3-10 shows that chromium has especially large γ' values. The small peak in $\gamma' = \gamma/\gamma_0$ near room temperature is due to the transition from the antiferromagnetic to the paramagnetic state at 311°K.

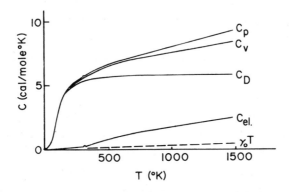

FIG. 3-11. Specific heat of chromium. The lattice specific heat c_D is the Debye specific heat where the Debye temperature is 460°K. The dashed line is the extrapolated low temperature electronic specific heat. $c_{el.}$ is obtained by subtracting c_D from c_v. After Ref. 98.

The interpretation of the temperature dependence of $\gamma(T)$ follows lines similar to those used in the interpretation of the Pauli susceptibility. The electronic specific heat is determined in an energy interval of the order of $k_B T$. If the density of states has a minimum at a given element, increasing temperatures should increase the average value of $N(E)$, $N(E)_{av.}$, in the energy interval $k_B T$, and with it $\gamma(T)$. Correspondingly, a maximum $N(E)$ value for a given element should lead with increasing temperature to a decrease in $\gamma(T)$ or $N(E)_{av.}$. One would therefore expect that $N(E)$ of chromium is at or near a minimum in the density of states curve.

For a more detailed computer calculation of the temperature dependence of $\gamma(T)$, one uses Eqs. 3-39 and 3-50 to determine the energy of the electron system, differentiates this energy with respect to temperature, and obtains in this way the electronic specific heat. This method requires, that $N(E)$ be known over an extended composition range, and that one knows the correlation between alloy composition and energy. Typically, one assumes that the rigid energy band concept can be used. Calculations of $\gamma(T)$ have been conducted by Shimizu et al. [98] on chromium, and by Hindley and Rhodes [151] on palladium. Both groups of authors obtained $N(E)$ as a function of E from liquid He temperature specific heat measurements of alloys. The calculation by Shimizu and coworkers may overestimate γ', since these authors used γ_o of Cr-V and Cr-Fe alloys and γ_o values of Cr-Fe alloys may be too high due to magnetic effects. High temperature specific heat measurements on Cr-Mn alloys by Giannuzzi et al. [152] show that the rigid band model cannot be used to analyze Cr-Mn and Cr-Fe alloys. If one uses γ_o of Cr-Mn alloys, which should be more appropriate for the study of the high temperature electronic specific heat of chromium since manganese is adjacent to chromium in the first long row of the periodic system, one would obtain lower γ_o values for some el./at. values than found for Cr-Fe alloys.

Hindley and Rhodes [151] used values of Pd-Ag and Pd-Rh alloys for their calculation of the electronic specific heat of palladium. Results of their calculation, together with experimental data, are

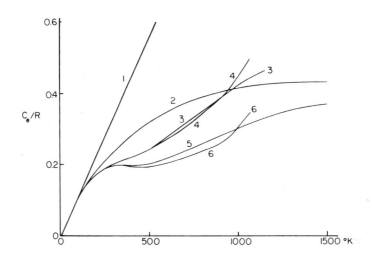

FIG. 3-12. Electronic specific heat of palladium. Linear extrapolation of low temperature data (1). Calculated from N(E) (2). Electronic specific heats obtained from experimental data are given in
(3 to 6). See Ref. 151 for more details. After Ref. 151, with permission.

given in Fig. 3-12. This figure shows that the rigid band model can
be a good approximation in the investigation of transition elements
and their alloys, especially if one keeps in mind that one does not
need adjustable parameters.

C. Alloys of Noble Metals and Non-transition Elements

Since the Fermi surfaces and the energy band structure of noble
metals are very similar, one would expect that γ_o values of solid
solutions with these elements are linear functions of composition.
Theories suggest that additional terms in the low temperature specific heat due to scattering exist, which would lead to an apparent
increase in $c_{el.}$, so that γ should increase more rapidly with the
addition of impurities than expected from a linear interpolation between elements. Figure 3-13a gives experimental data on silver-gold

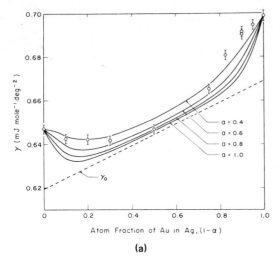

(a)

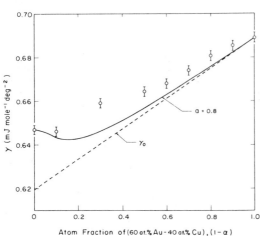

(b)

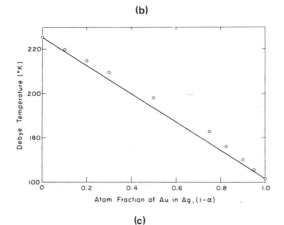

(c)

FIG. 3-13. Fit of phonon enhancement model to specific data for binary $Ag_\alpha Au_{1-\alpha}$ alloys (a). The parameter a depends on the anistropy of the electron-impurity scattering. Fit of phonon enhancement model to specific data for ternary $Ag_\alpha(60 \text{ at.\% Au-}40 \text{ at.\% Cu})_{1-\alpha}$ alloys (b). Variation of Debye temperature for $Ag_\alpha Au_{1-\alpha}$ binary disordered alloys (c). After Ref. 153, with permission.

alloys. It shows that γ decreases with the addition of small
amounts of gold to silver, disagreeing with the predictions. The
Debye temperature is a nearly linear function of composition (3-13c).
Similar measurements on copper-silver alloys again indicate that γ
decreases with alloying.

   N(E) is proportional to the area of the Fermi surface, weighted
by the reciprocal of the gradient of E in k-space. The decrease
of γ with alloying could therefore be explained in the following
way. The Fermi surface of noble metal elements is not spherical. It
touches the first Brillouin zone. These deviations from the spherical shape could increase γ above the value of the free electron
model even if the area of the Fermi surface not in contact with the
first Brillouin zone is smaller than the area of the perfectly spherical Fermi surface. Since alloying decreases γ, one could conclude
that the Fermi surface of the alloys is more spherical in noble metal
alloys. One may state it as: alloying leads to a "spheridization"
of the Fermi surface. This may be due to a reduction of the energy
gap across the first Brillouin zone in the < 1 1 1 > direction [58].

   Davis and Rayne [153] investigated binary silver-gold and ternary silver-copper-gold alloys. They interpreted their data with
the electron-phonon enhancement model used by Haga [154] and the virtual crystal model by Stern [155]. Electron-phonon enhancement is
especially strong for pure elements. It will decrease with increasing
impurity concentration. This means that γ of the pure elements is
not equal to γ from the energy band, but slightly larger. The enhancement effects depend also on the broadening of the energy levels.
Using this concept, γ was calculated for binary and ternary noble
metal alloys. Results are given in Fig. 3-13a and b as full lines.
a is an adjustable parameter. This calculation shows that the
electron-phonon enhancement leads to an increase of γ of the order
of 5%.

   These concepts should also be used in the interpretation of the
electronic specific heat of noble metals with higher valency B-metals
like zinc or cadmium. Figure 3-14 gives γ versus composition

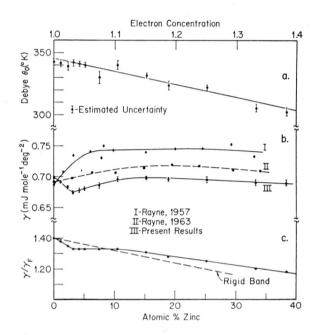

FIG. 3-14. Debye temperature Θ (a), electronic specific heat coefficient γ (b), and γ/γ_F (c) of CuZn alloys. Reproduced with permission from Ref. 156.

curves for these alloys as obtained by several authors (see Ref. 156 for details). The experimental results given in Fig. 3-14 show that some authors find an increase, some a decrease in γ as zinc is added to copper. These discrepancies cannot be explained with exper-imental uncertainties. One may suspect that short range order could influence the phonon-electron enhancement effect or the energy gap of the contact point of the Fermi surface with the first Brillouin zone. Obviously, specific heat measurements alone are not able to give con-clusive results on the changes in the Fermi surface of copper-rich alloys with increasing impurity concentration.

Figure 3-14c gives a comparison of the data by Isaaks and Mas-salski with a rigid band calculation based on a non-spherical Fermi surface. Ziman's phenomenological model [159] was used. It has as the adjustable parameter the energy gap at the first Brillouin zone.

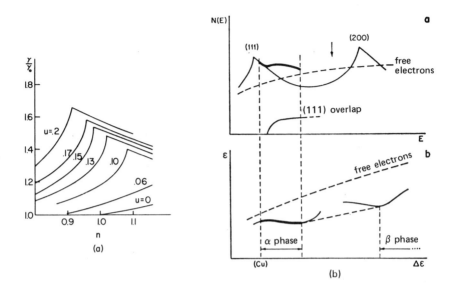

FIG. 3-15. γ/γ_0 as a function of n, the number of electrons per atom, and u, which is proportional to the energy gap and the effective mass (a, after Ref. 159, with permission). A diagramatic model of the trend in the density of states in Cu-Zn alloys; a and b illustrate the relationship between the density of states N(E) and the total electronic energy ε with energy (b, after Ref. 156, with permission).

Figure 3-15a shows γ/γ_0 as a function of electron concentration for this model. The peak in the curves corresponds to the first contact between the first Brillouin zone and the Fermi surface. The free electron model predicts that γ or N(E) should increase with increasing electron concentration. This result is given in (3-15a) for the case of u = 0, since u $\propto \Delta$E(gap)·m*. Figure 3-15b shows in section "a" as a dashed line N(E) as a function of E for the free electron model, and in solid lines the model proposed by Isaaks and Massalski from their γ data. N(E) for electrons in the 1st and 2nd Brillouin are given in light lines, and their sum is given as a heavy line. The first peak indicates the initial contact between the 1st Brillouin zone and the Fermi surface. The minimum in the heavy curve indicates the onset of the addition of electrons to the 2nd Brillouin zone. This would correspond to the minimum in γ close to $Cu_{96}Zn_4$ in

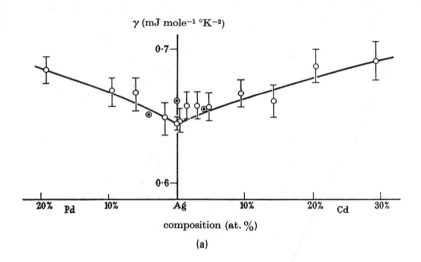

(a)

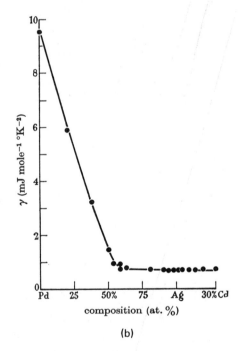

(b)

FIG. 3-16. Electronic specific heats of Ag-rich alloys (a). Circle with dot gives differential measurment. Circle with dot at pure silver represents $Ag_{92}Cd_4Pd_4$. Electronic specific heats of Ag-Pd and Ag-Cd alloys (b). After Ref. 160, with permission.

Fig. 3-14. Figure 3-15b gives the total energy $\varepsilon = \int E N(E) dE$ as a
function of energy. Since $N(E)$ obtained experimentally is larger
than $N(E)$ of the free electron model, E_F(exp.) $< E_F$(free electron
model). ε(free electron model) is larger than ε(nearly free electron model) for the same electron concentration.

The effect of small amounts of transition element impurities
on the electronic specific heat has been studied in copper and sil-
ver alloys. Figure 3-16 shows γ of $\underline{Ag}Cd$, $\underline{Ag}Pd$ and $Ag_{1-2c}Pd_cCd_c$
alloys. The addition of cadmium or palladium to silver leads to
an increase in γ. Also, γ of the ternary alloy is slightly larger
than γ of pure silver. This seems to imply that one cannot use
the rigid band model to explain the electronic specific heat coefficient of "pseudo silver" $Ag_{1-2c}Pd_cCd_c$, which has the same electron
concentration as pure silver. However, one may argue that pseudo
silver consists of very small volume elements of $\underline{Ag}Pd$ and $\underline{Ag}Cd$. The
sum of their electronic specific heats should be larger than the
electronic heat of pure silver. γ of $\underline{Ag}Pd$ alloys increases only
slightly with increasing palladium concentration, provided the palladium concentration is less than about 40 at.% higher palladium
concentrations lead to a rapid increase in γ. This is due to partially filled d-bands. This is discussed in more detail in the
following section.

D. Transition Element Alloys

Figure 3-17 gives examples of the electronic specific heat coefficients of extended solid solutions of transition elements with
each other, and of nickel-group and copper-group elements. The γ
versus composition curves for noble metal alloys are nearly horizontal for all alloys with more than 50 at.% noble metal constituents.
Below el./at. of 10.5 (here valency electrons and d-electrons in
outer shells are counted as number of outer electrons per atom), γ
increases suddenly with decreasing copper, silver, or gold concentration. The similarity for all curves of alloys in the first three

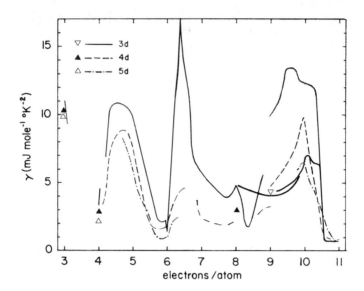

FIG. 3-17. Electronic specific heat coefficient γ of 3d-, 4d-, and 5d-alloys. Top curve above 9 outer electrons per atom shows the probable density of states curve for both spin orientations. Reproduced with permission from Ref. 161.

long periods is a strong support for the idea that the electronic structure of these alloys depends predominantly on the el./at. ratio, and that properties of these alloys can be described in first approximation with the rigid band model. The shape of the curves indicates that the d-band is filled at el./at. $\simeq$ 10.5. Calculated density of states per atom values from both specific heat and susceptibility measurements have been given by Gladstone et al. [162]. They are shown in Fig 3-18.

Figure 3-17 shows that γ has a sharp minimum at the chromium group, with el./at. = 6. This indicates that we have at least two separate d-electron subbands. Calculations and experiments indicate that the alloys slightly above el./at. = 6 contain predominantly localized d-electrons (ferromagnetic properties of iron are frequently attributed to localized electrons). Alloys with less than el./at. = 6 have electron wave functions similar to those of conduction electrons [163]. Calculations of the energy band structure of transition

elements as discussed in Chapter 1 show that the d-band has a high
density of states.

Alloys with el./at. ratios close to 6 have been investigated by
several authors who not only worked with binary but also with ternary alloys. Figure 3-19 gives a summary of experimental results,
which shows that nearly all alloy systems have a minimum at or close
to el./at. = 6. Several of these alloy systems also have a peak in
γ at el./at. = 6.36. Figure 3-20 shows that γ of ternary Ti-alloys
has a peak at el./at. $\simeq$ 6.36. However, data points are not as closely
spaced as in the other studies. It may be that part of the specific
heat of these samples is affected by magnetic contributions. This
would explain why the c/T versus T^2 plot for some of the alloys does
not give a straight line, as shown in Fig. 3-21. The fact that the
low temperature specific heat part usually attributed to electronic
effects is not associated with ferromagnetic ordering can be deduced
from high temperature specific heat measurements. Such measurements
on CrFe alloys above liquid nitrogen temperatures show a peak at
el./at. = 6.36, if c_p is plotted for a given temperature as a function of composition. This peak shows up above the Curie temperature
of this alloy which suggests it is electronic and not magnetic in
origin (3-22). $c_{el.}$ values calculated from these data give $\gamma(T)$
values between 100 and 600°K, which are nearly half the value of
$\gamma(4.2°K)$. A decrease of $\gamma(T)$ with temperature is to be expected
if the density of states curve has a maximum at el./at. = 6.36.

γ values of the isoelectric Cr-Mo and Cr-W alloy systems are
given in Fig. 3-23. In both systems, γ first   increases with decreasing chromium concentration, reaches a maximum, and then   decreases with decreasing chromium concentration. This result can be
explained if one keeps in mind that the Fermi surfaces of these three
elements are very similar in the paramagnetic state, but that antiferromagnetic ordering leads to a decrease in the area of the Fermi
surface. This should reduce γ. Therefore, the kink in the γ versus composition curves should indicate the onset of antiferromagnetic
ordering in these alloys. The γ curves extrapolated linearly from
tungsten or molybdenum to pure chromium should give γ of the

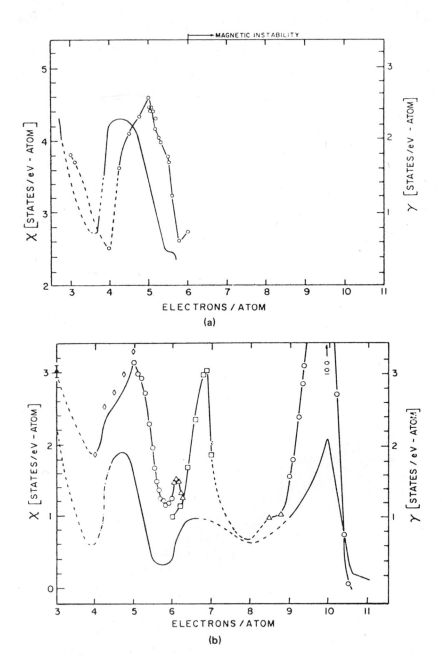

(a)

(b)

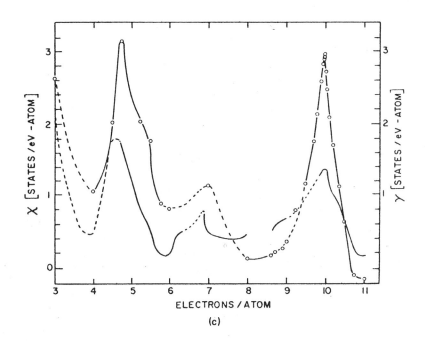

FIG. 3-18. Sections (a, opposite page), (b, opposite page), and (c) show respectively the variation of the magnetic susceptibility χ as one alloys along the 3d, 4d, and 5d transition metal series. The quantity (states/eV-atom) is actually the experimentally determined susceptibility divided by twice the square of the Bohr magneton. Superimposed on these curves are the "solid" curves (that is, those curves without data points) representing the specific heat coefficient γ. The susceptibility curves were plotted from the data using nearest-neighbor alloys whenever possible. Note the shift in scale for the 3d-elements in section (a). Reproduced with permission from Ref. 162.

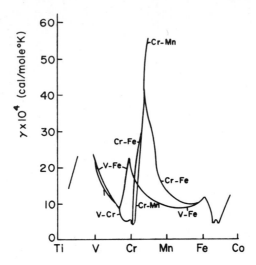

FIG. 3-19. γ versus electron concentration for bcc solid solutions. After. Ref. 164, with permission.

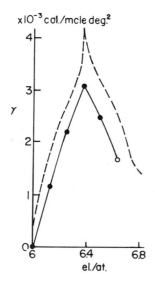

FIG. 3-20. γ versus average electron concentration for ternary Ti-alloys with CaCl type ordered structure. Full circles represent TiFe-TiCo alloys. Open circle stands for $(TiCo)_3TiNi$. Dashed line gives γ values for Cr-Fe alloys. After Ref. 165, with permission.

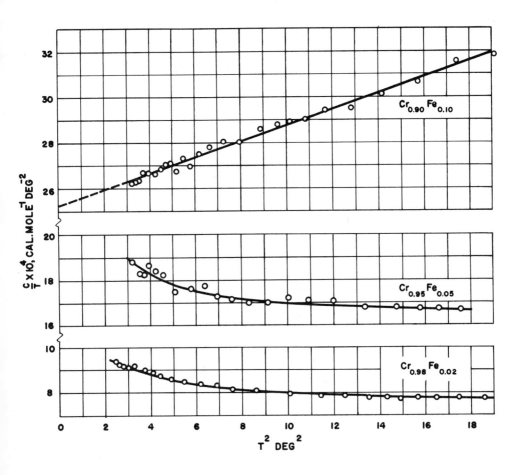

FIG. 3-21. Specific heat of bcc Cr-Fe solid solutions. Reproduced with permission from Ref. 164.

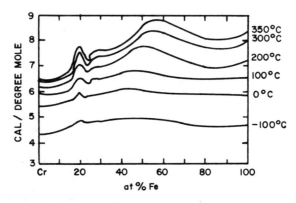

FIG. 3-22. Specific heat of Cr-Fe alloys as a function of composition for various temperatures. After Ref. 166, with permission.

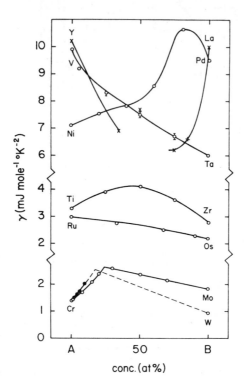

FIG. 3-23. Electronic specific heat coefficient γ of binary isoelectric alloys. Reproduced with permission from Ref. 161.

paramagnetic state of the alloys and chromium. c(T>100°K) confirm
that N(E) of these alloys are at a minimum [167].

IV. ATOMIC ORDERING

A. Energy of Atomic Ordering

The difference in the electronegativity between a large number
of metallic elements is small. Therefore, some alloys like $AuCu_3$,
$CuZn$, or Mg_3Li are disordered at high temperatures and show order
at low temperatures. This means that the chance of finding a

Cu-atom on a given lattice site in $AuCu_3$ is at high temperatures
3/4, or 1/2 in CuZn, independent of the type of atom which occupies
nearest neighbor or next nearest neighbor positions. At temperatures
below a critical temperature, provided the sample is at thermal equilibrium, the different types of atoms are arranged in a more regular
fashion since they attract each other more strongly than identical
atoms. Each type of atom will occupy preferably a position in a
well defined sublattice. In the case of $AuCu_3$, the Au-atoms prefer
the positions in the corner of an fcc unit cell, and the Cu-atoms
prefer face center positions. One would find perfect order under
equilibrium positions only at absolute zero temperature with all
Au-atoms in corner, and all Cu-atoms in face center positions. Perfect order is naturally only possible if the ratio of gold atoms to
copper atoms is 1:3.

The discussion of the energy bands and their effect on the stability of the lattice in Chapter 1 showed that the interaction of
atoms in a metal will extend much further than to the nearest neighbors. However, even a model which takes only nearest neighbor interaction into account to determine the energy changes of a crystal due
to atomic ordering is able to predict the essential features of the
temperature dependence in atomic ordering. In spite of its limitations, this model is therefore frequently used to analyze experimental results. In such a model, the binding energy between different
atoms i and j in a binary alloy with elements A and B in the
lattice are given by terms E_{AA}, E_{BB}, and E_{AB}. The binding energy
between atoms i and j is negative, since one assumes that the
potential of individual atoms at infinite separation is zero. The
potential energy of two atoms which attract each other decreases
with decreasing distance and reaches a minimum where attractive
forces between these two atoms are equal to repulsive forces.

It is naturally an oversimplification to assume that the binding
energy between two atoms in a solid is independent of the other nearest neighbor atoms these two atoms have. Each atom in an fcc or an
hcp crystal has twelve nearest neighbors, and they all interact with

each other directly or indirectly. Nevertheless, this model frequently leads to a reasonable description of experimental data. It
predicts that long range order is possible only below a critical
temperature, T_c, and that ordering increases with decreasing temperature below T_c.   The energy change due to the exchange of two different atoms in this model is:

$$E = E_{AA} + E_{BB} - 2E_{AB}. \tag{3-55}$$

All three terms, E_{AA}, E_{BB}, and E_{AB} are negative. If E is zero,
then the exchange of the two different atoms will not change the energy of the crystal. E < 0 means that two A–B bonds are stronger
than a pair of A–A and B–B bonds. This would therefore lead to
an ordered structure where different types of atoms attract each other.
E > 0 should lead to a system where A–atoms are surrounded by more
A–atoms then expected in a completly random atomic arrangement, and
naturally one will also have more B–atoms surrounding other B–atoms. This would lead to a segregation into A–atom rich and B–atom
rich volume elements.

     E = 0 is defined as an "ideal solution". The heat of mixing
in such a system is zero. If the heat of mixing is non–zero, but
one is able to assume that the atoms in the crystal are still nearly
as randomly distributed as in the ideal solution, then one has a
"regular solution". The configurational entropy of mixing per atom,
$\Delta_m S$, is given for both systems by:

$$\Delta_m S = - k_B (c_A \ln c_A + c_B \ln c_B). \tag{3-56}$$

This equation is derived in the same way as the entropy term in the
equation for the equilibrium concentration of defects. One again
uses the equation $\Delta_m S = k_B \ln W$, where W is equal to $N!/(N-n)!n!$,
with N the number of lattice sites, n the number of lattice sites
occupied by A–atoms, and N–n the number of lattice sites occupied
by B–atoms. Using the Stirling approximation, and keeping in
mind that c_A = n/N and c_B = (N-n)/N, gives again Eq. 3-56.

Let us assume that we have a lattice with N lattice sites, and each site has z nearest neighbors (nn). The total number of nn-bonds is $\frac{1}{2}zN$. The total number of bonds between different or equal atoms for such a system, if the atoms are randomly distributed, is [168]:

$$
\begin{aligned}
\text{Number of A-A bonds} &= zN_{AA} = \tfrac{1}{2}Nzc_A^2 \\
\text{Number of B-B bonds} &= zN_{BB} = \tfrac{1}{2}Nzc_B^2 \\
\text{Number of A-B bonds} &= zN_{AB} = Nzc_A c_B.
\end{aligned}
\tag{3-57}
$$

The number of A-A bonds can be determined in the following way. The number of A-atoms is Nc_A, and the number of nn A-atoms to each A-atom in the completely disordered system is $c_A z$. The product of Nc_A and $c_A z$ should give twice the number of A-A bonds, since each atom is counted twice. Similar arguments are used to determine the number of B-B bonds and A-B bonds. The potential energy for atomic ordering in a completely disordered system is therefore:

$$
\begin{aligned}
U &= -Nz(c_A^2 E_{AA} + c_B^2 E_{BB} + 2c_A c_B E_{AB}) \\
&= -Nz(c_A E_{AA} + c_B E_{BB} - c_A c_B E)
\end{aligned}
\tag{3-58}
$$

This gives for the enthalpy of mixing, $\Delta_m H$, or the energy of mixing, $\Delta_m U$ (both terms are very close to each other in solids):

$$
\Delta_m U \simeq \Delta_m H \simeq zNc_A c_B E.
\tag{3-59}
$$

The values for the components are $zNc_A E_{AA}$ for the A-atoms, and $zNc_B E_{BB}$ for the B-atoms. The free energy of mixing per atom, $\Delta_m G$, would be:

$$
\Delta_m G = \Delta_m H - T\Delta_m S
\tag{3-60}
$$

$$
= zc_A c_B E + k_B T(c_A \ln c_A + c_B \ln c_B).
\tag{3-61}
$$

B. Crystal Structure of Ordered Alloys

The most common types of superlattices or ordered binary alloys
are found in five different crystallographic structures [168]. Two
are derived from the bcc crystal, two from the fcc crystal, and only
one from the hcp crystal structure. Since the bcc unit cell contains
two atoms, one would expect that an ordered structure of the equiatomic composition AB could be constructed in the following way.
The center position of the unit cell is occupied by an A-atom, and
the corner positions by B-atoms. These two types of positions are
equivalent. Typical examples of this crystal structure (3-24c) are
CuZn, CuPd, and CoFe. A cell in the superlattice A_3B contains four
or a multiple of four atoms. One could put these four atoms into two
unit cells of the bcc lattice, but this would have no cubic symmetry.
Cubic symmetry requires a new cell with eight unit cells of two atoms
each (3-24d). This new large cell contains therefore sixteen atoms,
twelve of the A-type and four of the B-type. The new unit cell has
as lattice parameter 2a, if a is the lattice parameter of the
original bcc lattice.   The lattice sites [1/2 1/2 1/2],[3/2 3/2 1/2],
[1/2 3/2 3/2], and [3/2 1/2 3/2] are occupied by B-atoms, and the
rest by A-atoms. Typical examples of such structures are Fe_3Al and
Mg_3Li.

It is much easier to visualize an A_3B superlattice in the fcc
lattice (3-24a), since each unit cell contains four atoms. This
system was described before for Cu_3Au, where the Cu-atoms are placed
into face centered positions, and the Au-atoms into the corner positions. The AB superlattice could be constructed by placing A- and
B-atoms into alternating [1 0 0] planes, half the lattice parameter
apart. One can also use alternating [1 1 1] planes. The first arrangement is found more frequently. It is a slightly distorted lattice, since the distance between different atoms in adjacent planes
is slightly different from the distance of identical atoms in one plane.
Typical examples are AuCu and CoPt.

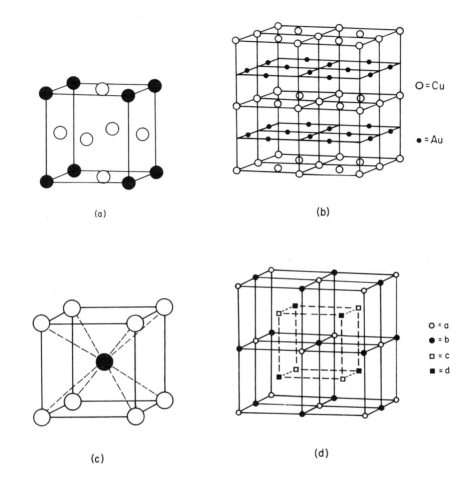

FIG. 3-24. Ordered fcc A_3B (a), fcc AB (b), bcc AB (c), and bcc A_3B structures (d).

Since the hcp lattice consists of closed packed planes in the sequence αβαβαβαβ parallel to the basal plane, it is easy to visu-alize an ordered AB structure. Adjacent alternating planes would be occupied by A- or B-atoms. However, one finds in nature only A_3B structures based on the hcp lattice. Again, since the unit cell of the hcp lattice contains only two atoms, the superlattice has to con-tain more than one unit cell.

C.   Short Range and Long Range Order

The ideally ordered lattice can exist at equilibrium only at
absolute zero temperature. This order will decrease with increasing
temperature and will either gradually or discontinuously disappear
at a critical temperature. The degree of order is characterized by
an order parameter, and one distinguishes between short and long
range order. In the first case, one considers only the arrangement
of atoms in a small volume element containing only a few atoms.
Long range order considers all atoms in a single crystal or a grain,
provided it contains no "antiphase boundaries". Antiphase boundaries
separate volume elements in a crystal. On one side of this antiphase
boundary all lattice points of one sublattice are occupied by one
type of atom, on the other side of the boundary by a different type
of atom (3-25). In a crystal with such boundaries, long range order
is determined in a volume element in a single crystal or grain bor-
dered by these antiphase domain boundaries. One can easily visualize
the creation of such antiphase domain boundaries, if one assumes that
one cools a single crystal from a temperature above the critical ordering temperature T_c to a temperature below T_c. It is usually
unavoidable that ordering starts in several different volume elements.
If ordering starts in two sections, then the sublattice for A-atoms
in one section may be the same sublattice which is occupied

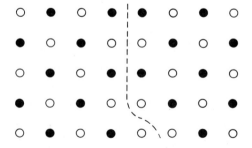

FIG. 3-25. Antiphase boundary.

by A-atoms at the second nucleation site. However, if this sublattice is occupied by A-atoms at one nucleation site, and by B-atoms
at the second nucleation site, then the two sections of the crystal
will not match if the two domains meet during further growth. The
domains are out of phase.   They are separated by an antiphase boundary, as shown in Fig. 3-25.

       We will consider in this section long range order only in a
single domain. If this domain is perfectly ordered in an AB crystal, all A-atoms will be on A-sites, and all B-atoms on B-sites,
where A-sites are all lattice positions on the A-sublattice, and all
B-sites positions on the B-sublattice. Let us assume that Nw_A A-atoms are on wrong sites. Then the same number of B-atoms are also on
wrong sites. Following Bragg and Williams [170, 171], one defines
the long range order parameter S with the following equation:

$$S = (r_A - c_A)/(1 - c_A) = (r_B - c_B)/(1 - c_B)$$
$$= 1 - c_W/c_A c_B,$$

(3-62)

where $r_A = 1 - c_w/c_A = 1 - w_A$ is the probability of finding an
A-atom in the A-sublattice, and $w_A = c_w/c_A$ the probability of
finding a B-atom in the A-sublattice. r_B and w_B are defined in
an analogous way. c_A would be the concentration of A-lattice sites,
and c_w would be the concentration of A-lattice sites occupied by
B-atoms (and vica versa). For a random system, S is zero. Perfect
ordering yields $S = 1$. Negative S values imply that more A-atoms
are found on the B-sublattice than in the A-sublattice. This means
that one should rename the two sublattices. In the AB alloy system,
one can set $w_A = w_B = w$ and $r_A = r_B = r$. There are $Nr/2$ A-atoms
and $Nw/2$ B-atoms on A-sublattice sites, since each sublattice has
$N/2$ lattice sites. This gives as numbers of pairs:

       Number of A-A pairs = $Nzrw/2$
       Number of B-B pairs = $Nzrw/2$ (3-63a)
       Number of A-B pairs = $Nz(r^2 + w^2)/2$

The internal energy is then [168]:

$$E = -Nzrw(E_{AA} + E_{BB}) - Nz(r^2 + w^2)E_{AB}$$
$$= -NzE_{AB} - NzrwE. \tag{3-63b}$$

The configurational free energy is then given by:

$$G \simeq F = -Nz(E_{AB} + rwE) + RT(r \cdot \ln r + w \cdot \ln w). \tag{3-64}$$

Equilibrium conditions require that $\partial G/\partial r = 0$, which gives:

$$\ln(r/w)/(2r - 1) = zE/k_B T \tag{3-65a}$$

or:

$$(2/S)\tanh^{-1}S = zE/k_B T. \tag{3-65b}$$

S is a function of E and T. Fig. 3-26 shows the temperature dependence of S, if one assumes that E is both temperature and order independent. The internal energy U as a function of S is given by:

$$E(S) = -Nz[E_{AB} + (1 - S^2)E/4]. \tag{3-66}$$

Figure 3-26b shows the temperature dependence of the configurational energy. dU/dT, the specific heat of atomic ordering, together with experimental results on ZnCu, are given in Fig. 3-26c.

The experimental results give a much steeper slope of the $c(T)$ curve below the critical temperature than predicted by the Bragg-William theory. This sharp peak in the specific heat is named "lambda" or λ-peak, since it resembles the shape of this Greek symbol. The experimentally determined specific heat of atomic ordering is not zero above T_c. This indicates that disordering is not complete at T_c.

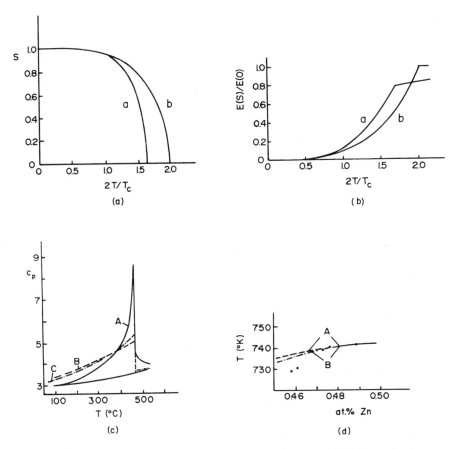

FIG. 3-26. Long-range order S as function of T for simple cubic lattice of AB alloy according to various methods of approximations: (a) Bethe model, (b) Bragg-Williams model (a). Configuration energy E(S) as function of T for simple cubic lattice of AB
alloy according to various methods of approximations: (a) Bethe
model, (b) Bragg-Williams model; E(0) = $Nk_bT_c/2$ (b). Molar heat
capacity in cal./mol. °C of β brass; A: experimental results;
B: Bragg-Williams; C: Bethe model (c). Dependence of the critical
temperature on atomic concentration of Zn for β brass: (A) Bragg-
Williams model; (B) Bethe model (d). After Ref. 172.

The Bragg-Williams theory has been applied also to A_3B ordered
structures. Figures 3-27a + b give the long range order parameter
and the configurational energy U of this system. It shows that
S changes discontinuously at T_c, indicating a first order transition.

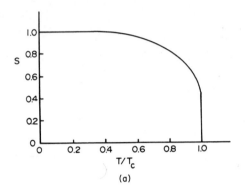

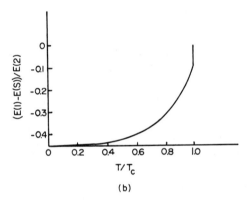

FIG. 3-27. Long-range order versus temperature for AB_3 alloy in the Bragg-Williams approximations (a). Configurational energy $E(S)$ of AB_3 alloy as function of T in the Bragg-Williams approximation. $E(2) = Nk_B T_c$ (b). After Ref. 172.

Experiments show that the specific heat of atomic ordering is not zero above the critical temperature. This indicates short range order above T_c. We mention here briefly the Bethe model [173]. This model is used to calculate the configurational energy of alloys as a function of temperature. It has been applied to determine the degree of ordering and associated energies both for AB and A_3B alloy systems. In the equiatomic AB case, the short range order parameter is defined as $\sigma = (q - q_r)/(q_m - q_r)$ where $q =$ (number of A-B pairs)/(number of all pairs). q_m is the maximum

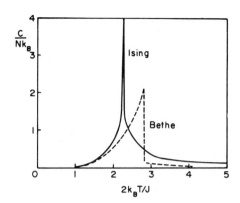

FIG. 3-28. Exact specific heat of the plane square Ising model
(full curve) compared with Bethe's approximation (dotted curve).
After Ref. 102.

value q can have in a given system, and q_r is the q-value of the
completely disordered state. q_m is equal to one for some simple
cases. σ is zero for the completely disordered state, and one for
the perfectly ordered state. Bethe then correlates the probability
of finding A or B-atoms as nearest neighbors for a given atom, and
then calculates the energy E and the short range order parameter as
a function of temperature. Results of these calculations are given
in Figs. 3-26 and 28. The data show that there exists a short range
order above T_c, and that the specific heat drops discontinuously by
Δc which is of the order of the Boltzmann constant k_B, but is
still finite above T_c. This gives better agreement with experiments than found in the Bragg-Williams theory. However, one should
keep in mind that even the Bethe theory is only an approximation.

The only rigorous solution to an ordering problem of physical
significance has been obtained for a two dimensional lattice with
the composition AB. It is equivalent to the two dimensional Ising
lattice, in which one assumes that the magnetic moments in each
lattice site are lined up in the up- or down-position. The up-
position is indicated by ' + ', the down-position by ' - '. Only
nearest neighbor interaction is taken into consideration, so that
the energy between nearest neighbors is either $E(++) = E(--)$ or

E(+ -) = E(- +). This is very similar to the Bragg-Williams model,
except that in the Bragg-Williams model a change in the atomic
arrangement and the energy can be obtained only by exchanging two
different atoms, whereas in the Ising model the system will change
if the orientation of only one spin is reversed. The specific heat
of this model, as calculated by Onsager [116] (see Fig. 3-28), also
shows a sharp λ-peak at T_c. However, there is no discontinuity of
c_{or} at T_c. The 2-dimensional Ising model predicts that c_{or}
should be for T near T_c a function of the form:

$$c_{or}/k_B = A^{\pm}\log |T/T_c - 1| .$$ (3-67a)

Here + indicates $T > T_c$, and - indicates $T < T_c$. A recent
calculation with the 3-dimensional Ising model by Sykes et al.
[174], gives for $T > T_c$ and T close to T_c the following
results:

$$c_{or}/R = A(1 - T_c/T)^{-1/8} + a \text{ for } T \to T_c .$$ (3-67b)

A is close to 1.1 and a is close to −1.24 for both fcc and bcc
lattices. This calculation also gives equations for larger temperature ranges. Computional difficulties in series calculations are
substantial. Future work may modify results presently reached.

Good agreement between Ising model calculation, taking the
compressibility into account, and the long range order parameter S
was found by Norwell and Als-Nielsen [175] from X-ray measurements
(3-29). However, Chipman and Walker [176] found that the experiments resulted in larger $S(T)$ values than the theoretical Ising
model predicted.

Ashman and Handler [177] obtain good agreement between their
data and this predicted power law. They plotted log $(dc_{or}/d\epsilon)$ with
$\epsilon = |T/T_c - 1|$ as a function of $\log|T/T_c - 1|$ and obtained
straight lines over several orders of magnitude of $|T/T_c - 1|$
(3-30). They calculated from the slope of their curve the

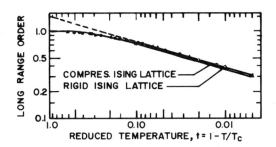

FIG. 3-29. Long range order parameter in β-brass as determined
by the intensity of the (1 0 0) and the (3 0 0) superlattice Bragg
reflections at constant pressure. The data were normalized to 0.996
at room temperature. Also shown are theoretical predictions for the
Ising model at constant volume and at constant pressure. After Ref.
175, with permission.

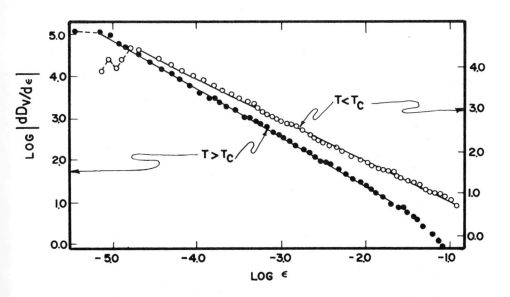

FIG. 3-30. Plot of $\log|dc_V/d\epsilon|$ versus $\log\epsilon$. The solid lines are
linear fits to the data points. The two vertical axes are slightly
displaced from one another but do have the same scale so that different slopes for $T > T_C$ and $T < T_C$ imply different critical ex-
ponents. $\epsilon = |1 - T/T_C|$. After Ref. 177, with permission.

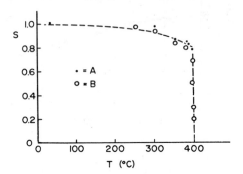

FIG. 3-31. Variation with temperature of the long-range order parameter S for Cu$_3$Au. Dots: Quenched sample. Open circles: Air temperature. After Ref. 179.

critical exponent α and obtained $\alpha = 0.13 \pm 0.01$, in good agreement with the theory. The exponent for $T < T$ was close to zero.

Figure 3-31 gives the variation of S for Cu$_3$Au as an example of an A$_3$B system [178]. These data were again obtained from X-ray measurements. They show the predicted first order transition at the critical temperature. The long range order parameter, S, drops discontinuously at this temperature, in agreement with the Bragg-William theory.

V. MAGNETIC ORDERING

A. Ferromagnetic Elements

The magnetization of ferromagnetic elements due to spin interaction is a function of temperature. This ordering process requires no atomic diffusion. Therefore, it is very rapid and reaches equilibrium very fast, even at low temperatures. This makes it possible to investigate the magnetic energy in much more detail in all temperature ranges than the energy due to atomic ordering. Several

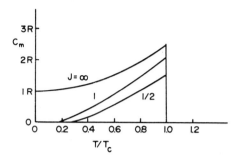

FIG. 3-32. Specific heat according to the Weiss molecular field theory. After Ref. 79.

model system have been worked out for magnetic ordering. Naturally, the Weiss molecular field theory has been investigated first. The magnetic energy due to a magnetic field is:

$$E_{mag.} = \int H dM. \tag{3-68}$$

Since the Weiss molecular field theory assumes for $T < T_c$ that the molecular field H_m is $H_W = W_m M$, one obtains for this model as the magnetic energy E_W:

$$E_W = \int H_m dM = \int W_m M dM = W_m M^2/2. \tag{3-69}$$

The magnetic specific heat in the Weiss approximation is $c_W = \partial E_W/\partial T$. This gives:

$$c_W = -(W_m/2)\partial M(T)^2/\partial T. \tag{3-70}$$

One can calculate c_W in this equation by using the temperature dependent Brillouin function for M (see Chapter 2). Results of such calculations are given in Fig. 3-32 for different quantum numbers. The Weiss theory predicts that c_W drops to zero at the Curie temperature, and this drop Δc_W is [79]:

$$\Delta c_W = \{5(J + 1)J/[J^2 + (J + 1)^2]\}n_o R \tag{3-71}$$

if one sets the gyromagnetic ratio equal to 2. n_o is the number of spins per atom, and J the total angular spin. Figure 3-28 gives c_{or} from calculations based on the Bethe model, which can be modified to incorporate magnetic fields (see Chapter 2) and these give the magnetic specific heat c_{BPW} of the Bethe-Peierls-Weiss model. This model also predicts a discontinuous drop of c_{BPW} at T_c and short range order above T_c which decreases with increasing temperatures. Therefore, c_{BPW} above T_c is not zero in agreement with experiments. Onsager was able to accurately calculate the specific heat of a two-dimensional Ising model. This model assumes that electron spins can line up only in the z-direction, both in the plus and minus direction. Only nearest neighbor interactions were taken into account. This model predicts for the magnetic ($c_{ma.}$) or ordering ($c_{or.}$) specific heats a logarithmic divergence:

$$c_{or}(T) = A^{\overset{+}{-}} \ln\left| (T - T_c)/T_c \right| + B^{\overset{+}{-}} \tag{3-72}$$

where $A^{\overset{+}{-}}$ and $B^{\overset{+}{-}}$ are constants, and where + indicates $T > T_c$, and − indicates $T < T_c$. Exact solutions cannot be obtained for the interesting 3-dimensional system; only some pathological cases can be solved. Presently, perturbation calculations of the "series expansion" type are used. The predict a type of curve for $c_{ma.}$ similar to that predicted for the 2-dimensional case. Near T_c, these calculations suggest in the case of atomic ordering:

$$c_{ma.}(T) \propto \left| T/T_c - 1 \right|^{-\alpha^{\overset{+}{-}}} \tag{3-73}$$

with $c_{ma.}(T^+ \to T_c) \to c_{ma.}(T^- \to T_c)$ for $T = T_c$. It is presently believed that $\alpha^{\overset{+}{-}}$ for several 3-dimensional Ising and Heisenber models should be 1/8.

The series expansion calculation assumes that the magnetic moments are localized. This is a good approximation for rare earth elements and also probably for iron, but cobalt and certainly nickel, should be described with a collective electron model.

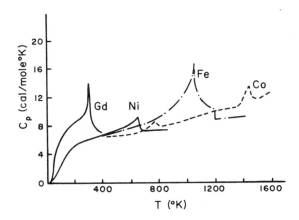

FIG. 3-33. Specific heat of iron, cobalt, nickel and gadolinium. After Ref. 114.

Katsuki [103] tried to calculate $c_{ma.}$ with the Stoner (collective
electron) model for nickel, but he was not able to determine $c_{ma.}$
near the critical point. Therefore, the magnetic specific heat is
usually analyzed with models of localized electrons, and the polarization of conduction electrons is sometimes taken into account as
a correction. This can be done by splitting the conduction band
into spin up and spin down sections, and filling each band to
$E_F - (1/2)\mu_B H_w$ or $E_F + (1/2)\mu_B H_W$. Since $H_m = W_m M$ is temperature dependent, the filling of the conduction band is also temperature dependent. One can show that the slope of $c_{el.} (T \to T_e^-)$ can
be divergent at T_c. The divergency depends on the radius of curvature of the density of states curve, a property which cannot be
measured accurately.

Figure 3-33 shows the experimentally determined specific heat
of ferromagnetic elements. A detailed analysis of the specific
heat of nickel near the Curie temperature has been given by Connelly
et al. [179]. They used a calorimeter in which the temperature of
a small single crystal sample fluctuated periodically in a furnace
since it also absorbed heat by illumination with a periodic (square
wave) light source. Their experimental technique allowed a

resolution of the specific heat in temperature intervals of about 0.01K. Results are shown in Fig. 3-34. Agreement with previously published data is good. The data were analyzed with an equation of a form similar to Eq. 3-67b: $c = (A/\alpha)|1 - T/T_c|^{-\alpha} + K$ for $T > T_c$. This equation yields: $\log_{10}(\partial c/\partial T) = -(\alpha + 1) \log_{10}|1 - T/T_c| + $ const. α is replaced by α' for $T < T_c$. Experiments show that $\log |1 - T_c/T|$ is a linear function of $\log_{10} |1 - T_c/T|$. For $T < T_c$, α, A and K are replaced by α', A' and K'. The calculation showed that the values of the critical exponents depend markedly on the value selected for the Curie temperature. A good fit to the experimental data can be obtained, if one uses $T_c = 631.58K$. This gives for $T > T_c$ as α value: $\alpha = -0.10 \pm 0.03$, and for $T < T_c$: $\alpha' = -0.10 \pm 0.03$. A straight line fit is obtained for experimental data with these exponents over more than two decades of $\epsilon = |1 - T/T_c|$. However, near the Curie temperature, some rounding occurs, which cannot be described with this power law. Further, $T_c = 631.58K$ is not the maximum value of c_p. This maximum is found about 0.13K below T_c. Selecting 631.45K as Curie temperature would lead to a poor fit of the power curve. The main disadvantage of this lower T_c value would be that $\alpha = 0.00$ for $T > T_c$ and $\alpha' = 0.18$ for $T < T_c$. Theoretical arguments require that $\alpha = \alpha'$.

The rounding of the specific heat peak over several degrees K can be related to inhomgeneities in the sample. The single crystal used in these experiments has small amounts of impurities. Major contaminants were 58 ppm copper, 793 ppm cobalt, 114 ppm iron, 990 ppm silicon, 140 ppm aluminum, and 256 ppm sodium. This means the major metallic impurities were close to 0.3 at.%. Further, one should suspect additional non-metallic impurities in the sample. A specific heat run in the same system on a 99.999% pure polycrystalline nickel foil showed a similar rounding to that of the single crystal except that the position of the peak was shifted to slightly lower temperatures. A much broader rounding was found (about 1K) for a deformed single crystal. The largest amount of rounding was found for the single crystal exposed to a magnetic field.

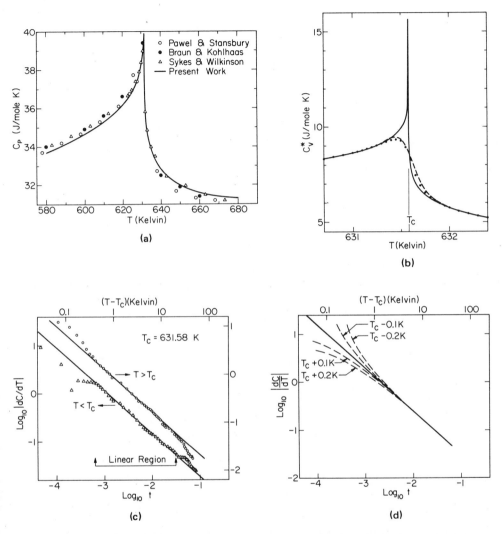

FIG. 3-34. Comparison of experimental specific heat data for Ni
(see Ref. 179 for details) (a). Temperature variation of the magnetic
specific heat very close to the Curie temperature (b). Solid curve
shows an analytic fit with the Curie temperature $T_C = 631.58°K$ and
$\alpha = \alpha' = -0.10$. The dashed curve shows a calculation using a Gaussian
distribution function. $\log|\partial c/\partial T|$ vs $\log(t)$ of Ni (c). Straight lines
indicate the exact power law with $\alpha' = 0.10$. The origin in the ordinate has been displaced to distinguish data measured above and below T_C. Effect of uncertainty in T_C on the apparent linearity (d).
Dashed curves show the effect of shifting the value of T_C by the
indicated amounts. Reproduced with permission from Ref. 179.

240 Oe were sufficient to produce a peak broadening over several
degrees Kelvin.

Since broadening effects can be attributed to internal strains
due to dislocations and to internal magnetic fields, it is not
surprising that even c_p of the underformed single crystal shows
some broadening. Even close to the Curie temperature, internal
magnetic fields may lead to energy variations in the sample, especially since the sample contained a relatively large amount of
impurites with localized moments.

Figure 3-35 shows experimental data for the specific heat of
rare earth elements. There are further peaks below $7^{\circ}K$ (not visible
in Fig. 3-35a). They are due to superconducting transitions of α-La
at $4.9^{\circ}K$ and β-La at $5.85^{\circ}K$. The peak at $8^{\circ}K$ in neodymium is due to
antiferromagnetic ordering. The heat capacity of cerium (3-35b) shows
one anomaly at $12.5^{\circ}K$, associated with a paramagnetic-antiferromag-
netic transition of β-Ce, whereas the peak at $170^{\circ}K$ is due to the
α – γ transition. The pikes near $13^{\circ}K$ and $105^{\circ}K$ in samarium are
probably of magnetic origin. This is also true for the peaks in
c_p shown in Fig. 3-35c. The reader should consult the original
reference for more details. Since in these elements the magnetism
is due to localized electrons, the theoretical predictions on the
temperature dependence should be closely followed. However, rounding is found near T_c, just as for transition elements. Some of
the rare earth elements show not only ferromagnetic transitions,
but also antiferromagnetic phase changes.

In spite of extensive theoretical and experimental studies, our
understanding of the magnetic specific heat of ferromagnetic elements is not very good. One can predict the general temperature
dependence of c_{ma}, but it is not possible to predict the absolute
values of critical exponents with sufficiently high accuracy to
calculate c_{ma} from first principles. An accurate determination of
the magnetic specific heat is only possible for spin waves at very
low temperatures, since the temperature dependence of the magnetization is known.

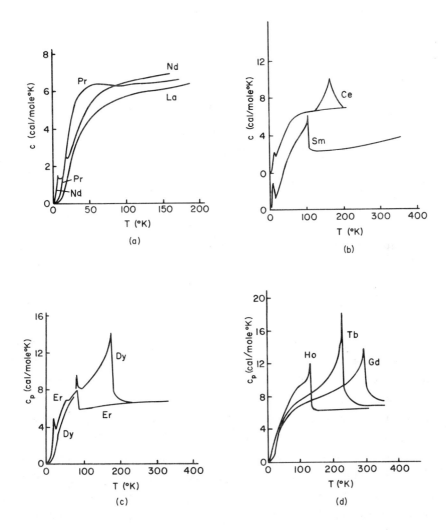

FIG. 3-35. Specific heats of lanthanum, neodynium and praseo-
dium as functions of temperature (a). Specific heats of cerium and
samarium as functions of temperature (b). Specific heats of dysprosium and erbium as functions of temperature (c). Specific heats of
gadolinium, holmium and terbium as functions of temperature (d).
Scale is shifted in section (b). After Ref. 180, with permission.

B. Ferromagnetic Alloys and Compounds

Several models have been used to describe the magnetization of
alloys. Some have been discussed in Chapter 2. The simplest model
would assume that the saturation magnetization M_o and the molecular field H_m at zero degrees, or the Curie temperature, are linear
functions of composition. All the parameters are zero at the non-
magnetic element. This may be a reasonable first approximation for
the nickel-palladium system. One more frequently finds for alloy
systems, in which one element is non-magnetic, that the Curie temperature of the magnetization is already zero for c(ferromagnetic
element)$_o$ > 0. The onset of ferromagnetism occurs for the copper
nickel alloy system at c_o = c_{Ni} ≃ 0.55 at.% Ni, and in the iron-
vanadium system for c_o = c_{Fe} ≃ 30 at.% Fe. In alloy systems in
which both T_c and the saturation magnetization are linear functions
of composition, and where the onset of ferromagnetixm occurs at c_o,
the magnetic energy is proportional to $(c - c_o)^2$, since the energy
per Bohr magneton is proportional to H_m and T_c. If one knows the
quantum number associated with the magnetic moment of the atom, one
can use the Brillouin function to determine the specific heat of the
sample in the same way as for the pure element. This would give
for the alloy a magnetic specific heat with a sharp λ-peak. However, it seems reasonable to assume that the sample has short range
fluctuations in its chemical composition. This means that the Curie
temperature in separate volume elements fluctuates around an average
Curie temperature. This should lead to a rounding of the peak in
alloys.

Such a model has been used successfully by Takahashi and
Shimizu to describe $c_{ma.}$ of PdFe and PdCo alloys [181]. These
authors assumed that the 4d-electrons are itinerant and the magnetic
moments of iron and cobalt are localized. The sample is divided
into small volume elements, in which the composition is assumed to
be constant. The composition in different volume elements is given

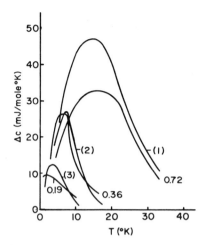

FIG. 3-36. Experimental magnetic specific heat of Pd-Fe alloys and calculated results at very low temperature. Curves (1), (2) and (3) are calculated values for c = 0.72 at.% Fe, c = 0.36 at.% Fe, and c = 0.19 at.% Fe, respectively. Experimental curves are characterized by numbers giving the impurity concentration in at.%. After Ref. 181, with permission.

by a Gaussion distribution function, with the maximum at the nominal composition. The magnetic specific heat depends on both the spin of the magnetic impurity, which is assumed to be either 1 or 3/2, and on the half width of the Gaussian distribution function. The magnetic specific heat in each region of uniform composition is a function with a λ-peak. Its Curie temperature is proportional to the impurity concentration and $S(S + 1)$, where S is the quantum number of the impurity. The sum of the magnetic specific heats of all volume elements gives the total magnetic specific heat. Figure 3-36 shows calculated and experimental c_{ma}. curves. Agreement between theory and experiments is very good. Therefore the PdFe and PdCo alloys may be regarded as examples in which the model of magnetic independent volume elements may be justified.

It is much more difficult to visualize magnetically independent volume elements in alloys far below the Curie temperature if these volume elements are large compared with the volume of one atom. Thermal excitations of the magnetic structure in such systems lead

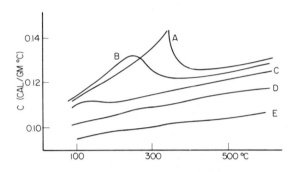

FIG. 3-37. Specific heat of Cu-Ni alloys. The specific heat of copper (E), nickel (A) and copper-nickel alloys containing (B) 90.05, (c) 75.07 and (D) 50.04 wt.% Ni. After Ref. 182, with permission.

to spin waves, which travel through the crystal and will extend over volume elements much larger than chemical fluctuation ranges. Their specific heat contribution follows the same laws as spin waves of pure elements. Variations in chemical composition will not affect spin waves at low temperature. A model of collective excitations may not be appropriately close to the Curie temperature. There magnetic moments show short range order. In that temperature range one should therefore envision that the crystal consists of magnetic clusters. The magnetic cluster diameter decreases with increasing temperature. It should be influenced by variations in the chemical composition of the sample as soon as the "magnetic" cluster diameter is of the same order or smaller than the "chemical" cluster diameter. Since the Curie temperature of each magnetic cluster depends on its chemical composition, one would expect to find a range of Curie temperatures. This should lead, just as in PdCo or PdFe alloys discussed above, to a rounded peak in c_{ma}. near the average Curie temperature. Specific heat measurements on nickel-copper alloys near the Curie temperature seem to support this view. A few percent copper in nickel broadens the magnetic specific heat peak considerably (3-37). At 15 at.% copper, the peak in the specific heat curve has practically disappeared. A

quantitative analysis of $c_{ma.}$ of these alloys with up to 25 at.%
copper is not available on the basis of localized electron models.
However, since the magnetization and specific heats of copper-
nickel alloys at the onset of ferromagnetism in the liquid He-temperature range can be described with models using the cluster concept, it seems reasonable to assume that this atomic ordering model
should be also appropriate for nickel-rich alloys.

Figure 3-38 gives the specific heats of fcc Co-Ni alloys.   It
shows that the sharp λ-peak is retained for all alloys. Similar
results were obtained for fcc Fe-Ni alloys. The λ-peak is not as
sharp as found for the pure nickel experiment. This is not unexpected, since the experimental resolution of c_p in the alloy
experiments is of the order of $\Delta T/T_c \simeq 2 \times 10^{-3}$. The data in these
experiments (3-38b and c) were analyzed in the following way: the
total experimental specific heat $c_p(T)$ at temperature T is assumed
to be composed predominantly of dilation $(c_p - c_v)$, lattice $c_{1a.}(T)$,
electronic $c_{el.}(T)$, and magnetic $c_{ma.}(T)$, contributions: $c_p(T) =$
$[(c_p(T) - c_v(T)] + c_{1a.}(T) + c_{el.}(T) + c_{ma.}(T)$. At each temperature, the lattice dilation, $(c_p - c_v)$, and the electronic and lattice specific heat background was estimated and subtracted from the
experimental c_p values to obtain the magnetic specific heat, $c_{ma.}(T)$.

The dilation correction $(c_p - c_v)$ can be described usually by
a function linear in temperature. $c_p - c_v$ is estimated to be only
about 2% of $c_p(T)$ although thermodynamic derivatives like the thermal expansion and compressibility do exhibit anomalous behavior in
the neighborhood of the critical point.

The harmonic lattice specific heat can be determined quite
easily since the Debye model specific heat is expected to give an
adequate representation of the lattice specific heat. The characteristic Debye temperature, T_D, may be taken from liquid He temperature measurements. The anharmonic contribution to $c_{1a.}$ com-
prises a positive linear function of temperature.

The electronic contribution at temperatures below T_c was
calculated from the relationship $c_{el.} = \gamma_0 T$, where the γ_0

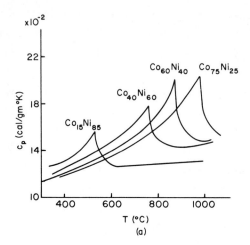

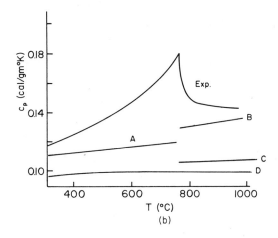

FIG. 3-38. Specific heat of Co-Ni alloys (a). Specific heat of
Co$_{40}$Ni$_{60}$ (b). Full lines give calculated lattice (D) and lattice
plus electronic specific heats (A, B, and C). The electronic term is
adjusted in such a way that $\int c_m dT$ has the right value for the Ising
(B) or Heisenberg model (C). Semilogarithmic plot of c_m vs temperature for Co$_{40}$Ni$_{60}$ (c, opposite page). Open circles and left hand
scale correspond to $T > T_c$, full circles and right hand scale to
$T < T_c$. The Ising model was used. Critical parameter A^+ is defined
with Eq. 3-72 for Ni-alloys (d, opposite page). After Ref. 183,
with permission.

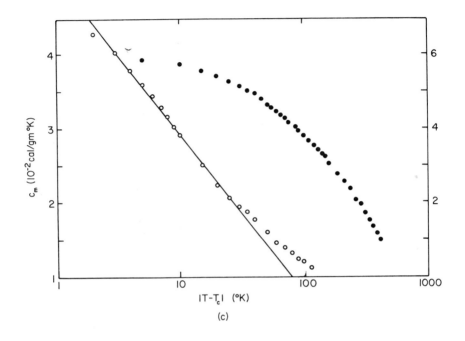

(c)

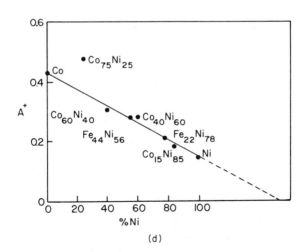

(d)

coefficient was taken from liquid He temperature specific heat
measurements. (γ_o should include those anharmonic lattice specific
heat terms which are linear functions of temperature.) $c_{el.}(T) =$
$\gamma_o T$ was added to the previously calculated lattice contribution.
One should keep in mind that the approximation $\gamma_o T = c_{el.}$ is
probably incorrect close to T_c, since the magnetization in this
temperature range changes rapidly. Therefore, $N(E)$ should depend
on T, and $c_{el.} \propto N(E)T$ will not be a linear function of T.

The data above T_c cannot be calculated in this way, since
$c_{el.}$ above T_c is not known. However, the c_p data above T_c
can be used to obtain a crude approximation of the density of states
for each alloy in the paramagnetic region above T_c. The analysis
requires two major assumptions: first, that the lattice specific
heat is given by the temperature-independent Dulong-Petit value and
second, that the electron specific heat is a linear function of temperature $c_{el.} = \gamma T$ for $T > T_c$, where γ is proportional to the
paramagnetic density of states. $c_{la.} + c_{el.}$ above T_c was constructed in such a way (γ was selected as an adjustable parameter) that
the enclosed area between it and the experimental specific heat
curve $c_p - (c_L + c_E)$ equals the magnetic energy associated with
the theoretical Heisenberg model. In exactly the same manner,
another background was constructed for each alloy from the energy
predictions of the Ising model. Both the constructed Heisenberg
and Ising background are presented in Fig. 3-38b for the $Co_{40}Ni_{60}$
alloy. Usually only the Heisenberg or the Ising model matches up
reasonably well with the experimental data. Figure 3-38c gives a
plot of $c_{ma.}$ vs log T for this alloy. The experimental data
follow a straight line except for some rounding near T_c.

The A^+ values of this alloy were found to be independent of
the background choice for the $Co_{40}Ni_{60}$ alloy where both the Ising
and Heisenberg models give a reasonable match with the high temperature experimental data. This indicates that A^+ is most likely independent of the background calculation for alloys in this investigation.

The compositional dependency of A^+ is presented in Fig. 3-38d. The A^+ values for the four fcc materials, high temperature Co, Ni, Co Ni alloys and NiFe alloys, are linear functions of nickel concentration. The linear plot extrapolated to A^+ (fcc) = 0 intercepts the abscissa at approximately the composition $Ni_{40}Cu_{60}$, where the d band is filled and the spontaneous magnetization is zero. It is presently not possible to explain why A^+ should be a linear function of composition.

Iron base alloys with non-ferromagnetic impurities should be good examples of alloy systems in which localized electrons are responsible for most of the ferromagnetism. Figure 3-39a gives a plot of c_p vs T for an FeCr alloy. It shows a λ-peak which is much sharper than those found for similar impurity levels in NiCu alloys. Figures 3-39b and c give c_p as a function of $\log|1 - T/T_c|$ for pure iron and two FeCr alloys. Again, the fit of the data to a straight line is much better above the Curie temperature than below.

The feature that the $c_{ma.}(T)$ data of all the alloys display a logarithmic temperature dependence above T_c and, in general, not below T_c is similar to the behavior found in the insulating materials (3-40). The non-logarithmic behavior found both above and below T_c in EuS has been attributed to intrinsic effects. McCoy and Wu [187] calculated the specific heat for this system using the Ising model with impurities, They obtained good agreement between theory and experiments for $T > T_c$ if they assumed that coupling energies in individual lattice planes were constant, but found that coupling energies between adjacent planes could vary in a given energy range. This model could explain the observed rounding of the specific heat curve close to the Curie temperature.

C. Antiferromagnetic and Ferrimagnetic Alloys and Compounds

Specific heat anomalies due to antiferromagnetic ordering sometimes show the same λ-type anomaly as found for ferromagnetic transitions. This was mentioned in the discussion of Fig. 3-35a and b.

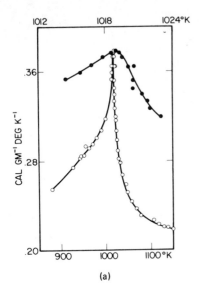

(a)

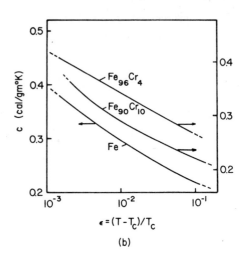

$$\epsilon = (T - T_c)/T_c$$

(b)

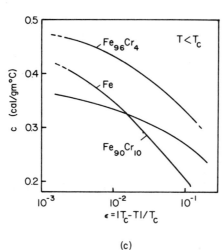

$$\epsilon = |T_c - T|/T_c$$

(c)

FIG. 3-39. Specific heat of $Fe_{90}Cr_{10}$ near its magnetic critical temperature (a). Open circles correspond to the bottom scale, full circles to the top scale of temperature. After Ref. 184. Specific heat data of iron and iron-chromium alloys for temperatures above the Curie temperature (b). Specific heat data for iron and iron-chromium alloys below the Curie temperature (c). The Curie temperature was determined from measurements above the Curie temperature both for (b) and (c). After Ref. 185.

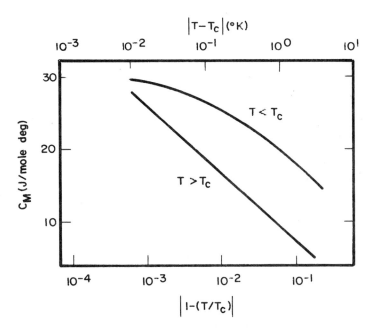

FIG. 3-40. Specific heat of EuS. After Ref. 186.

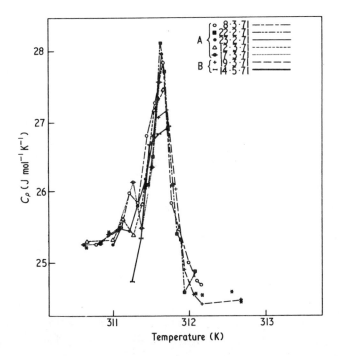

FIG. 3-41. Specific heat of polycrystalline chromium near the
Néel temperature. The data designated group A were taken after the
annealed sample had relaxed, the group of curves B were taken in
the unrelaxed state. Reproduced with permission from Ref. 188.

Chromium and manganese are antiferromagnetic. The moment per atom is about 0.4 Bohr magneton per atom at 0^OK. Chromium shows a very small specific heat anomaly near its Néel temperature, $T_N =$ 312^OK (3-41). The transition from the antiferromagnetic to the paramagnetic state is of first order. The associated energy change is very small. The specific heat change near the Néel temperature of chromium is less than 1% of the lattice specific heat. This has been attributed to the fact that the magnetization is due to spin density waves of the electron gas. The addition of small amounts of manganese increases the number of Bohr magnetons drastically and makes the spin density wave commensurate with the lattice. This, however, does not produce a noticeable magnetic specific peak. Figure 3-42 shows c_p of chromium and chromium alloys with up to 50% Mn. There is a slight maximum in these curves near the Curie temperature. Only the addition of quite large amounts of manganese leads to peaks in the specific heat curves near the Néel temperature. The shape of this peak is not similar to the λ-anomaly found for the typical second order magnetic transitions. It may be possible that these changes in the specific heat near the magnetic transition temperature should be attributed to changes in the electronic specific heat of conduction electrons. The electronic specific heat coefficient for Cr-Mn alloys changes noticeably with temperature, even well below the Néel temperature. The liquid He temperature data indicate that γ_o increases rapidly with increasing manganese concentration, reaching values of 23 mJ/K^o for the alloy with 50 at.% Mn, whereas the room temperature data give, except for pure chromium, γ values of 8 mJ/K, independent of manganese concentration [152].

D. Spin Waves and Magnetic Clusters

The magnetic moments of electrons responsible for ferromagnetism and also antiferromagnetism and ferrimagnetism are lined up in the perfectly ordered state in well defined crystallographic

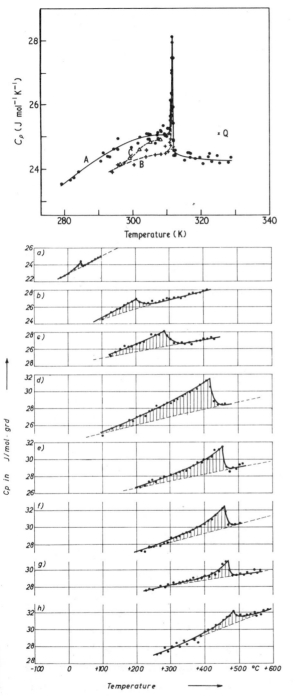

FIG. 3-42. Specific heat of chromium from 280 to 330°K (a). Reproduced with permission from Ref. 188. Specific heats of Cr-Mn alloys (b).

0 at.% Mn: a,
1.18 at.% Mn: b,
2.48 at.% Mn: c,
8.11 at.% Mn: d,
14.65 at.% Mn: e,
23.30 at.% Mn: f,
29.84 at.% Mn: g,
50.13 at.% Mn: h.

Reproduced with permission from Ref. 189.

directions. Due to thermal excitations, travelling spin waves (or magnons) are created in the crystal.

Figure 2-41 gives a model of a travelling spin wave. In the lowest energy state, the spins are lined up in a ferromagnetic system in only one direction. Some of the spins will deviate from the direction of easy magnetization. They will precess around the axis of easy magnetization. The precession movement of one spin will influence the neighbors, which will then follow the movement of the other electrons with a phase shift. The angular displacement of the spin from the easy magnetization increases with increasing thermal excitation. The temperature dependence of M is given by the Bloch Law in Eq. (2-113). From M(T), one again determines the magnetic energy. By differentiation one obtains the specific heat of the ferromagnetic spin wave [106]:

$$c_B(\text{ferromagnetic spin wave})_v = 0.113 \ k_B^{5/2} T^{3/2} (N/nJ) \qquad (3\text{-}74)$$

n = 1, 2, or 4 respectively for sc, bcc or fcc crystals. An analysis by Van Kranendonk and Van Vleck [190] shows that the specific heat due to the thermal excitation of antiferromagnetic spin waves is:

$$c_{KV}(\text{antiferromagnet spin wave}) = (4R/15)(k_B/2JS(2z)^{1/2})^3 \cdot T^3$$
$$(3\text{-}75)$$

for cubic lattices. z is the number of nearest neighbors, S is the total spin number, and J is the exchange integral. The specific heat due to these spin waves follows the same temperature dependence as the lattice specific heat. Therefore, it is not possible to separate these two terms, if one develops the measured specific heat in a power series. On the other hand, the lattice specific heat of metals is so well known, that one can determine c_{1a} from the Debye theory using either reasonable Debye temperatures determined from high temperature elastic constants or from specific heat data above the liquid N_2 temperature. If one assumes, for instance, in

some chromium alloys that the lattice specific heat is essentially
independent of composition, then one finds extra T^3 power terms
for alloys with 10 to 16 at.% Fe, or 1 and 2 at.% Ni. These alloys
have all Néel temperatures above liquid He temperature but below
room temperatures. This can lead, according to an analysis by Baum
and Schroder [191], to noticeable antiferromagnetic spin wave contributions to the specific heat.

A similar specific heat contribution, again due to the crystal
energy which lines up the spins of electrons, has been associated
with small ferromagnetic volume elements in a non-ferromagnetic
matrix. These groups of ferromagnetic atoms, which may also include
non-ferromagnetic atoms as long as the group together is a magnetic
domain, are called clusters. Due to thermal excitations, the magnetic moment of the cluster is not lined up in the direction of
easy magnetization. For small angular deflections the crystal
energy of the cluster is $E_{crystal} = KV\alpha^2/2$, where K is the crystal
anisotropy energy, V is the cluster volume, and α is the angle
between the direction of easy magnetization and the magnetization
vector of the cluster. The deflection leads to a torque on the
magnetic cluster, equal to $KV\alpha$. Since this torque is proportional
to the deflection, the system corresponds to a harmonic oscillator.
The magnetization vector will rotate around the vector of easy magnetization in a precession, only the projection of the magnetization vector in a plane containing the direction of easy magnetization will oscillate like a harmonic oscillator. This is similar to
the case of an electron circling an atom, where only the projection
of the electron movement in a given direction on the plane of movement will behave like a simple harmonic oscillator. The energy
levels of the oscillating (or rotating) magnetization vector of the
cluster are evenly spaced energy levels, just as in the Einstein
specific heat model. Therefore, the specific heat of each spinning
cluster should be given by the Einstein function, $c_{cl.} \propto E(T/T_S)$,
with T_S equivalent to the Einstein temperature. $T_S = 4K/k_B \, n_o$,
with K the crystal energy per volume and n_o the number of spins

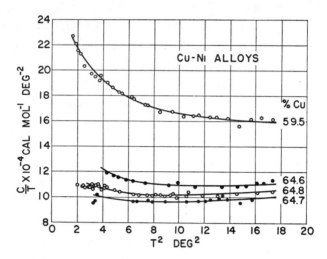

FIG. 3-43. The heat capacities of Cu-Ni alloys. The solid curves represent fits of the form $C = \gamma T + \beta T^3 + C_M$. Reproduced with permission from Ref. 192.

per volume of the cluster [192]. T_S should be in the range from 0.01°K to 0.1°K for pure iron clusters.

This equation for c_S can be used only for small α values. Livingston and Bean [193] calculated the specific heat of a spinning cluster for large α values, and for more complex crystal energy systems in the classical case. They showed that the cluster specific heat decreases at higher temperatures. Therefore, $c_S(T)$ follows for $T \ll T_S$ an exponential function, c_S is nearly constant between T_S and $T \ll KV/k_B$, and for $k_B T > KV$ c_S decreases to zero with increasing temperature.

It is possible to determine the cluster concentration from low temperature specific heat measurements. Typically, one analyzes the specific heat with a function of the form $c_p = \gamma T + \beta T^3 + C_M$, where the first term represents the electronic specific heat, the second the lattice specific heat, and the third term is due to clustering. C_M is proportional to the cluster concentration. β can be obtained from the Debye temperature. Therefore, the equation has only two adjustable parameters, namely γ and the cluster concentration. Figure 3-43 shows experimental data as circles, and

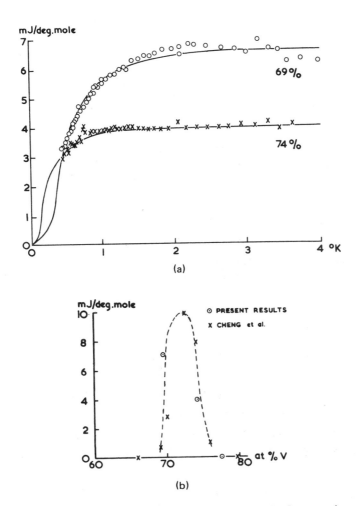

FIG. 3-44. Specific heat minus lattice and electronic contributions vs temperature for a $Fe_{69}V_{31}$ alloy. Solid line is a fitted Einstein function (a). Constant specific heat parameter C_M vs vanadium concentration (b). After Ref. 194.

the full lines were calculated with γ and C_M as adjustable parameters. The curves fit the data very well.

Scurlock and Wray [194] measured the specific heat of Fe-V alloys to temperatures below $1^{\circ}K$. They subtracted from the measured c_p values both lattice and electronic contributions. The results

are plotted in Fig. 3-44a. The points obtained this way closely
follow an Einstein function. The measurements indicate that the
Einstein temperature is of the order of $1^{\circ}K$. This indicates that
the crystal anisotropy constant of the clusters is not that of pure
iron, but of Fe-V alloys.

Figure 3-44b gives C_M as a function of alloy concentration.
C_M is proportional to the number of Einstein oscillators, which
should be again proportional to the number of clusters. It shows
that the cluster concentration follows a bell-shaped curve. Its
maximum is close to the onset of ferromagnetism. This would be expected if the clusters are independent ferromagnetic groups of atoms.

Measurements by Robbins et al. [195] showed that C_M or the
cluster concentration increased if the sample was deformed. This
would be expected, since plastic deformation leads to a break-up of
clusters due to slip, and each cluster fragment acts as a new independent cluster.

The interaction of magnetic impurities with each other and with
a magnetic or non-magnetic matrix can be very complex. A review
on these subjects has been given by Phillips [134], who
discusses exchange enhancement effects, localized states of impurities in a simple metal matrix, impurity-conduction electron interaction, and also the nuclear heat capacity due to the interaction of
nuclei with a magnetic moment with fields in the crystal.

Chapter 4

ELECTRICAL AND THERMAL TRANSPORT PROPERTIES

I. INTRODUCTION

Electrical units are named after men like Volta, Ampere, Faraday, Ohm, Coulomb, and Henry, who lived in the 18th and 19th centuries. In this period, the first systematic investigations of electrical phenomena began. This late interest in electrical studies is due to the fact that these phenomena usually produce small effects in nature. The only large electrical effect known to men for a long period of time was lightning.

The first techniques for measuring electrical effects were developed around 1800. Correlations between voltages and currents in metals were investigated shortly afterwards. Seebeck showed that temperature differences between different metal junctions produce voltages. Oersted discovered that an electric current produces a force on magnets. Faraday showed that moving a magnet induces a voltage in a wire. His measurements of the ratio of gas developed at an electrode to the electric current, together with the already used concept that atoms are the building blocks of the material, should have led to the concept of quantized electric charges – electric units like the electron. Its discovery, however, came many years later. In his research of the electrical conductivity of silver sulphide, Faraday found that the electrical resistivity decreased with increasing temperature. This type of temperature dependence is generally regarded as typical for semiconductors.

The electron was discovered shortly before 1900. Experiments
by Lennard and Thomson made it possible to investigate the properties of individual electrons. The ratio of its mass to its charge
could be determined from its movement in electrical and magnetic
fields. Millikan measured the charge of individual electrons through
the forces on ionized oil droplets. It was proposed from these experiments that the electrical conductivity of a metal could be explained with the assumption that electrons move freely through a
system of ions. This model, sometimes referred to as the Drude or
"classical" free electron model, gives the right ratio of the electrical to thermal conductivity of metals.

It is now possible to calculate the essential features of the
electron transport process in solids with concepts of quantized
energy states. However, it sometimes takes a rather long time to
answer seemingly simple questions. For instance, only a few years
ago it became possible to calculate the correct sign of the thermopower in copper. The theoreticians predicted for years that the
Seebeck coefficient should be negative, whereas experimentalists
were adamant in their claim that it was positive. This example indicates that there is still a long way to go before the electron
transport process in alloys can be calculated from first principles.

An electrical current with the current density j will flow
through a metal if an electric field (V/ℓ) is applied; a temperature gradient, $\nabla_r T$, leads to a heat flow. The ratios of $j/(V/\ell)$
and of $\dot{q}/\nabla_r T$, where $\dot{q}$ is the amount of heat which flows per unit
time through a unit cross sectional area, are materials properties.
The theory of semiconductors, metals,and alloys tries to show how
these and other properties can be correlated with each other and
how they depend on such parameters as the crystal structure, crystal defects, density of states, Fermi surface, alloy composition,
temperature, and magnetic state. Typically, one initially uses
models like the Drude theory, in which one assumes that conduction
electrons behave like particles of an ideal gas with kinetic energies
of $(3/2)k_B T$. This classical model gives the right order of magnitude of the electrical conductivity at room temperature for metals.

Naturally, it fails at low temperatures where quantum mechanical concepts have to be taken into consideration. Similar to the analysis
of the thermal properties of an electron gas, only electrons close
to the Fermi energy level give important contributions to the electronic transport process. This is partly due to the fact that energies due to applied electric and magnetic fields are small compared
with the Fermi energy of electrons. Far below the Fermi surface,
one finds for each electron with the velocity v or momentum $\hbar k$
another electron with $-v$ or $-\hbar k$, and their combined contribution to the electrical current is zero. Only electrons near the
Fermi energy level can be excited to higher states, leaving "holes"
below the Fermi surface. Therefore, one should be able to calculate
$j/(\Delta V/\ell)$ or $\dot{q}/\nabla_r T$ if one knows the shape of the Fermi surface and
the scattering mechanism of electrons. The results of such calcula-
tions have some similarities with results from classical theories.
Therefore, one frequently describes transport processes with classical parameters as the electron mass, but then replaces this mass by
an "effective mass" m^* to accomodate quantum mechanical results.

In this chapter, we will deal with the following properties:
a) the electric conductivity σ (or its reciprocal, the electrical
resistivity $\rho = 1/\sigma$), b) the thermal conductivity κ, c) the Hall
coefficient R_H, d) the thermoelectric power S (and its relation
to the Thomson coefficient and Peltier coefficient).

The electrical resistivity is easily determined on wires or
rods with known length between potential leads and given cross sectional area A. Electrical currents I and the voltage drop V_R between the potential leads are related by (see Fig. 4-1):

$$V_R/\ell = F = \rho \, I/A = \rho j \, . \tag{4-1}$$

j is the current density I/A, with A the cross sectional area
of the sample; and $F = V_R/\ell$ is the electric field where V_R is
the voltage drop between the two potential leads, and ℓ is their
distance. The electrical conductivity may depend on an applied magnetic field, or on the magnetization of magnetically ordered samples.

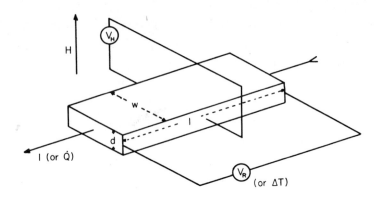

FIG. 4-1. Schematic diagram for Hall voltage and thermal and electrical conductivity (κ and σ) measurements. V_H: Hall voltage; I: electrical currents; V_R: voltage drop along the sample of length 1; $\dot{Q}$: heat flow through the sample under a temperature gradient $\Delta T/1$. W is the width of the sample, and d is the thickness.

The "magneto resistance" is usually measured with the magnetic field either parallel or perpendicular to the electrical current.

The thermal conductivity, κ, can be measured very similarly to the electrical conductivity. An experimental problem is that thermal insulation is much more difficult to obtain than electrical insulation. Heat will flow through the rod in Fig. 4-1 if a temperature gradient $\nabla_r T = \Delta T/\ell$ is applied. The amount of heat which flows per second at steady state condition through this rod, $\dot{Q}$, is proportional to the temperature gradient $\nabla_r T$. This gives the correlation:

$$\Delta T/\ell = \nabla_r T = \dot{Q}/A\kappa = \dot{q}/\kappa \quad . \qquad (4\text{-}2)$$

If a magnetic field is applied perpendicular to a rod which carries an electrical current with current density j, one finds that an electric field (or voltage) is induced perpendicular to both current flow direction and magnetic field direction. This is the Hall effect, which produces the Hall field F_H or Hall voltage V_H. Figure 4-1 shows schematically the experimental arrangement to

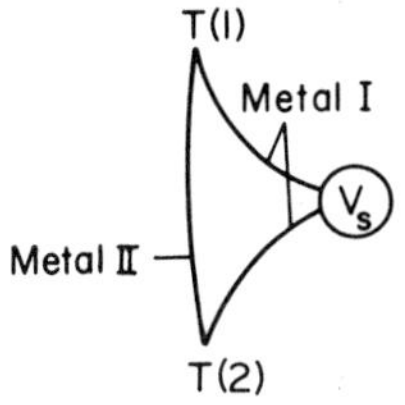

FIG. 4-2. Seebeck effect measurement. V_s: Seebeck voltage.
T(i): temperature at junction i of metal I and II.

measure the Hall effect. One finds frequently that the Hall field
is proportional to the current density and the polarization B (experimentalists use sometimes H instead of B):

$$V_H/W = F_H = R_H jB \quad . \tag{4-3}$$

W is the width of the sample, and R_H the Hall coefficient.

The Seebeck coefficient is measured in an experimental arrangement given schematically in Fig. 4-2. It consists of two wires I
and II, connected at junctions 1 and 2 with temperatures T(1)
and T(2), respectively. The voltage generated in this circuit,
V_s, is for T(1) → T(2) proportional to the temperature difference
T(1) − T(2). The proportionality constant is the Seebeck coefficient S_{I-II}:

$$V_s = S_{I-II}[T(1) - T(2)], \quad T(1) \to T(2) \quad . \tag{4-4}$$

S_{I-II} depends on the properties of both wires I and II. Since
this seems to indicate that the Seebeck coefficient is defined for
a pair of two different metals, one may suspect that S_{I-II} is a
characteristic property of the interface between the two wires.
This is incorrect; one can show that S_{I-II} can be separated into
two terms, each characteristic only of metal I or metal II.

$$S_{I-II} = S_I - S_{II} \quad . \tag{4-5}$$

Experiments show that the Seebeck coefficient of a non-supercon-
ductor I with respect to a superconducting wire SC, S_{I-SC}, is
independent of the superconducting material. It is therefore not
unreasonable to assume that the Seebeck coefficient of all superconductors is equal to zero. Therefore, S_{I-SC} is called the absolute
Seebeck coefficient of the metal I, S_I. The transition temperatures
from the normal to the superconducting state are very low. S_{I-SC} =
S_I can be determined only in a very small temperature range. However, one can determine S_I for all temperatures with the equation:

$$S_I(T) = \int_{T=0}^{T} (\mu/T)dT \qquad (4-6)$$

where μ is the Thomson coefficient. Thomson suggested that aside
from the irreversible heat generated per unit time in a unit volume,
$\dot{q}_{ir.} = \rho j^2$, a second reversible heat contribution per unit time in
a unit volume, $\dot{q}_{rev.}$ is absorbed or emitted from this wire. $\dot{q}_{rev.}$
depends linearly on the temperature gradient and on the current
density, and the proportionality constant is the Thomson coefficient
μ:

$$\dot{q}_{rev.} = \mu j (dT/dx) \qquad . \qquad (4-7)$$

The heat generated by the irreversible process, ρj^2, is independent
of the temperature gradient. One can visualize the reversible heat
generation process in the following way. Electrons in a volume element at $\vec{r}(1)$ with temperature $T(\vec{r}(1))$ have a certain energy distribution at this temperature, which depends on T and the energy band
structure. If these electrons are moved by an electric current into
another volume element at a different temperature, their energy distribution changes, and they will absorb or emit reversibly the energy
difference of the electrons in these two places. This model and
Eq. 4-6 show clearly that S_I is a property of the bulk of the metal
I, and not an effect at the interface between two metals I and II.

The same is true for the Peltier effect. If a current with the current density j passes from metal A to metal B, the energy distribution of the electrons changes. Both metals have at the interface the same temperature, but the density of states curve and with it the total energy of the electron is different in both metals. Therefore, the electrons will absorb or emit energy reversibly in this flow process, and the energy change per time interval per unit area of the interface is:

$$\dot{q}_p = \pi_{A-B} j = (\pi_A - \pi_B) j \qquad (4-8)$$

where π_{A-B} is the Peltier coefficient. π_A is related to the absolute Seebeck coefficient by:

$$\pi_A = T S_A \qquad (4-9)$$

and can therefore be determined from the bulk property S_A.

II.   ELECTRICAL CONDUCTIVITY

A.   Free Conduction Electron Model

After the discovery of the electron, it was proposed that the electrical current should be described in the following way in a metal. The electron with charge e is accelerated by an applied electric field F in the x_1-direction, with $\partial^2 x_1/\partial t^2 = eF/m$. m is the mass of the electron. After an average time interval τ, it collides with an ion or a crystal defect, and its drift velocity is reduced to zero. The mean drift velocity v_{dr} is eF τ/m, since:

$$v_{dr} = \int_0^\tau (e/m)F dt = (eF/m)\tau. \qquad (4-10)$$

The mean free path, which is the average path length of an electron between collisions, is equal to:

$$\ell = \tau v_{dr}. \tag{4-11}$$

The electrons in this classical model have an average thermal kinetic energy of $(3/2)k_B T$. The acceleration due to an outside field is superimposed on the kinetic velocity, $v_{therm.} = (3k_B T/m)^{1/2}$. This model gives the right order of magnitude for the resistivity of metals at room temperature. It is convenient to use to describe electronic transport phenomena, because results of quantum mechanical calculations can be incorporated into this model by using an "effective" electron mass or by incorporating scattering of s-electrons into the d-band into the calculation of the mean free path.

The descriptions of the electron transport process in the classical model and in quantum mechanics follow to a certain degree similar lines. Instead of assuming that the drift velocity is superimposed on the thermal velocity of electrons, it is superimposed on the Fermi velocity of the electrons. The Fermi velocity is the velocity of electrons with the Fermi energy. Let us assume that the initial energy distribution of electrons is given in Fig. 4-3 by the solid line for a 1-dimensional system at finite temperature, or in Fig. 4-4 for a two-dimensional system at 0°K. Applying an electric field for a short time interval δt will change the wave vector k of electrons by:

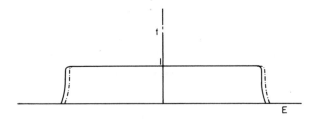

FIG. 4-3. Shift of electron distribution in the energy space due to an electric field.

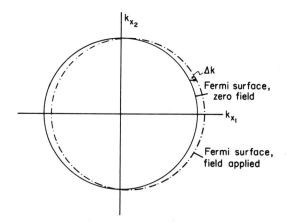

FIG. 4-4. Shift of Fermi surface due to an electric field.

$$\delta k = eF \, \delta t / \hbar \qquad\qquad (4\text{-}12)$$

Together with:

$$\delta E = (\partial E / \partial k) \delta k \qquad\qquad (4\text{-}13)$$

and the conservation of energy:

$$\delta E = eFv \, \delta t \qquad\qquad (4\text{-}14)$$

for the electron acceleration in the one dimensional case this gives:

$$\dot{v} = (1/\hbar)(\partial / \partial t)(\partial E / \partial k) = (\partial^2 E / \partial k^2) eF / \hbar^2. \qquad\qquad (4\text{-}15a)$$

$\dot{v}$ is again proportional to eF, just as in Eq. 4-10. Therefore, one writes:

$$\dot{v} = (1/m^*) eF \qquad\qquad (4\text{-}15b)$$

with m* defined as the effective mass:

$$m* = \hbar^2/(\partial^2 E/\partial k^2). \qquad\qquad (4-16a)$$

One obtains for the 3-dimensional case:

$$m*_{ij} = \hbar^2/(\partial^2 E/\partial k_i \partial k_j). \qquad\qquad (4-16b)$$

These equations show that the effective mass depends on the shape of the energy bands as discussed in detail in Chapter 1. $m* = m$ for the free electrons, since $E = \hbar^2 k^2/2m$. Electrons in a narrow band, such as the d-band with localized electrons states in transition elements, have a high effective mass $m*$. Their acceleration in an applied field is much smaller than that of conduction electrons in the s-band.

Wave functions are not only able to describe the average properties of a large number of electrons. Wave functions can be used to analyze the behavior of localized electrons. One describes such an electron with a "wave packet". This is a superposition of a large number of plane waves with a range of wave numbers. Wave numbers, amplitudes and spacing of these plane waves are selected in such a way that their superposition leads to a maximum of the electron density in one volume element. This maximum moves with the velocity of the localized individual electron. The volume for this electron spreads out with time.

The electron system will return to the equilibrium state if the electric field F is turned off. The redistribution of the electron momentum follows an exponential decay process. This process has the same relaxation time τ which determines the mean free path. This leads therefore to:

$$\partial k/\partial t = -\, \delta k/\tau. \qquad\qquad (4-17a)$$

This process opposes the acceleration of electrons. In equilibrium, both processes balance and one obtains with Eq. 4-12:

$\delta k = e\tau F/\hbar$. (4-17b)

An applied electric field will displace the Fermi surface by $\delta k =$ $e\tau F/\hbar$. The current associated with an electron of velocity v will usually be compensated by the current contribution of an electron with velocity -v. The macroscopic current due to the displacement δk of the Fermi surface by an applied field F will therefore come only from sections near the Fermi surface, which, due to $F \neq 0$, are not compensated by an electron with opposite k value.

The current in the classical model can be given in the following ways :

$$J = n \cdot v_{dr.,av.} e = \sum_{el.} v_{dr.} e = \int_{el.} v_{dr.} e \, dn \qquad (4\text{-}18)$$

where n is the number of electrons per unit volume, and $v_{dr.,av.}$ is their average drift velocity. The term $n \cdot v_{dr.} \cdot e$ would be appropriate if all electrons have the same drift velocity. The expression $\sum_{el.} v_{dr.} e$ would be used for arbitrary drift velocity distributions, and $\int dn \, v_{dr} \cdot e$ should be used if the drift velocity is a smooth function of the electron concentration. In the quantum mechanical model, one determines the current in a similar way, except one sums the electrons per volume, n, not in real space, but in k-space. One keeps in mind that (summary or integrating only over occupied states):

$$n = \int dn = 2 \int (d\vec{k}/2\pi)^3 = 2 \int dA_F \partial k/8\pi^3 \qquad (4\text{-}19)$$

where the factor 2 in front of the integral is due to the fact that 2 electrons occupy each electron state. Since there is space only for one electron with one spin direction per atom in each Brillouin zone, which extends in one dimension from $k = -\pi/a$ to $k = \pi/a$, an integration over this span should give one electron per atom per spin orientation for a completely filled Brillouin zone. Therefore the integral $\int \partial k$ in this equation has to be normalized by the factor $a/2\pi$. The factor "a" in $a/2\pi$ is already taken care of by calculating "n" as number of electrons per unit volume in a 3-dimensional case, or by calculating the number of electrons per unit length in a one-dimensional model. Since, as mentioned above, only electron states at the Fermi surface contribute to the measured macroscopic current, one obtains from:

$$j = \int dn\ v_{dr.}e = (1/4\pi^3)\int (dk)^3 v_{dr.}e = (e/4\pi^3)\int d\vec{A}_F \cdot d\vec{k}_\perp v_{dr.}$$

$$(4\text{-}20a)$$

with Eq. 4-17, which gives the displacement of k by an applied field:

$$j = (e^2\tau/4\pi^3\hbar)\int \vec{v}_{dr.} \cdot d\vec{A}_F\ F \qquad\qquad (4\text{-}20b)$$

or:

$$\sigma = (e^2\tau/4\pi^3\hbar)\int \vec{v}_{dr.} \cdot d\vec{A}_F. \qquad\qquad (4\text{-}20c)$$

This gives for a cubic system:

$$\sigma = (e^2/12\ \pi^3\hbar)\ell A_F = e^2\tau v_{dr.}A_F/12\hbar\pi^3. \qquad\qquad (4\text{-}20d)$$

with A_F the area of the Fermi surface, and ℓ the mean free path.

These equations show that the conductivity of a metal depends on the geometry of the Fermi surface and on the scattering process, characterized by a relaxation time τ. Equivalent to τ in the determination of σ is the use of the mean free path, or the scattering probability P. P is inversely proportional to τ.

B. Temperature Dependence of the
Metallic Conductivity

The calculation of the temperature dependence of the resistivity follows essentially the following arguments. Deviations from the perfect lattice are caused by thermal oscillations of ions. One expects that the scattering cross section of an ion should be proportional to the square of the ion displacement. This was Wien's hypothesis. One can show that this assumption gives the essential features of the temperature dependence of the resistivity of a pure metal even for the simple Einstein model of a harmonic oscillator. In such an oscillator $M(\partial^2 X/\partial t^2) + \alpha X = 0$, where M is the mass of the atom, X the displacement of the atom, and α is the force constant. This force constant is equal to $\alpha = M \omega^2$. The mean potential energy of the atom above the Einstein temperature T_E is $(1/2) \alpha X^2 = (3/2)k_B T$; below T_E it should be $h\nu/[\exp(h\nu/k_B T)-1]$. With Wien's hypothesis, this leads to $\rho = C^* \overline{X^2} = C^* 3 h^2 T/MT_E^2 k_B$ for $T > T_E$, and for $T < T_E$ $\rho = C^*3 \hbar^2/[(\exp[h\nu/k_B T]-1)MT_E^2 k_B^2]$. C^* is a constant. This gives the right temperature dependence of ρ above T_E, and the right trend of $\rho(T)$ below T_E. Naturally, this model predicts incorrect results at very low temperatures, since the Einstein model is an oversimplification. Instead, one should use the Debye model, which leads to $\rho \propto T^5$.

Bloch showed with a quantum mechanical model that the scattering process in metals leads to $\rho = C\overline{X^2}$. He calculated the scattering probability of an electron by a periodic lattice, in which atoms are displaced from their equilibrium position by a vector $\vec{R} = \{X_1 X_2 X_3\}$. The perturbation potential (e.g. see Ref. 29) for such a system would be given by $U(R) = [(\text{actual potential}) - (\text{perfect lattice potential})]$. The scattering process has to be described with the time dependent Schrödinger equation:

$$H\psi = i\hbar(\partial\psi/\partial t).\qquad\qquad(4\text{-}21)$$

The Schrödinger equation, its Hamiltonian, potential energy, and wave functions for the perfect lattice with $R = 0$ will be indicated by the superscript ^{o}. Therefore, $H^o\psi^o = E^o\psi^o$ for the unperturbed system. On can write $H = H^o + U(R)$. The scattering process of an electron with wave vector $\vec{k}$ leads to a wave function which is given after the scattering process by:

$$\psi = \Omega^{\frac{1}{2}} \sum_{\vec{k}'} a_{k'}(t)\psi_{k'}(r)\ e^{-iE_{k'}t/\hbar} \tag{4-22}$$

where $|a_{k'}(t)|^2$ is the probability that the electron is in state k'. Ω is the volume per atom. The boundary conditions would be:

$$a_k(t=0) = 1, \quad a_{k'}(t=0) = 0 \quad \text{for} \quad k \neq k' \tag{4-23}$$

Inserting Eq. 4-22 into 4-21 and neglecting higher terms like $U a_{k'}$ gives a differential equation for $a_{k'}(t)$:

$$i\hbar(\partial a_{k'}/\partial t) = \Omega^{-1}U_{kk'} \exp[i(E_{k'} - E_k)t/\hbar] \tag{4-24}$$

with $U_{kk'} = \int\psi_{k'}^* U \psi_k\ dx_1\ dx_2\ dx_3$. This leads to solutions of the form:

$$|a_{k'}(t)|^2 = 2|U_{kk'}|^2[(1 - \cos x\, t)/x^2]/\hbar^2\Omega^2 \tag{4-25}$$

with $x = (E_{k'} - E_k)/t\hbar$. $|a_k(t)|^2$ has for large values of "t" a maximum for $x \to 0$. This shows that electrons are predominantly scattered into states close to their original state, since small x values correspond to $E_{k'}$ close to E_k. The probability of scattering P should be proportional to the change of the number of electrons in the given state. This change is equal to $d|a_{k'}(t)|^2/dt$. Therefore one obtains finally for the scattering of electrons into a given area of the Fermi surface, PdA_F:

$$P(kk') \ dA_F = (dA_F/8\pi^3\hbar^2\Omega)[\partial\{\int|U_{kk'}|^2 \ 2[(1-\cos x \ t)/x^2]dk_\perp\}/\partial t].$$

$$(4-26)$$

This gives:

$$P(kk') \ dA_F = (dA_F/4\pi^2\Omega|V_{kk'}|^2/(\partial E/\partial k_n). \tag{4-27}$$

$U_{kk'}$ can be calculated in the following way: One assumes that the potential can be developed into a power series. This gives in first approximation:

$$U = V(r + R) - V(r)$$
$$= -[X_1(\partial V/\partial x_1) + X_2(\partial V/\partial x_2) + X_3(\partial V/\partial x_3)]. \tag{4-28}$$

One obtains, since terms of the form X_iX_j should be zero for $i \neq j$ (for each X_iX_j combination one finds a $-X_iX_j$ combination):

$$|U_{kk'}|^2 = \sum_{i=1}^{3} \overline{X_i^2} \ |\int\psi_{k'}^* \ \text{grad} \ V \ \psi_k \ dx_1 dx_2 dx_3|. \tag{4-29}$$

Again using the harmonic oscillator model, which gives for $T > T_E$ $x^2 = k_B T h^2/MT_E^2 k_B^2$, one obtains:

$$|U_{kk'}|^2 = (Th^2/MT_E^2 k_B^2)|\int\psi_{k'}^* \ \text{grad} \ V \ \psi_k \ dx_1 dx_2 dx_3|. \tag{4-30}$$

The volume integral can be changed to a surface integral. This then gives for the volume integral a surface integral of the form $(h^2/2M)\int[\psi_k^* \ (\partial/\partial n) \ \text{grad} \ \psi_k \ -\text{grad} \ \psi_k \ (\partial\psi_k/\partial n)]dA_F$. This type of integral may explain why the details of the potential near the nucleus are unimportant for the description of the scattering process of electrons in a metal. Similar calculations below the Einstein temperature show that the resistivity decreases rapidly with temperature.

The numerical calculation of the electrical resistivity of metals from transition probabilities is possible only for certain

particularly simple cases. The experimental results for metals can
be described by the Grüneisen-Bloch function. For low temperatures,
ρ follows a T^5 power law [29]:

$$\rho \simeq 497.6 \; (T/T_D)^5 \rho \, (T_D) \qquad \text{for} \qquad T << T_D \qquad (4\text{-}31)$$

For high temperatures, one obtains:

$$\rho \simeq \rho \, (T_D) \cdot T/T_D. \qquad (4\text{-}32)$$

The low temperature T^5 law is as characteristic a quantum effect
as the T^3 law for the lattice specific heat. Figure 4-5 shows the
good fit between experimental data and the Grüneisen-Bloch curve derived under certain simplifying assumptions. The best prediction of
the absolute value of the resistivity, restricting ourselves to the
high temperature limit $T >> T_D$, as obtained from the Bloch calculation, can be written as with n = electrons/volumen as:

$$(\hbar k_F \sqrt{6} \; \pi \, /ne^2 48)(k_B T/\tfrac{1}{2} Mv^2)(1/a) = \rho \qquad (4\text{-}33)$$

where "v" is the velocity of sound at the Debye temperature and
"a" the lattice parameter. One can deduce from this equation a relation for the mean free path:

$$\ell \simeq (2 \; Mv^2/k_B T) \cdot a. \qquad (4\text{-}34)$$

This equation shows that the mean free path in pure metals is the
ratio of the kinetic energy of the ion, $\tfrac{1}{2} Mv^2$, and the thermal energy
$k_B T$ times the lattice parameter. This equation can be even more
simplified by using the Lindemann melting formula [197] which states
that the melting temperature of a metal T_m is given by $T_m \simeq T_D^2 V_o^{2/3} B$,
with $B \simeq A/120^2$ (A = atomic weight, V_o = atomic volume). This gives:

$$\ell \sim 50 \; (T_m/T) \, a. \qquad (4\text{-}35)$$

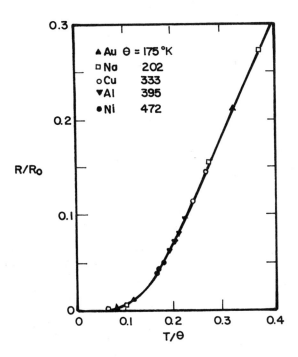

FIG. 4-5. The Grüneisen-Bloch formula fitted to experimental data on the resistivity of metals. After Ref. 196.

C. Electronic Conduction in Semiconductors

The temperature dependence of ρ is rather complex for semi-conductors. In these materials, the bottom or top of some energy bands are close to the Fermi energy level. Energy differences $|E_i - E_F|$, where "i" stands for bottom or top of the energy band, are of the order of or smaller than $k_B T$. Therefore, one cannot re-place the Fermi-Dirac statistic by a step function. Not only is the scattering process of electrons and holes a function of tempera-ture, but also the current carrier concentration. Further, the electron transport has to be considered in several bands.

The conductivity in multiband systems can be described by the parallel movement of the different carriers in the applied field. One obtains therefore:

$$\sigma = \sum_i \sigma_i \qquad\qquad (4\text{-}36)$$

where σ_i is the conductivity of carriers in band "i". Since one likes to visualize the electron transport process by the movement of individual electrons or holes, one writes:

$$\sigma = \sum_i n_i |e| \mu_i \qquad\qquad (4\text{-}37)$$

where μ_i is the mobility of the carrier in band "i". This may be either an electron or a hole. μ_i is always positive, since the mobility is defined by $\mu_i = |v_i/F|$. Therefore, if one replaces the velocity v by the mobility times the electric field, one also has to replace the charge of a carrier by the absolute value of the charge.

As described in Chapter 1, three types of carriers have to be taken into account in the discussion of the electronic transport phenomena in germanium and silicon and in III-IV and II-V semiconductors. Therefore, one should write for conductivity:

$$\sigma = |e| \sum_{i=1}^{3} n_i \mu_i. \qquad\qquad (4\text{-}38)$$

n_1 would be the concentration of electrons, $n_2 + n_3$ is the concentration of holes. Electrical neutrality requires that $n_1 = n_2 + n_3$ for intrinsic semiconductors. Equation 4-34 is not easily handled since one has five temperature dependent parameters. One can sometimes estimate the relative values of the mobility from effective mass data, since:

$$\mu_i = |e| \tau_i / m_i^*. \qquad\qquad (4\text{-}39)$$

Frequently, one makes some simplifying assumptions for an analysis of the electrical conductivity. One may assume that only one type of hole and one type of electron is important. This gives:

$$\sigma = |e| (n_{el.} \mu_{el.} + n_{ho.} \mu_{ho.}).$$ \hfill (4-40)

$n_{el.} = n_{ho.}$ for an intrinsic semiconductor. If one type of the carrier has a much lower mass than the other, it carries most of the current and even a one band model can be used as first approximation.

D. Temperature Dependence of the Electrical Conductivity in Intrinsic and Extrinsic Semiconductors

As discussed in Chapter 3, the carrier concentration in intrinsic semiconductors is essentially an exponential function of temperature. The change of mobility with temperature is not as strong; therefore, if the mobility of one carrier is much larger than the mobility of the second carrier, the electrical conductivity in the intrinsic temperature range (usually at high temperatures) is a simple exponential function of the temperature.

The temperature dependence of the resistivity in the extrinsic range of semiconductors is quite complicated (detailed discussions of the problem of this section and section E are given by Blatt [198] and Ziman [2]). Measurements show that ρ can both increase and decrease with increasing temperature. Starting at very low temperatures, ρ may first decrease, then increase, and finally decrease. The last stage is associated with intrinsic conduction, the first and second stage with impurity electron and hole (extrinsic) conduction. The general features of the temperature dependence of ρ can be explained in the following way.

At very low temperatures the carriers are predominantly due to electrons from donor states, and holes from acceptor states, if

the activation energies associated with these states are markedly
smaller than the activation energy for transitions from the valency
to the conduction band. At very low temperatures, ρ is inversely
proportional to the concentration of carriers, n, times their scattering probability. Since n is an exponential function of T, ρ
decreases with increasing temperature and log ρ is essentially
proportional to 1/T.

Above the very low temperature range, the carrier concentration
may become constant if the intrinsic carrier concentration is much
smaller than the carrier concentration from impurity states. Essentially, all impurity states are ionized in this range. Since lattice scattering should increase with increasing temperature, and should
be the essential factor which affects the temperature dependence of the
resistivity, the electrical resistivity increases in this temperature.
A further increase in temperature leads again to a decrease in ρ,
since the carrier concentration for the intrinsic current increases
exponentially.

Energy changes of electrons in the scattering processes with
phonons are very small. The average electron at room temperature
does not gain or lose more than one tenth of its energy in each
collision. A detailed calculation for such a system gives for each
event a relaxation time, which is proportional to the square of
the amplitude of thermal fluctuations and the density of states into which an electron can be scattered (this is similar to that found
in metals). The first term should be proportional to $k_B T$. The
mean free path is for a parabolic band with $N(E) \sim E^{\frac{1}{2}}$ independent
of the energy of the carrier, and the mobility in the lattice scattering range is proportional to $T^{-3/2}$.

More important than lattice scattering is scattering on ionic
impurities for lower temperatures. Then in this case the drift
mobility is proportional to $T^{3/2}$. The mobility due to scattering
on neutral impurities is independent of temperature for low energy
electrons, because the scattering cross section of electrically
neutral impurities is inversely proportional to the electron velocity, and the relaxation time is independent of the carrier energy.

Lattice scattering processes in semiconductors are separated
into scattering by longitudinal and transverse waves with   both
acoustical and optical modes. Acoustical waves which distort
the lattice periodically naturally also distort the reciprocal lattice and with it the Brillouin zone structure. Changes in the
energy band sections very close to the Brillouin zone boundary lead
to changes in the electron and hole concentration and to changes in
the effective masses of these carriers. Bardeen and Shockley
showed that one can calculate the effect of these lattice distortions on the transport process with the concept of the "deformation
potential". Essentially periodic variations of the lattice parameter due to elastic deformations have the same effect as a periodic
potential on the interaction of electrons with the periodicity of
the lattice.

Electron scattering can be of the intervalley and the intravalley types. Figure 1-67 shows the Fermi surface of germanium.
The electrons are found in four ellipsoidal pockets. If an electron stays in the same pocket after a collision in which it was before the collision, then the scattering process is an intravalley
scattering process. In this case, the change in $\vec{k}$ can be very
small. In intervalley scattering, electrons are scattered into a
different electron pocket. The change in $\vec{k}$ is large, and the
phonon involved in the scattering process (either created or annihilated) has a q-vector which is a large fraction of the Brillouin
zone diameter. The theory of the lattice scattering in a semiconductor can be treated in simplified form if one assumes that the
carriers are concentrated near a single minimum in $\vec{k}$-space. Then
the changes in the electron wave vectors and the phonon wave vectors
are very small. One may expect that the phonon occupancy number
above the Debye temperature will be given by $\langle n \rangle = k_B T / \hbar \omega$.

The relaxation time in a system in which both lattice and impurity scattering has to be taken into account should be determined
with an equation of the form:

$$1/\tau(E) \simeq 1/\tau_{1a.}(E) + 1/\tau_{im.}(E) \qquad\qquad (4-41)$$

where $\tau_{1a.}$ is the lattice, and $\tau_{im.}$ the impurity relaxation time. These relaxation times depend in a different way on energy. One would be tempted to propose $1/\mu = 1/\mu_{1a.} + 1/\mu_{im.}$. However, this equation is not accurate. Large deviations from it have been found, especially if $\mu_{1a.}$ and $\mu_{im.}$ are close.

E. Resistivity Changes in Semiconductors due to Static Elastic Deformations

The electrical resistivity of semiconductors can change drastically by static elastic deformations. ρ can be doubled or halved in some semiconductors, if the sample is strained by only one percent. This resistance change may be due to several effects. In an intrinsic semiconductor, changes in the lattice parameter will lead to changes in the energy gaps, and hence to changes in carrier concentrations. One may write:

$$\partial \ln \rho/\partial \ln V \sim \partial \ln n/\partial \ln V \sim (1/k_B T)(\partial E_g/\partial\Delta) \qquad (4-42)$$

where Δ is the dilatation. The change in carrier concentration in an extrinsic semiconductor due to pressure can in most cases be neglected. In those systems, only the change in mobility due to lattice distortions are important. An elastic strain will remove the cubic symmetry of the sample, and the resistivity will become a tensor of high rank. One would expect changes in the effective mass tensor, and the scattering process will be modified. For instance,if one looks at the intervalley scattering process, the symmetry of the valleys is lost in a strained sample. A hydrostatic pressure, of course, causes no change in symmetry. The contours of constant energy may shift to higher positions in one direction and to lower positions in another one.

The electrical resistivity change with stress is called the "piezoelectric resistance". It is found in III-V and II-VI compounds with partial ionic bonding. There stresses can introduce unsymmetric charge distributions or a polarization. This leads to scattering of carriers.

F. Metallic Elements

The room temperature values of the resistivity of metallic elements are given in Table 4-1. The lattice resistivity $\rho_{1a.}$ depends on the characteristic resistivity parameter R, which in turn is a function of the Fermi surface area A and the number of conduction electrons n [2]:

$$\rho_{1a.} = (\pi^3 \hbar^3 q_0 T/4e^2 M k_B T_D^3 2^{1/3})R \tag{4-43a}$$

with:

$$R \sim n^{3/2}(A_{fr.\ el.}/A)^2 \cdot C. \tag{4-43b}$$

$A_{fr.\ el.}$ is the area of the Fermi surface in the free electron model. C represents a double integral. $C = 1$ for the Bloch model, and $C \sim 1.6$ for the Bardeen model. Fig. 4-6 gives R values as a function of the atomic number. R was calculated with the Eq. 4-43a from experimental data of $\rho_{1a.}$. Table 4-2 shows results of a calculation in which the area of the real Fermi surface as obtained from $\rho_{1a.}$ is compared with the Fermi surface area obtained from the anomalous skin effect. The last row in this table should be constant if R is inversely proportional to the square of the ratio of the real Fermi surface area to the free electron Fermi surface area. Within the uncertainty of the argument and the values of the parameters, agreement between theory and experiment is satisfactory [2]. It shows that it is in principle possible to determine the

TABLE 4-1

Thermal Conductivity, κ, Electrical Resistivity, ρ, and its
Temperature Dependence, $\partial\rho/\rho\partial T$, Near Room Temperature

	ρ $\mu\Omega\text{cm}$	$\partial\rho/\rho\partial T$ $^\circ K^{-1}$	κ $\dfrac{\text{cal}}{^\circ C \text{ cm sec}}$		ρ $\mu\Omega\text{cm}$	$\partial\rho/\rho\partial T$ $^\circ K^{-1}$	κ $\dfrac{\text{cal}}{^\circ C \text{ cm sec}}$
Li	8.55	4.35	0.17	Fe	9.71	6.57	0.18
Na	4.2	5.5	0.32	Co	6.24	6.58	0.165
K	6.15	5.7	0.24	Ni	6.84	6.75	0.22
Rb	12.5	4.8	–	Re	–	4.63	–
Cs	20	5.0	–	Ru	7.6	–	–
Be	4	10	0.35	Os	9.5	4.45	–
Mg	4.46	4.2	0.38	Rh	4.5	4.43	0.21
Ca	3.43	3.8	0.3	Ir	5.3	4.11	0.14
Sr	23	3.5	–	Pd	10.8	3.77	0.17
Ba	–	–	–	Pt	9.83	3.92	0.17
Al	2.65	4.67	0.53	Cu	1.673	(4.31)	0.94
La	58	–	0.033	Ag	1.59	4.10	1
Ti	42	5.46	6.6	Au	2.19	3.98	0.71
Zr	40	4.4	0.211	Zn	5.91	4.20	0.27
Hf	35	–	0.22	Cd	6.83	4.26	0.22
Th	19	–	0.09	Hg	94.1	–	0.02
V	26	–	0.07	In	8.37	5.1	0.057
Nb	13.1	–	–	Tl	18	5.2	0.093
Ta	12.4	3.47	0.13	Sn	11	4.63	0.16
Cr	13	–	0.16	Pb	20.65	4.22	0.083
Mo	5.2	4.73	0.34	As	33.3	–	–
W	5.6	(4.82)	0.4	Sb	39	5.4	0.045
Mn	185	–	–	Bi	106.8	4.45	0.02

[9, 199]

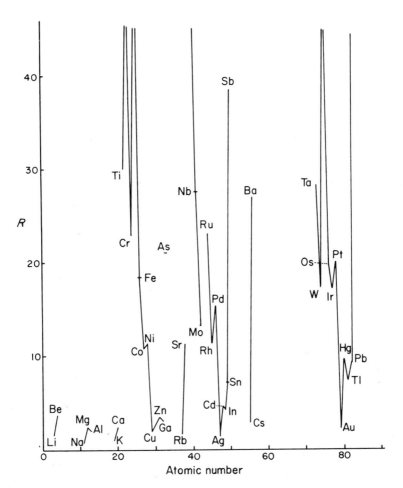

FIG. 4-6. Reduced resistance R (see Eq. 4-43). After Ref. 2, with permission.

TABLE 4-2

Correlation of the Reduced Resistance R with the Area of the Fermi Surface A obtained from the Anomalous Skin Effect [2]

Element	Cu	Ag	Au	Al	Sn
$A/A_{fr.\ el.}$	1.00	0.72	0.71	0.81	0.43
$R \sim (A/A_{fr.\ el.})^2 n^{2/3}$	1.67	0.82	0.38	0.62	0.52

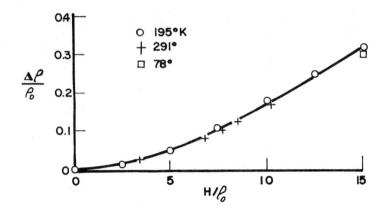

FIG. 4-7. Magnetoresistance data on magnesium at various temperatures, fitted to Kohler's rule. After Ref. 201.

resistivity of non-transition elements from parameters like the Debye temperature and the area of the Fermi surface.

The resistivity of elements is markedly affected by magnetic ordering. Experimental results for polycrystalline materials can be frequently described with an equation of the form:

$$\Delta\rho/\rho_o = f(H/\rho_o) \qquad\qquad (4\text{-}44)$$

where ρ_o is the resistivity without an external field and $\Delta\rho$ is the resistivity change due to the field. This equation represents "Kohler's rule". It shows that magnetoresistance effects should be more noticeable at low temperatures, since ρ_o usually decreases with decreasing temperature (4-7). Experimental results of magnetoresistance measurements are sometimes plotted in the "reduced Kohler diagram", where $\Delta\rho/\rho_o$ is plotted as a function of sH, with $s = \rho_o(T)/\rho_o(T_o)$. T_o is a standard reference temperature. It may be the room temperature or the Debye temperature. Figure 4-8 shows results. Typically, $\Delta\rho/\rho_o$ is proportional to H^2. Transverse or longitudinal magnetoresistivities ($H_\perp$ or $H_{||}$ to wire axis) are usually of the same order of magnitude.

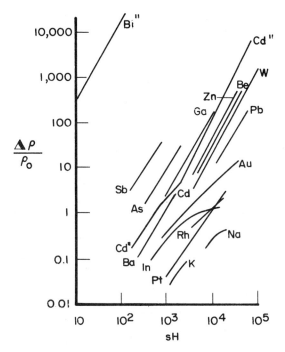

FIG. 4-8. Reduced Kohler diagram. After Ref. 202.

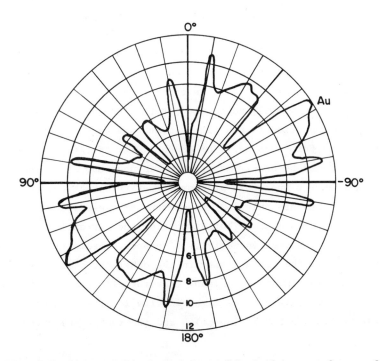

FIG. 4-9. Magnetoresistance of a gold single crystal as a function of orientation in repect to a magnetic field. After Ref. 203.

The magnetoresistance of single crystals is, even for cubic
crystals,markedly orientation dependent. Figure 4-9 shows results
obtained on gold. The lack of cubic symmetry in this diagram may
be due to experimental alignment problems.

Ferromagnetic ordering in iron, cobalt and nickel decreases
the resistivity to less than the extrapolated paramagnetic resistivity. This can be seen in Fig. 4-10 which shows the resistivity
of both nickel and palladium. The change in the slope of ρ, $d\rho/dT$,
at the Curie temperature has to be due to magnetic ordering because
palladium, which has a similar band structure to nickel in its paramagnetic state, shows no anomaly. Figure 4-11 shows schematically
a band model for nickel in its paramagnetic and ferromagnetic states.
One may explain the decrease of the resistivity due to ferromagnetic
ordering in the following very simplified way. The conduction band
electrons (about 1/2 conduction electron per atom) occupy bands
with spin-up or spin-down. If the spin direction is largely con-
served in the scattering process, then electrons with spin-up will
be scattered preferredly into the d-subband with spin-up. The d-
band in nickel is nearly filled. The change from the paramagnetic
to the ferromagnetic state will lead to an increase in the density
of states in one subband (probably not very much if the top of the
d-band is nearly parabolic). The other d-subband, however, will be
filled in the completely magnetized state. s-electrons with the
same spin orientation as the filled d-subband cannot be scattered
any more into this subband. The conductivity of these conduction
electrons should be nearly as great as in one subband of copper.
One should expect therefore at room temperature that the conductivity
of nickel (which has only 1/2 the number of s-electrons, and only
those in one subband contributing most of the current) should be
about one quarter the conductivity of copper, neglecting the small
contribution to the conductivity in the other s-subband. This
agrees with experiments which show that for nickel the room temperature conductivity (where the magnetization is already close to
saturation) is about 0.22 $(\mu\Omega \text{ cm})^{-1}$ and the room temperature

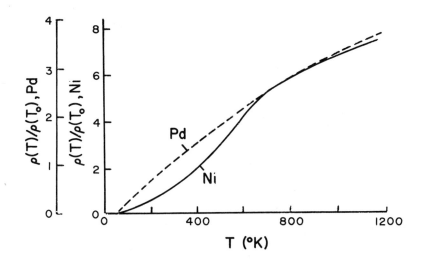

FIG. 4-10. Relative temperature dependence of the resistivity of palladium and nickel. The scales were selected in such a way that the curves coincide just above the Curie temperature. After Ref. 204.

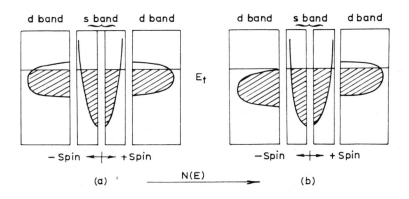

FIG. 4-11. Relative positions of the s- and d-bands for a metal in the paramagnetic (a) or ferromagnetic state (b). After Ref. 205, with permission.

conductivity of copper is about 0.98 $(\mu\Omega\ \text{cm})^{-1}$. The resistivity ρ
and $\partial\rho/\partial T$ for iron, cobalt and nickel is given in Fig. 4-12a and
4-12b. $\partial\rho/\partial T$ shows for iron and nickel a sharp peak. Cobalt, in
between these two elements in the periodic table, shows only a
rather broad maximum. Surprisingly, ρ and $\partial\rho/\partial T$ do not change much
during the first order transition from hcp to fcc.

The temperature gradient of the resistivity near the critical
point (see Chapter 2 about critical phenomena) is frequently described with the equation of the form:

$$\partial\rho/\partial T = (A/\lambda)(\{T/T_c - 1\}^{-\lambda} - 1) + B \tag{4-45}$$

where A, B are constants, or are only weakly temperature dependent.
λ is the exponent which characterizes the critical resistance behavior. This equation leads to:

$$\partial^2\rho/\partial T^2 = -(A/T_c)[T/T_c - 1]^{-\lambda - 1} \tag{4-46}$$

$\log \partial^2\rho/\partial T^2$ should be essentially a straight line if plotted as a
function of $\log|1 - T/T_c|$. Log $(\partial^2\rho/\partial T^2)$ is given as a function
of $\log|1 - T/T_c|$ in Fig. 4-12c. It shows the predicted straight
line behavior only over a limited temperature range. A detailed
discussion of these data has been given by Kawatra and Budnick [200].

The model used to analyze the resistivity of nickel cannot be
used to explain the high resistivity values of iron. Scattering of
conduction electrons into the d-band with empty states for spin-up
and spin-down has to be taken into account both above and below the
Curie temperature for this metal. For iron, $\rho \simeq 110\ \mu\Omega$ cm at
1300°K. A linear extrapolation of the paramagnetic resistivity to
0°K yields $\sim 70\ \mu\Omega$ cm. This high resistivity value cannot be associated with high density of d-states values. It is most likely due
to spin scattering of conduction electrons on localized spins [208].
The spin-scattering mechanism will not explain the high ρ values of
β-titanium in the high temperature bcc phase. ρ is $\sim 140\ \mu\Omega$ cm at

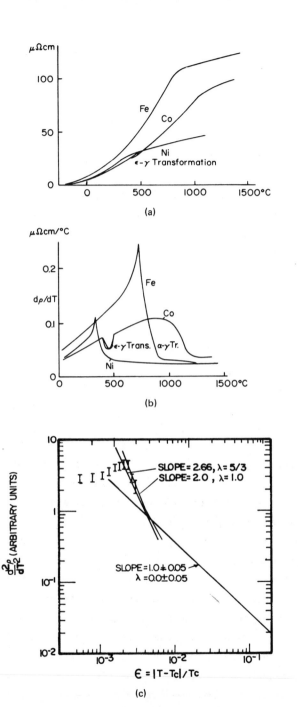

FIG. 4-12. Temperature dependence of the electrical resistivity of iron, cobalt and nickel (a). Temperature dependence of $\partial\rho/\partial T$ of iron, cobalt and nickel (b). After Ref. 206. Log $\partial\rho^2/\partial T^2$ as a function of ε for nickel above the Curie temperature (c). After Ref. 207.

$1200^\circ K$ and $d\rho/dT$ is very small. The to $0^\circ K$ extrapolated resistivity is about 130 $\mu\Omega$ cm. The susceptibility of β-titanium is not
unusually high for a transition element. Therefore, the high values
of the high temperature resistivity cannot be due to localized spins.
High temperature specific heat measurements rule out high density
of d-states values as explanation of high ρ-values for this bcc
phase. All elements in the titanium and scandium group in the
periodic system have high resistivity values, as seen in the survey
of the resistivity of 3d-, 4d- and 5d-transition elements given
in Fig. 4-13. The peculiar $\rho(T)$ curve of manganese should be
associated with antiferromagnetism in this element.

G. Semiconducting Elements

The temperature dependence of the resistivity of p- and n-type
silicon (4-14 and 15) is given as a plot of log R = log ρ vs 1/T.
The n-type semiconductor has phosphorus impurities, and the p-type
has boron impurities. Only alloys with the smallest amounts of dopants (0.0013 at.% B or 0.00091 at.% P) showed the intrinsic resistivity range at high temperatures. These figures show that log R
is a nearly linear function of 1/T at high temperatures. The
slope in this straight region should be proportional to $\frac{1}{2}E_g$. It
cannot be a perfectly straight line even at high temperature for an
intrinsic semiconductor, since $n \propto kT^{3/2} \exp[-E_g/2k_B T]$. Marked deviations from straight lines are found only at low temperatures.
There the samples are extrinsic conductors and the current contribution due to impurity states cannot be neglected. Table 4-3 gives
energy gap values as determined from optical and electrical resistivity measurements. The last can be smaller than gaps obtained from
optical measurements since indirect transitions for electrons are
important for the electron movement in applied electric fields. Optical excitations are usually direct with no change in the wave number. The scattering process in an electric field makes it possible

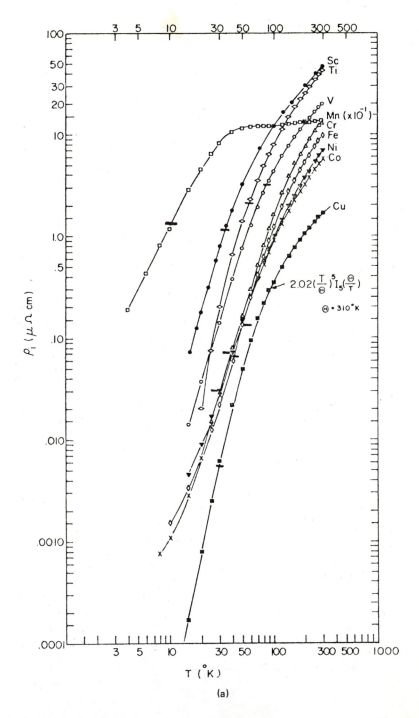

FIG. 4-13a. Electrical resistivity of 3d-transition elements.
After Ref. 162, reproduced with permission.

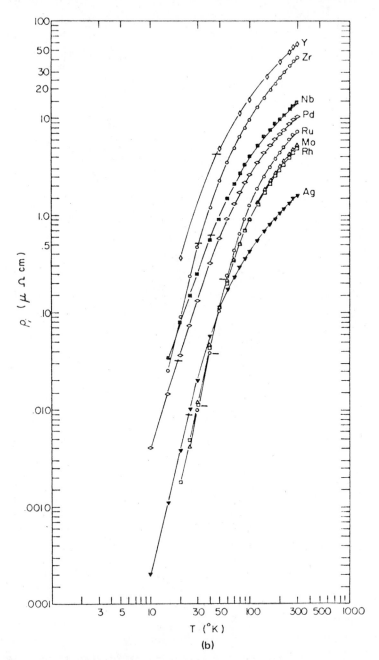

FIG. 4-13b. Electrical resistivity of 4d-transition elements.
Reproduced with permission from Ref. 162.

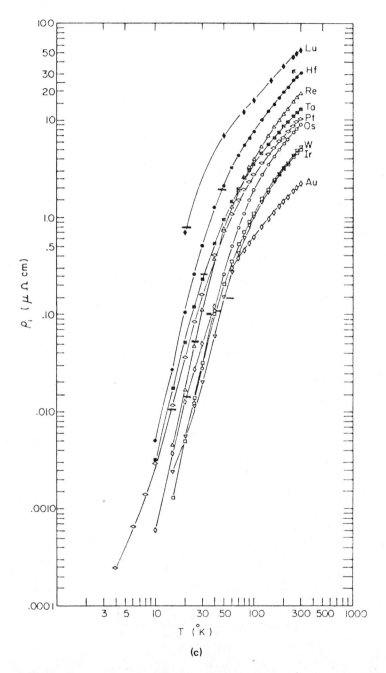

FIG. 4-13c. Electrical resistivity of 5d-transition elements.
Reproduced with permission from Ref. 162.

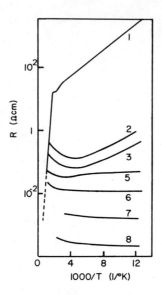

FIG. 4-14. Resistivity of p-type silicon. Sample 1: pure Si;
sample 2: 0.0013 at.% B; sample 3: 0.0026 at.% B; sample 4: 0.0052
at.%B; sample 5: 0.013 at.% B, sample 6: 0.026 at.% B; sample 7:
0.26 at.% B; sample 8: 2.6 at.% B. After Ref. 210, with permission.

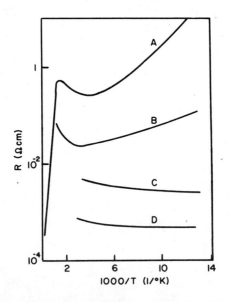

FIG. 4-15. Resistivity of n-type silicon. Sample A: 0.00091 at.%
P; sample B: 0.0072 at.% P. Sample C: 0.091 at.% P; Sample D: 0.91
at.% P. After Ref. 210, with permission.

TABLE 4-3

Values of the Energy Gap Between the Valence and Conduction
Bands in Semiconductors, at Room Temperature

Crystal	E_g, eV	Crystal	E_g, eV
Diamond	5.3	PbS	0.34-0.37
Si	1.1	PbSe	0.27
Ge	0.72	PbTe	0.30
InSb	0.23	CdS	2.4
InAs	0.33	CdSe	1.7
InP	1.25	CdTe	1.4
GaAs	1.4	ZnO	3.2
AlSb	1.6-1.7	ZnS	3.6
GaP	2.2	ZnSe	2.6
SiC	3	AgCl	3.2
Te	0.33	AgI	2.8
ZnSb	0.56	Cu_2O	2.1
GaSb	0.78	TiO_2	3

for electrons to change their wave number, if they jump from the
valency to the conduction band, since they may absorb or create a
phonon. Since the energy exchange of electrons with the lattice
is usually of the same order of magnitude as the energy absorbed
from the electric field responsible for the current, indirect transitions are much more likely in the electron scattering process in an
electric field than electron excitation by photons. Therefore, the
energy gap found in resistivity measurements is usually smaller than
that found in optical experiments.

The calculation of mobilities from the electrical conductivity
requires that the carrier concentration be known. Data for the
mobility have been obtained directly for n-type germanium by Prince
[211]. The principle of the experiment is simple. At one point of

the sample minority carriers are injected into the sample. These
carriers drift under the effect of an applied field through the
sample and after a time t^0 reach a second contact where some of
the extra charges are absorbed. This leads to an electric signal
which makes it possible to measure t^0. The drift mobility would
then be calculated from $\mu_{drift} = |v/F| = L^2/Vt^0$, where L is the
distance between the emitter and collector probe pressed against the
sample, and V is the voltage drop between these two probes. Prince
[211,212] measured both the hole and the electron mobility. The
experiments give $\mu_{de.} = 3.5 \cdot 10^7 T^{-1.6} cm^2/Vs$ for the electron mobility, and $\mu_{dh.} = 9.1 \times 10^8 T^{-2.3} cm^2/Vs$ for the holes. More recent
experiments by Ludwig and Watters [213] gave on purer samples $\mu_{de.}$
$\propto T^{-2.7}$ and $\mu_{dh.} \propto T^{-2.5}$. Surprisingly, their room temperature data
agree well with those of Prince. A review of the experimental data
has been given by Smith [50]. Theories predict that electron and
hole mobilities are proportional to $T^{-1.5}$ in the lattice scattering
range. This deviates from the experimental results. It may be due
to the fact that the assumption of pure lattice scattering is a
simplification. Lattice defects may have to be considered. It has
been suggested that one has to take impurity interaction into account.
In germanium, the Bohr radius for the impurity electron (radius of
the outer electron in its lowest energy state) is about 80A. Therefore, even for impurity concentrations of only $10^{15}/cm^3$, impurity
electrons interact and one should discuss their properties with the
concept of an impurity band. Such a model leads to deviations from
the $T^{-3/2}$ power prediction for electron mobilities. Figure 4-16
shows that μ depends on the sample conductivity, which in turn depends
on the sample purity. This confirms that μ is not influenced by
lattice scattering only.

The discussion of electron transport phenomena assumed up to
now that the electrons returned to equilibrium conditions after each
collision process. This is correct for metals, where electric fields
are always small since the electrical conductivity is high. However,
in semiconductors fields of the order of $10^5 V/m$ can be obtained

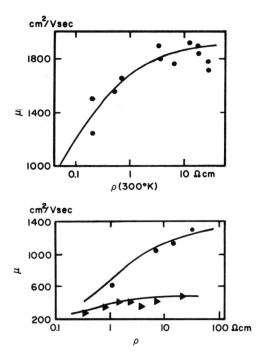

FIG. 4-16. Variation of drift mobility with resistivity for electrons in p-type germanium (top diagram), for electrons in p-type
silicon (circles, bottom diagram), and for holes in n-type silicon
(triangles, bottom diagram). After Refs. 211 and 212.

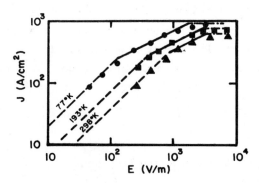

FIG. 4-17. Current density in n-type germanium as a function of
the electric field. After Ref. 214.

TABLE 4-4

Carrier Mobilities at Room Temperature

	Diamond	Si	Ge	GaP	GaAs	InP	
Electron mobility	0.17	0.14	0.39	0.05	0.85	0.34	m^2/V sec.
Hole mobility	0.12	0.05	0.19	0.002	0.45	0.06	m^2/V sec

without an electrical breakdown. At such high fields, the current
density may not be proportional to the applied field F. In other
words, Ohm's law is not satisfied. Figure 4-17 shows experimental
results obtained on n-type germanium, in which log j is plotted as
a function of log F. It shows three distinctive ranges. For low
fields, j is proportional to F (or: log j = log F + const). In the
middle range j is proportional to $F^{1/2}$. For high fields, saturation
is observed and the current density is constant. These data can be
explained in the following way. At low fields, the electron-lattice
interaction is sufficient to bring the electron during a scattering
event back to the equilibrium condition. At higher fields, the
electron energy increase due to the field is much larger than the
phonon energy. The energy transfer to the lattice becomes too slow
to establish equilibrium conditions. The electrons become "hot
electrons". The problem of energy transfer for this case is similar
to that in a gas discharge where electrons of the plasma interact
with ions and molecules. The drift mobility for this case is given
by $\mu \propto \sqrt{F}$ for $F > F_{crit}$.

Current saturation is reached for very high fields when the
electron energy is sufficiently high to create optical phonons. Their
energy is much higher than the phonon energy in the acoustical branch.
As soon as an electron has a very high energy, it will create an optical phonon and return to thermal equilibrium. This process leads
to a maximum drift velocity. The average drift velocity should be
half this maximum drift velocity.

H. Semiconducting Compounds

The conductivity of some semiconducting compounds is shown in Fig.
4-18. The analysis of the data follows the same approach used for
germanium and silicon. However, the analysis of the mobility is more
difficult in the compounds than in the elements. The III-V semiconductors contain two types of atoms, with different electric charges.
The optical frequency spectrum of the lattice is due to the relative

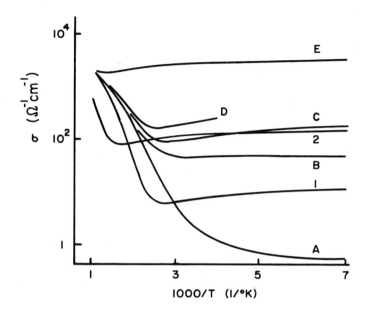

FIG. 4-18. Temperature dependence of the conductivity in InAs for five n-type (A-E) and two p-type (1,2) samples. After Ref. 215.

oscillation of these two different types of charges with respect to each other. The polar interaction of electrons with optical lattice vibrations is a very important scattering mechanism in III-V and II-VI semiconducting compounds. In contrast to acoustical scattering, no relaxation time can be defined for scattering in the optical lattice vibrational spectrum [107]. Ehrenreich investigated the theoretical formulas for the mobility, the Hall coefficient, and the thermoelectric power with a variational principle. The results can be formally reproduced as a solution of the Boltzmann equation, where the energy dependence of the relaxation time is given by $\tau \propto T^r$. Ehrenreich found that r = 1/2 for all three properties at elevated temperatures. r decreases with decreasing temperatures and is zero for $h\nu_{la.} = k_B T$. $\nu_{la.}$ is the frequency associated with optical modes of the lattice. The complete range of expected r values according to Ehrenreich is given in Fig. 4-19.

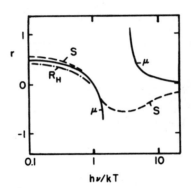

FIG. 4-19. Exponent r of the energy dependence of the effective relaxation τ for polar scattering. Since no true relaxation
time can be defined, different effective relaxation times apply for
the mobility μ, the thermoelectric power S, and the Hall coefficient R_H. After Ref. 217.

Since Hall coefficient and electrical conductivity measurements are used for most of the investigations on the mobility in
semiconductors, the discussion of this subject will be continued
in Section IV of this chapter.

I.   Noble Metal Element Alloys

The resistivity of metallic alloys can be separated into a temperature dependent, and a temperature independent part. The last
term is the "residual resistivity" which is frequently measured at
the temperature of boiling helium. This value is usually so close
to the to $0°K$ extrapolated ρ value that one can neglect the difference between $\rho(4.2°K)$ and $\rho(0°K)$. In the case of small amounts
of impurities, typically of the order of a few percent, one finds
for ρ.

$$\rho = \rho_{1a.}(T) + \rho_{re.} \qquad\qquad (4\text{-}47a)$$

where $\rho_{re.}$ is the residual resistivity, and $\rho_{1a.}(T)$ is the resistivity of the lattice without impurities. Experimentally determined values of the resistance increase of base alloys due to the addition of 1 at.% impurities are given in Table 4-5. $\rho_{re.}$ is frequently proportional to the impurity concentration c:

$$\rho_{re.} = c(\partial\rho/\partial c). \qquad\qquad (4\text{-}47b)$$

Equation 4-47 gives Mathiessen's rule. It is one of the first results of experimental studies on alloy series where the alloy composition was systematically varied   (deviations from this rule are discussed by Bass [218].) R(room temp.)/R(4.2°K) is the residual resistance ratio (RRR), which is frequently used to characterize the impurity resistance contribution. This ratio for small amounts of impurities in the sample is inversely proportional to the impurity concentration, since RRR = $([\rho_{1a.}$ (room temp.)$/\rho_{re.}$ (4.2°K)] + 1) $\simeq \rho_{1a.}$ (room temp.)/$\rho_{re.} \propto$ c. RRR can be measured easily. Since it characterizes impurities by their scattering power in electron transport processes, it is a very useful indication of the impurity concentration in a sample and of the impurity effect on all types of electron conduction processes.

The residual resistivity of some extended binary solid solutions with randomly distributed elements is given by:

$$\rho_{re.} = K\, c_A c_B = K\, c_A\, (1-c_A) \qquad\qquad (4\text{-}48)$$

This is Nordheim's rule. It is applicable to alloys of elements which have a similar electron structure. For small impurity concentrations ($c_A \ll 1$), one can write it as $\rho_{re.} = K\, c_A$.

TABLE 4-5

The Resistivity Increase in the Noble Metals Due to
Small Concentrations of Alloying Agents

Alloying Agent	Base Metal			Alloying Agent	Base Metal		
	Cu	Ag	Au		Cu	Ag	Au
Be	0.62			As	6.8	8.5	8.0
Mg	0.65	0.50	1.30	Rh	4.40		4.15
Al	1.25	1.95	1.87	Pd	0.89	0.44	0.41
Si	3.95			Ag	0.14		0.36
P	6.7			Cd	0.30	0.38	0.63
V				In	1.06	1.78	1.39
Ti			12.9	Sn	2.88	4.36	3.36
Cr	3.6		4.25	Sb	5.4	7.25	6.8
Mn	2.90	1.60	2.41	Ir	5.7		
Fe	9.3		7.9	Pt	2.1	1.6	1.01
Co	6.35		6.1	Au	0.55	0.36	
Ni	1.25		0.79	Hg	1.0	0.79	0.44
Cu		0.077	0.45	Tl		2.27	1.9
Zn	0.32	0.64	0.95	Pb		4.65	3.9
Ga	1.42	2.36	2.2	Bi		7.3	6.5
Ge	3.79	5.5	5.2				

The values of $\Delta\rho$ are in μohm-cm/atomic per cent of alloying agent.
[209]

One can explain Nordheim's rule with the following arguments.
The average potential V for ions in the crystal is:

$$V = c_A V_A + c_B V_B \qquad\qquad (4\text{-}49)$$

where V_A is the potential of atom 'A', and V_B the potential of atom
'B'. The deviation from this potential for 'A' or 'B' atoms is:

$$V - V_A = c_B \ (V_B - V_A) \qquad\qquad (4\text{-}50)$$

$$V - V_B = c_A \ (V_A - V_B)$$

The probability of scattering for an electron on an A-atom will
be proportional to $V - V_A$:

$$P \propto \left| c_B \int \psi_{k'}^* \ (V_B - V_A) \ \psi_k \ dx_1 dx_2 dx_3 \right|^2 . \qquad\qquad (4\text{-}51)$$

The total scattering at A-atoms would be equal to the product of
this term with c_A. Together with a similar expression for scattering by B atoms, one obtains Eq. 4-48.

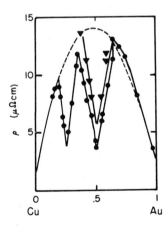

FIG. 4-20. Resistivity of ordered (full lines) and disordered
binary alloys (dashed line). After Ref. 219.

The resistivity decreases sharply below this value in these alloys if ordering takes place. Figure 4-20 gives as examples the copper-gold system in the ordered state. This decrease in the resistivity due to ordering seems to be surprising, but one has to keep in mind that a perfectly ordered lattice with two or more types of atoms is just as perfect in the sense of a periodic structure as a lattice with only one type of atom, and it should have zero residual resistivity.

Resistivities of noble metals with small amounts of B-group metals have been investigated in detail. Norbury pointed out that there exists a connection between ρ_{re} and the difference in valency Z between host and impurity. The larger the difference in valency, the larger is ρ_{re} for 1% impurity concentration. Linde [220-222] gave a qualitative correlation (Linde's rule) for small c values:

$$\rho_{re} = c(a+b\Delta Z^2) \qquad (4-52)$$

where c is the impurity concentration, and ΔZ the difference in valency electrons of impurity atom and host. a and b are constants, which depend only on the solvent metal and the specific row of the periodic system for the solute atom. Figure 4-21 shows that this correlation describes residual resistivities very well. It was already pointed out by Mott and Jones [29] that such a result could be understood only if one assumes that the extra charge of the impurity atom is screened, with the screening potential:

$$V = \frac{\Delta Z e^2}{r} \exp[- r/r_{sc.}] \qquad (4-53)$$

$r_{sc.}$ is the screening radius.

Equation 4-53 shows that most interactions between impurity atoms and conduction electrons decrease rapidly with distance since the screening radius is of the order of 10^{-10} m.

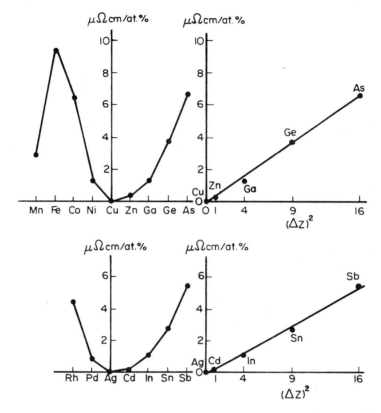

FIG. 4-21. Increase in the resistance of copper (top row) or silver (bottom row) due to one atomic percent of various metals in solid solution. ΔZ denotes the differences in the number of electrons outside an inert gas shell. After Ref. 220.

J.   Noble Metal and Transition Element Alloys

Figure 4-21 shows that the resistance increase due to one percent of transition element impurities in copper cannot be described with Linde's rule. The curve in Fig. 4-21 has a peak at iron. It does not increase with higher ΔZ values. Similar measurements on aluminum and nickel base alloys show that a maximum increase of $\Delta\rho/c$ is found if 3d-transition elements from the center of the transition element

group are dissolved in the matrix. This type of composition dependence is not unexpected if one keeps in mind that the number of un-
paired d-electrons has, according to Hund's rule, its maximum at
chromium in its free state. In alloys, discussed above, one should
expect this maximum in the general area of the middle of the 3d-
transition element group. Figure 2-13 shows schematically the virtual states of transition metal impurities in a copper matrix, as
proposed by Friedel, which could explain these high $\Delta\rho/c$ values.

Impurity states are frequently described with the concept of
virtual energy states. The calculation of their scattering cross-
section shows that the phase shift η_ℓ of the scattered wave is related
to the excess charge of the impurity atom with respect to the matrix
by $\Delta Z = (2/\pi) \sum_\ell (2\ell + 1) \eta_\ell$, where ℓ is an integer. This is Friedel's
sum rule which is derived essentially from two principles: the
electric charge of the impurity atom has to be neutralized, and the
Fermi wave vector at large distances away from the impurity must be
the same as in the pure crystal. This leads to an oscillating charge
density, which tends to zero with $1/r^3$ (see Fig. 2-14).

It is usually not possible to describe the residual resistivity
of extended solid solutions in which at least one component is a
transition element with as simple rules as are applicable to noble
metal alloys. This is due to the fact that the resistivity is not
only influenced by the concentration of conduction electrons and
scattering centers, but it is also important to evaluate the composition dependence of the number of states into which s- electrons
can be scattered. $N(E)$ can be a rather complex function of alloy
concentration. This means that one cannot use Nordheim's rule for
Cu-Ni or Ag-Pd alloys.

Another complication in the evaluation of the impurity resistivity in these alloys is due to the fact that conduction electrons
and impurity electrons may be "polarized". This effect is probably
responsible for the "resistance minimum phenomena". Older experiments on "pure" elements showed at low temperature a minimum in $\rho(T)$
and a maximum at still lower temperatures. ρ_{max}/ρ_{min} is usually only
of the order of a few percent. One realizes now that this resistance

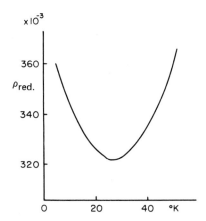

FIG. 4-22. The resistance minimum in Cu + 0.044% Fe.   After Ref. 223.

minimum is frequently associated with small amounts of ferromagnetic impurities in a non-magnetic matrix.   Figure 4-22 shows that even 0.044% iron can produce a resistance minimum in copper.   It is not necessary for the transition element impurity in its bulk state to be ferromagnetic.   Chromium is anti-ferromagnetic, but is responsible for a resistance minimum in gold alloys [224].   Figure 4-23 shows that 0.14% at.% chromium in gold yields a pronounced minimum.   Such effects are noticeable even in the alloy with 0.56 at.%   chromium. Table 4-6 lists some of the alloy systems which show a resistance minimum.   It seems difficult to explain the data systematically and it is possible that more than one scattering mechanism is responsible for these low temperature resistance anomalies.   Reasonable agreement between theory and experiments seems to exist for the analysis of $\underline{Au}Fe$ alloys with the Kondo model [225] in which the resistivity due to s-d electron interaction leads to a logarithmic term $\rho \propto 1 + a \cdot \ln T$.   The coefficient "a" depends on the exchange energy between conduction electrons and the localized d-electrons of impurities. The temperature at the minimum of the resistance minimum, T_{min}, is proportional to $c^{1/5}$.   Kondo found experimentally for the

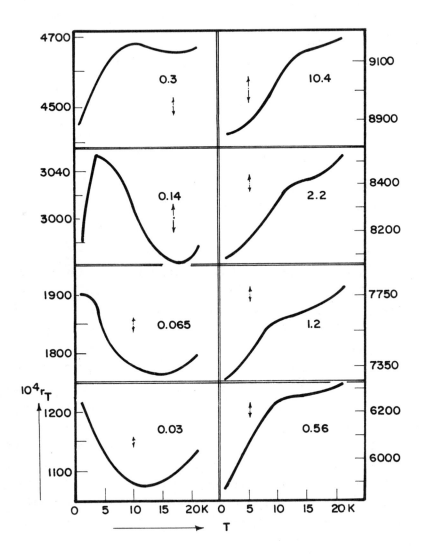

FIG. 4-23. Resistance maxima and minima in AuCr alloys as a function of temperature. Numbers give the percentage of chromium. After Ref. 224.

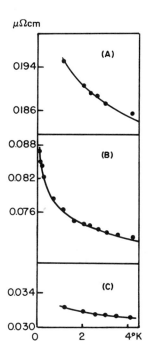

FIG. 4-24. Measured and calculated resistivity curves for gold rich Au-Fe alloys. Points give measurements, full lines are calculated. (A): 0.02 at.% Fe; (B): 0.006 at.% Fe; (C): 0.002 at.% Fe. After Ref. 225.

TABLE 4-6

Examples of Alloys with a Resistance Minima

Solvent	Solute [2,232]
Cu	Fe, Mn, Cr, Co, Ga, In, C, Ge, Sn, Pb, Bi.
Ag	Mn.
Au	Fe, Mn, Cr, Ni.
Cr	Co

AuFe system $T_{min} \propto c^{1/5.3}$. Figure 4-24 shows the good fit between the experimental data and the logarithmic function. Exchange energies between s- and d-electrons, as deduced from the logarithmic term, are equal to -0.15 eV. This agrees reasonably well with other calculations. The resistance minimum phenomena is sometimes referred to as the "Kondo-effect".

Noble metals form complete series of solid solutions with palladium. Low temperature specific heat measurements and susceptibility measurements show that the variation of N(E) is given approximately by $C^*(p - c)^2$, if c < 0.6. C^* is a constant, c is the concentration of the palladium, and p = 0.6. At c = 0.6, the d-band is filled. As shown before, the deviation from the average periodic field is proportional to c. The scattering probability from the s- to the d-state is proportional to [29]:

$$(1 - c)c^2 \ N(E) \left| \int \psi_d^* \ (V_A - V_B)\psi_s \ (dx)^3 \right| \propto P \ . \qquad (4-54)$$

This yields as the scattering probability P for s-d transitions if one assumes that $N(E) \propto (p - c)^2$:

$$P \propto (p - c)^2(1 - c)c^2, \qquad\qquad c \leq 9.6 \qquad (4-55)$$

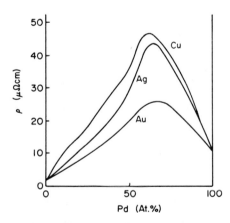

FIG. 4-25. Resistance of alloys where a continuous range of solid solutions is formed. After Ref. 29.

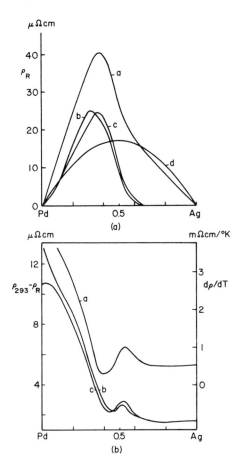

FIG. 4-26. Residual resistivities $\rho_R = \rho_{re}$ of palladium-silver alloys (a). (a) measured values; (b) calculated values of $\rho_R(s-d)$; (c) $\rho_R-Kc(1-c) = \rho_R(s-d)$; (d) $Kc(1-c)$ fitted to ρ_R at $c = 0.65$. Temperature dependence of resistivity in palladium-silver alloys (b). (a) $d\rho/dT$ at 273°K, (right-hand scale); (b) calculated $\rho_{293} - \rho_R$ (left-hand scale); (c) $\rho_{293} - \rho_R$. After Ref. 226.

Figure 4-25 shows experimental resistivities at 0°C of Pd–Cu, Pd–Ag, and Pd–Au alloys. These curves are not as symmetrical as the resistivity curve of Cu–Au alloys. The shift in the resistivity peak in the Pd-alloys in the direction of pure palladium should be associated with s–d scattering.

Adding the two curves for s-s and s-d scattering yields functions similar to experimentally found data (Fig. 4-25). The multiplying factor for the resistance contributions due to s-d scattering has the largest value for the copper, and the smallest for the gold alloy.

The resistivity of these alloys is frequently separated into two terms, $\rho_{th.} + \rho_{re.}$. If one can neglect magnetic ordering and spin density wave scattering effects, it is possible to obtain good agreement between theory and experimental values not only for the residual resistivities, but also for values for $\rho - \rho_{re.} = \rho_{la.}$ or $\partial\rho/\partial T$ with only a few assumptions for Au - Pd. For instance, Coles and Taylor [216] calculated the resistivity curves of Ag-Pd alloys by assuming that only s-electrons carry the current, and that s-s and s-d scattering is responsible for the resistivity, with s-d scattering proportional to $N_d(E)$. Following the Mott model, they obtain:

$$\rho_{la.} = (1 + A)\{N_d(E)\pi^2\hbar\, mT/n_s\, Me^2 k_B T_D\}$$
$$\int_o^\pi |\psi_d^*(\partial V/\partial x)\psi_s\, dr|^2 \sin\theta d\theta$$

(4-56)

with A equal to

$$A = (\pi^2 k_B^2/6)\{3[(\partial N_d(E)/\partial E)/N_d(E)]^2 - [\partial^2 N_d(E)/\partial E^2]/N_d(E)\}$$

(4-57)

Coles and Taylor assumed in their calculation the rigid band model, with dn = 2 N(E) dE. Their results are given in Fig. 4-26.

K. Resistivity of Transition Element Alloys

A similar investigation as for Pd-Ag has been made for non-ferromagnetic Fe-V alloys by Yessik [228]. The density of states curve of this alloy system is given in Fig. 3-19. The resistivity at constant temperature of V-Fe alloys initially increases with increasing iron concentration and reaches a maximum for alloys with 33 at.% iron.

The large resistivity values are essentially due to high residual resistivities. The temperature coefficient is strongly composition dependent. Both $\rho(300^{\circ}K) - \rho(100^{\circ}K)$ and $\rho(200^{\circ}K) - \rho(100^{\circ}K)$ have a minimum at about 30 at.% iron and a maximum at about 33 at.% iron. The qualitative features of the resistivity data can be explained from the density of states curve $N(E)$ as determined from low temperature specific heat measurements. $N(E)$ has a minimum at about 25 at.% iron and a maximum at about 31 at.% iron. This explains the peak in $\rho_{re.}$ near 33 at.% iron. The temperature dependent part of the resistivity is also proportional to $N(E)$. The minimum or maximum of $\rho(300^{\circ}K) - \rho(100^{\circ}K)$ should therefore be close to the minimum or maximum or $N(E)$, respectively.

L. Resistivity of Alloys with Magnetic Ordering

The isothermal resistivity versus composition diagrams for a series of Ni-Cu and Pd-Ag alloys are plotted in Fig. 4-27. They show that the $\rho(c, T = const.)$ curves for these two alloy systems are quite

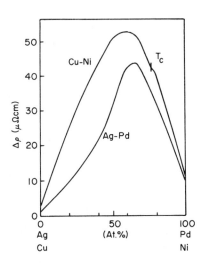

FIG. 4-27. Room temperature resistance of alloys of transition and noble metals. After Ref. 229.

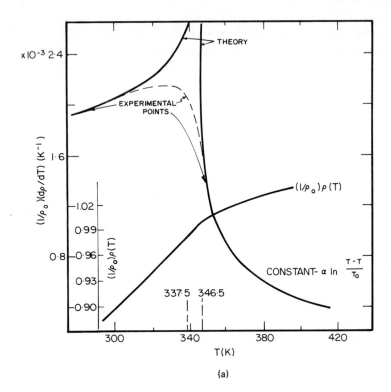

FIG. 4-28a. The temperature dependence of the electrical re-
duced resistivity $\rho(T)/\rho(T_o)$ and its derivative $(1/\rho_o)(d\rho/dT)$ near
the Curie temperature T_g for $Cu_{30}Ni_{70}$. The full line for
$(1/\rho_o)(\partial\rho/\partial T)$ correspond to an analytical fit with $T_o = 346.5K$.
$(1/\rho_o)(\partial\rho/\partial T)$ data have a maximum at $T_{max.} = 337.5K$. After Ref. 230.

similar. However, it is interesting that the $Cu_{30}Ni_{70}$ alloy (Fig.
4-28) shows a sharp change in the slope of the resistivity-tempera-
ture curve near the Curie temperature, similar to that found for
pure nickel. $(1/\rho_o)(\partial\rho/\partial T)$ shows a sharp λ-peak (see Fig. 4-28b),
and follows above T_o a function of the form:

$$\left|(1/\rho_o)(\partial\rho/\partial T)\right| = (A/\lambda)(\left|\varepsilon\right|^{-\lambda} - 1) + B \qquad (4\text{-}58)$$

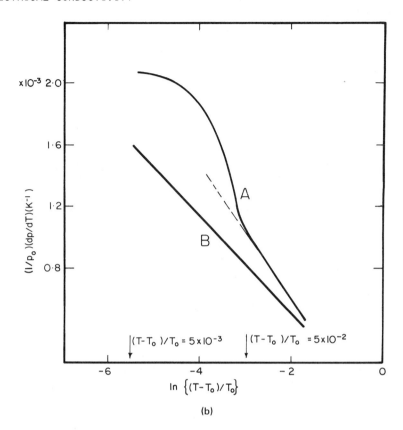

FIG. 4-28b. $(1/\rho_o)$ $(d\rho/dT)$ against $\ln (T - T_o)/T_o$ for $T > T_o$ with $T_o = T_{max}$ (curve A) and $T_o = 346.5K$ (curve B) for a $Cu_{30}Ni_{70}$ alloy. After Ref. 230.

With $\varepsilon = |1 - T_o/T|$, λ the critical exponent, and A and B constants. Similar results were found below T_o. This type of logarithmic divergence is the same as found in pure nickel.

As discussed before, chromium has a resistance anomaly near the Néel temperature, with a minimum near $T_N = 311°K$ and a maximum near $308°K$. The resistance minimum is associated with the onset of anti-ferromagnetic ordering. Figures 4-29 to 31 show the resistivity of several chromium alloys. The resistivity sometimes increases very rapidly with anti-ferromagnetic ordering. The resistance anomaly in

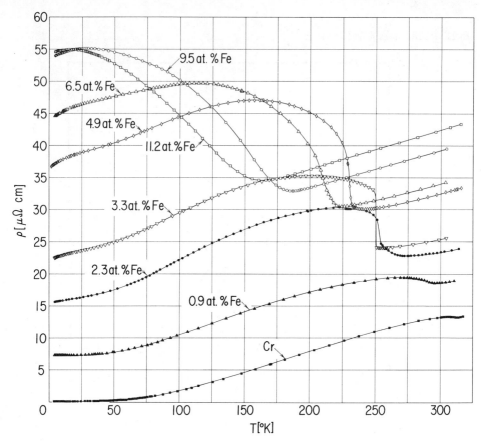

FIG. 4-29. ρ of CrFe. Reproduced with permission from Ref. 231.

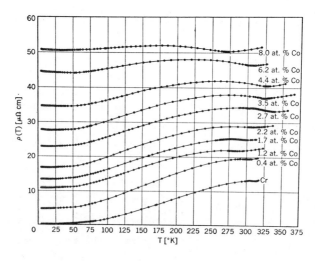

FIG. 4-30. ρ of CrCo. Reproduced with permission from Ref. 232

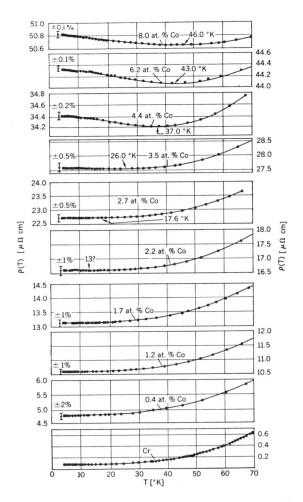

FIG. 4-31. Electrical resistivity of chromium-cobalt alloys below 70°K showing the resistivity minimum. After Ref. 232.

some CrFe alloys is much sharper than in pure chromium. Most alloying elements reduce T_N. Exceptions are found for some impurity atoms with higher outer electron concentration than chromium, such as Mn, Re, Ru and Rh. It is difficult to give a systematic description of the results. In some cases the composition dependence of T_N can be given with a rigid band model as in the case of ternary CrMnV (see Fig. 4-32) alloys. Figure 4-32a gives $\rho(T)$ of these alloys. T_N

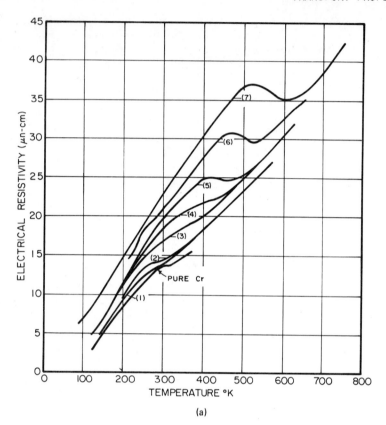

(a)

FIG. 4-32a. Resistivity of Cr-Mn-V alloys. Impurity concentration in at.%. (1): 0.4%V, 0.34% Mn; (2): 0.54%V, 0.86% Mn; (3):
0.51%V, 1.07% Mn; (4): 0.59%V, 1.18% Mn; (5): 0.54%V, 1.66% Mn;
(6): 0.57%V, 2.47% Mn; (7): 0.52%V, 3.60% Mn. After Ref. 233.

obtained from these curves is given in Fig. 4-32b. It is surprising that all data follow the same line. The average magnetic moment of the ternary alloys is given in Fig. 4-32c for comparison.
These magnetic moment values were obtained from neutron diffraction
data.

It is peculiar that the addition of iron and nickel to chromium
reduces T_N markedly, whereas cobalt with an electron concentration
just between these two elements has only a small effect on T_N. There
exists a second resistance minimum close to liquid He temperatures
far below T_N in CrCo (Fig. 4-31). This low temperature minimum may

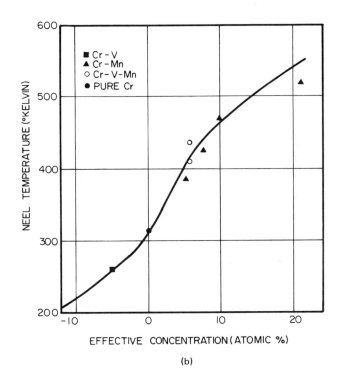

(b)

FIG. 4-32b. Concentration dependence of Néel temperature T_N for Cr-V-Mn alloys with those for pure Cr, Cr-V, and Cr-Mn alloys for comparison. After Ref. 233.

be related to the resistance anomaly (Kondo effect) found for noble metal alloys with transition element impurities since it occurs in the same temperature range. However, the Kondo theory in its original form for the minimum in ρ is probably not applicable to the Cr-Co system because this theory uses a positive exchange energy. Such an exchange energy is difficult to visualize in an anti-ferromagnetic matrix.

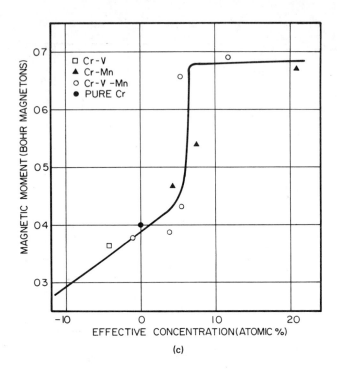

FIG. 4-32c. Concentration dependence of average magnetic moment μ for Cr-V-Mn alloys with those for pure chromium and binary chromium alloys for comparison. After Ref. 233.

M.   Resistivity of Semimetal and Semiconductor Alloys

The electrical resistivity of pure bismuth is markedly orientation dependent at low temperatures. H. Thompson [234] showed that
$\rho(\alpha) = \rho_{||} \cos^2\alpha. + \rho_{\perp} \sin^2\alpha.$ α is the angle between dcurrent direction and principal axis.

Bismuth-antimony alloys form a complete series of solid solutions.
Both elements are semimetals. The electrical resistivity of some
Bi-Sb alloys is much more complicated than that of elements. Figure
4-33a shows that the typical temperature dependence of ρ for metals
and semimetals (namely that ρ decreases with $1/T$) is found only for

bismuth and the alloy with 4.8 at.% antimony. Higher antimony concentrations lead to a temperature dependence of the resistivity which
is typical for semiconductors: ρ increases in some temperature intervals with decreasing temperature. This is found for alloys with 5.8
to 41 at.% Sb. The resistivity of Bi-Sb alloy at liquid He temperatures is given in Figure 4-33b. It shows a rapid increase of ρ_{re} with
composition for alloys with more than 5.8 at.% Sb. The largest resistivity value is found for the alloy with 12.5 at.% Sb. The ρ_{re} value
is still very high for the alloy with 40 at.% Sb.

The electrical resistivity at room temperature for these alloys
is nearly a linear function of composition and changes by less than
a factor of two over the composition range. This would mean that the
product of electron concentration times the mobility is essentially
constant at room temperature. The activation energy associated with
electron transitions in these alloys, as determined from the temperature dependence of ρ, is given in Fig. 4-33c. It shows energy gaps
for alloys with 5 to 40 at.% Sb, and band overlaps for lower Sb-con-
centrations and pure bismuth.

The addition of small amounts of lead to bismuth increases ρ,
and even leads to a maximum in ρ for some alloys, similar to that
found in semiconductors (Fig. 4-34a). One should therefore suspect
that lead impurities, just as antimony impurities, lead to shifts of
energy bands and the Fermi level. This is also found for BiSn, BiIn
and BiZn alloys (Fig. 4-34b and c). In all cases, the outer electron
concentration of the impurity atom is less than that of bismuth. BiSb
shows also a resistance maximum. Antimony has the same outer electrons
as bismuth. One should therefore conclude that these elements modify
the band structure and electron concentration in such a way that the
samples become semiconducting in their properties. It is easily seen
that the concepts of changes in the electron per atom ratio alone
cannot explain the data, since orientation effects can have a much
larger effect on the resistivity than alloying. This can be seen
in Figs. 4-34d and e. 0.59 at.% Se reduces ρ from 140 $\mu\Omega$ cm to 100 $\mu\Omega$cm
at 300°K if ρ is measured with the current parallel to the principal
axis, whereas ρ changes by less than 10 $\mu\Omega$ cm if up to 2.77 %Se is

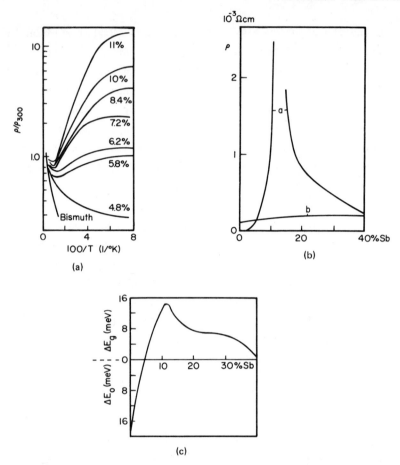

FIG. 4-33. $\rho(T)/\rho(300^\circ K)$ versus $100/T$ for samples with up to 11 % Sb (a). Resistivity versus concentration of Sb (b). Section a corresponds to data at 4.2°K, and section b to data at 300°K. Activation energy as a function of Sb-concentration (c). The energy gap ΔE_g corresponds to the semiconducting state, the energy overlap ΔE_0 to the semimetallic state. After Ref. 235.

FIG. 4-34 (opposite page). $\rho(T)$ of Bi-Pb alloys (a). $\rho(T)$ of Bi-base alloys. (b). $\rho(T)$ of Bi-Sn alloys (c). $\rho(T)$ of Bi-Se alloys (d and e). Measurements parallel to the principal axis for data in (b), (c), and (d), except for Cd alloys in (b) which were twinned. Measurements perpendicular to the principal axis in (e). In (e), the origin is shifted upwards by $10^{-5}\Omega cm$ for each curve . The higher Bi-curve in (e) represents a previous measurement by Shubnicov. After Ref. 234.

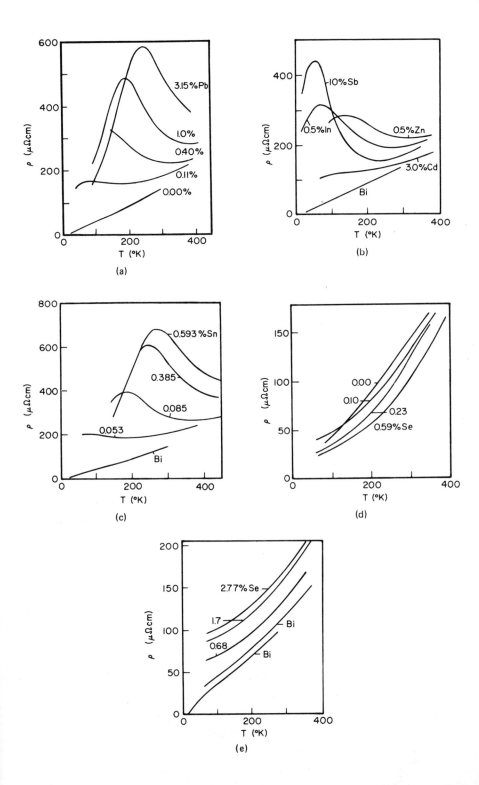

(a)

(b)

(c)

(d)

(e)

added when the resistance is measured with the current perpendicular
to the principal axis.

Germanium and silicon form a complete series of solid solutions.
Figure 4-35a shows log σ as a function of composition for their alloys
and for the pure elements. The temperature dependence of σ for all
samples is typical of semiconductors in these tests. Above room temperature σ decreases with 1/T. For most alloys, log σ is a nearly
linear function of 1/T in the temperature range under investigation.
These alloys show intrinsic conduction. Only silicon and silicon-
rich alloys show a change in slope, typical for the case where impurity conduction becomes noticeable. It is not surprising that extrinsic conductivity is more noticeable in SiGe alloys than in GeSi
alloys, since the energy gap in the former alloys is higher than in
the latter ones. Therefore, σ(intrinsic) is much smaller in SiGe
alloys, and conduction by holes and electrons from donors and accept-
ors is much more noticeable in samples with low intrinsic conduction.

The energy gap obtained from the slope of log σ versus 1/T in
the intrinsic conduction range E_g(cond) is given in Fig. 4-35b together with the energy gap obtained from optical data, E_g(opt). As
mentioned before E_g(opt) $\geq E_g$(cond), since the electron transitions
in conduction experiments are frequently phonon assisted. Therefore,
the dashed line in Fig. 4-35b should always be at or above the full
line. Within experimental error, both curves are essentially the
same. The E_g-composition curve shows a change in slope near $Ge_{80}Si_{20}$.
This would be the composition where the energy gap from $\Gamma_{25'}$ to $\Gamma_{2'}$
(this is the lowest energy gap in germanium) and the energy gap from
$\Gamma_{25'}$ to Δ (this is the lowest energy gap in silicon) have the same
values. Figure 1-77 shows the composition dependence of the different
energy gaps in Ge-Si alloys. Log σ at 1000°C, which is essentially
equal to log σ(intrinsic),changes smoothly as a function of composition (see Fig. 4-35c). The data, if extrapolated to room temperature,
show a change in slope between 15 and 20 at.% silicon (see Fig. 4-35d).
Such a change in slope would be expected from the composition dependence of E_g shown in Fig. 4-35b. The electron and hole mobilities
calculated from these data are given in Figs. 4-35e and f. The shape

of the curve follows a similar composition dependence to that of
the conductivity in metallic alloys like Cu-Au.

N. Electron Scattering by Point and Line Defects

Lattice defects like vacancies, impurity atoms, or dislocations
increase the electrical resistivity because they lead to deviations
from perfect periodicity. The calculation of the resistivity of a
metal due to vacancies is in principle very similar to the calculation of ρ due to a substitutional impurity with a different valency
than the matrix. In both cases, one has a screened Coulomb charge
and some lattice distortion in the matrix. Since the resistivity
increase in a copper alloy with 1 at.% impurities is of the
order of 0.1 $\mu\Omega$cm, one would expect similar values for one percent
of vacancies. This estimate agrees with more advanced calculations,
which give values of 0.4 to 1.7 $\mu\Omega$cm/at.% vacancies for copper.
Results of calculations by different authors are given by Blatt [198].
The calculated $\Delta\rho$ values are slightly larger than experimental $\Delta\rho$
values on most substitutional impurities. The lattice distortion and
resistance change due to interstitials are frequently larger than that
due to vacancies. One finds $\Delta\rho$ values from 0.6 to 10.5 $\mu\Omega$cm/at.%
interstitials in copper [198].

It can be shown that an edge dislocation corresponds to an electric dipole. The crystal section on top of the dislocation line is
compressed, and below it is expanded. This leads to both changes in
the lattice parameter with modifications in the electron energy, and
to changes in the concentration of positive ions per unit volume.
Both effects lead to a dipole moment. In copper one finds for randomly oriented edge dislocations $\Delta\rho$ to be of the order of $10^{-20}\Omega$cm
for a unit dislocation/cm^2. The resistivity due to a screw dislocation should, according to this model, be zero, since a screw dislocation leads to no elastic volume change in an elastic medium.
Higher order terms are of the same order as for the case of the
edge dislocation.

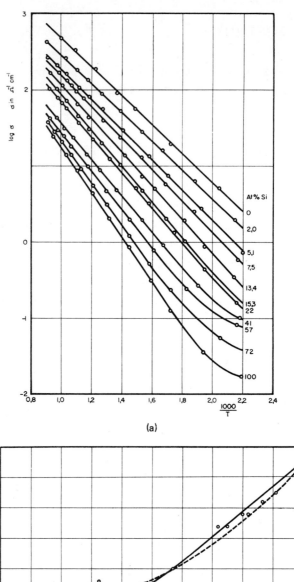

(a)

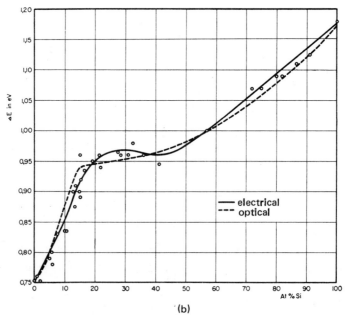

(b)

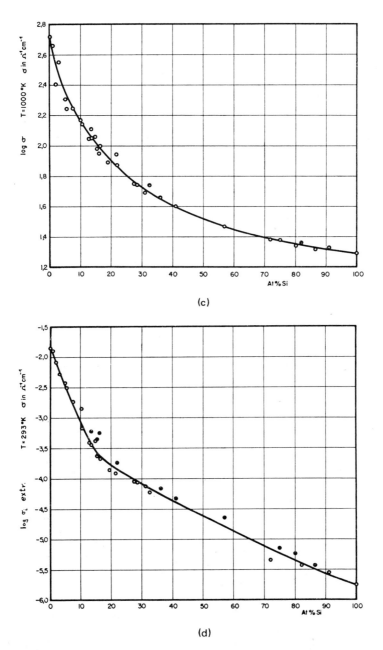

FIG. 4-35 a to d. Log σ as a function of 1000/T for Ge-Si alloys (a, opposite page). Optical and electrical energy gap in Ge-Si alloys (b, opposite page). Log σ at 1000°K for Ge-Si alloys as a function of Si-concentration in Ge-Si (c). To room temperature extrapolated log σ values of Ge-Si alloys (d). After Ref. 236, with permission.

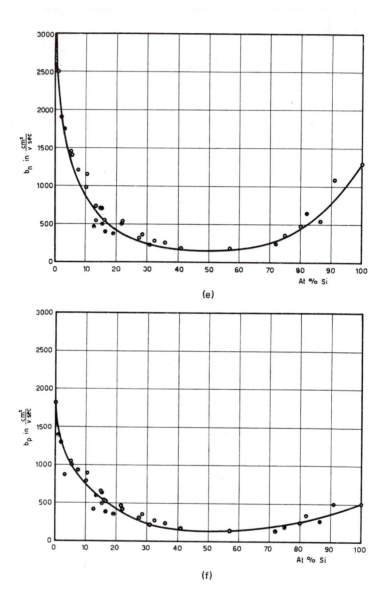

FIG. 4-35 e and f. Electron mobility $b_n = \mu$ as a function of composition of Ge-Si alloys (e). Hole mobility $b_p = \mu$ as a function of concentration in Ge-Si alloys (f). After Ref. 236, with permission.

The theory predicts that the resistivity of a dislocation split into partial dislocations can be noticeably larger than that of a perfect dislocation. However, experiments show that stacking faults contribute very little to the resistivity. Little work has been done on the study of scattering by grain boundaries.

Experimental values of the resistivity due to vacancies and dislocations are difficult to determine accurately, since the resistivity change due to one vancancy or one dislocation is very small. Therefore, one usually measures resistivity changes due to large changes in the number of vacancies and dislocations. Figure 3-7 gives an example. However, as for example stored energy measurements show, interaction of vacancies with impurities will modify the vacancy concentration. One should expect that the electrical resistivity will not only depend on vacancy concentration and dislocation density but also on their spacial distribution and on the interaction with other defects.

III. THERMAL CONDUCTIVITY

A. Heat Transport by Phonons

The thermal energy transport through a metal is due to two processes. Part of the heat flow is carried by electrons and the other part by phonons. One can visualize the mechanism of heat transfer in the following way: particles, either electrons or phonons, at thermal equilibrium in the sample at $\vec{r}_1$ at temperature $T(\vec{r}_1)$, will diffuse through the crystal and interact with the lattice in another section at $\vec{r}_2$ in such a way that the particle will reach the equilibrium temperature $T(\vec{r}_2)$. This means that the particle will lose or gain some energy, leading to an energy or heat flow. This model assumes that the heat transfer is essentially a diffusion process. The energy transfer from one unit volume ΔQ can be calculated in the following way [3]. Particles in a volume element at $\vec{r}_1$ with the temperature $T(\vec{r}_1)$ move a distance of the mean free path in the x_1 direction, collide and thereby reach the temperature $T(\vec{r}_2)$. They lose or gain the energy:

$$\Delta Q = c \; V_a [(T(\vec{r}_1) - T(\vec{r}_2))] = c \; V_a \; v_{x_1} \; \tau (\partial T/\partial x) \; . \qquad (4\text{-}59)$$

c is the heat capacity per mole, and V_a is the number of moles per unit volume. v_{x_1} is the velocity of particles in the x_1-direction, and τ is the relaxation time. The energy flow is:

$$\partial q/\partial t = \Delta Q v_{x_1} = c(\partial T/\partial r) < v_{x_1}^{\;2} > V_A \tau$$

$$= c \; V_A \; v\ell \, (\partial T/\partial r)/3 \qquad (4\text{-}60)$$

v is the particle velocity, and $\ell = v\tau$ is the mean free path. This gives for the thermal conductivity $\kappa = \dot{q}/\nabla_r T$:

$$\kappa = (1/3) \; c \; v\ell \; V_A \qquad (4\text{-}61)$$

The thermal conductivity due to phonons can be obtained from this equation, if one uses for c the lattice specific heat, $c_{1a.}$, and for v the phonon velocity. If κ is known, the mean free path can be calculated. This equation cannot be tested easily on metals, since the heat is only partially carried by phonons. However, quartz is a good electrical insulator. Therefore, no heat will be carried by electrons. Only phonons transport heat in this process. By using the sound velocity in quartz $v = 5 \cdot 10^5$ cm/sec, and setting $c_{exp.} = c_{1a.}$, one obtains for the mean free path of phonons $\ell_{ph.} = 40\text{Å}$ at room temperature, and 540Å at $-190°C$, the liquid air temperature. These mean free path values are surprisingly large. This is due to the fact, that phonon-phonon interaction leads to an energy transfer, only if a three phonon interaction process of the form:

$$q_1 + q_2 = q_3 + G \qquad\qquad G \neq 0 \qquad (4\text{-}62)$$

takes place. $\vec{q}_i$ is the wave vector of the phonon, and $\vec{G}$ is a lattice
vector in reciprocal space. A three phonon process with $\vec{G} = 0$ will
not change the sum of the momenta of the three phonons, and therefore not contribute to the heat flow. $G = 0$ are normal (or N-) processes, whereas $G \neq 0$ leads to an Umklapp (or U-) process (Umklapp
stands for "flipping over"). At high temperatures ($T > T_D$), all
elastic modes are excited. Then both N- and U- processes are possible.
At these temperatures the phonon concentration is proportional to
T. The mean free path ℓ should be inversely proportional to T. Both
the lattice specific heat and the sound velocity are constant. Therefore, the thermal conductivity due to phonons should in first approximation be inversely proportional to the absolute temperature. On
the other hand, for very low temperatures and very pure samples,
the mean free path can reach the dimensions of the sample. Naturally,
it cannot reach larger values. Therefore, ℓ_{ph} stays constant below
a critical temperature. c_{1a} for $T \ll T_D$ is proportional to T^3, and
the sound velocity is constant. Therefore:

$$\kappa \propto T^3, \qquad T \ll T_D . \qquad\qquad\qquad (4\text{-}63)$$

A more detailed analysis for κ of infinitely large samples at low
temperatures shows that the thermal conductivity should depend essentially on the number of possible U-processes which is given by an
exponential function of the temperature. The theory predicts [2]
that:

$$\kappa = a(T/T_D)^d \exp(T_D/bT) \qquad\qquad\qquad (4\text{-}64)$$

should give a reasonable description of κ, where "a", "b" and "d"
have to be determined experimentally. Figure 4-36 gives the thermal
conductivity of Al_2O_3 where one finds the predicted correlation κ
$\propto T^3$ at low temperatures, $\kappa \propto T^n \exp(b/T)$ at immediate temperatures,
and $\kappa \propto 1/T$ at higher temperature.

The thermal conductivity is not only markedly impurity dependent
but can be also influenced by isotopes. The scattering of elastic

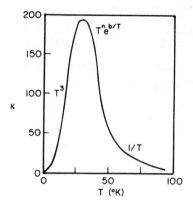

FIG. 4-36. Thermal conductivity of Al_2O_3 (κ in Watt units). After Ref. 237.

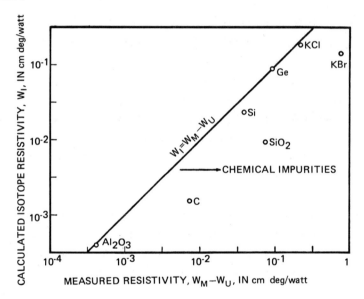

FIG. 4-37. Peak thermal resistivity against calculated isotope resistivity. After Ref. 238.

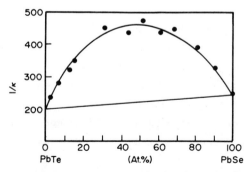

FIG. 4-38. Thermal resistance of mixed crystals. After Ref. 241

waves on atoms in a solid depends only on the spring constants and
the mass of an atom at a given lattice site. Therefore, an isotope
with a different mass than most other atoms in a pure element crystal
acts similar to an impurity atom in the lattice for electrical
resistivity measurements. Isotopes can be very strong scattering
centers for lattice waves and one can show that isotope scattering
in some materials is of the same order of magnitude as scattering by
thermal vibrations, even at room temperature. Figure 4-37 gives a
plot of the measured peak thermal resistivity versus calculated isotope resistivity. Experimental points should lie on a straight line
if samples are chemically pure. The points would be shifted to the
right if chemical impurities are present. Good agreement between
most experimental points and the isotope calculation indicates that
the isotope effect is well understood.

The effects of chemical impurities on the thermal conductivity
of insulators has been studied both experimentally and theoretically.
Calculations, just as in the case of isotopes, start with the problem
of scattering of elastic waves on a sphere. This gives for long
wavelength the Rayleigh scattering, where the scattering cross-section
is inversely proportional to the fourth power of the wavelength. Experiments by Eucken and Kuhn [239] showed that the extra thermal
resistivity above the thermal resistivity of the pure lattice (the
thermal resistivity is the reciprocal of κ) in the KCl-KBr mixtures
is proportional to the impurity concentration. Similar results are
reported on KCl containing F-center [240]. Solid solutions of semiconductors show a composition dependence of the thermal resistivity
which is similar to the Nordheim's rule for electrical resistivities
(Fig. 4-38). The marked effect of lattice defects in quartz on the
thermal conductivity is shown in Fig. 4-39. The defects were created
by neutron irradiation. Neutron irradiation first creates point
defects in clusters. They finally interact and lead to a general
structural disordering. The sample changes from a crystal to a glass
(4-39). The peak in κ is gradually suppressed. Experiments on
polycrystalline Al_2O_3, BeO and graphite also show scattering on
grain boundaries [244].

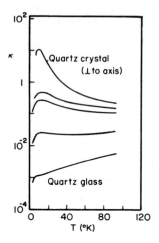

FIG. 4-39. The effect of irradiation on the thermal conductivity of SiO$_2$. After Ref. 243.

B.   Heat Transfer by Electrons

The thermal conductivity of some very pure single crystal insulators can be very high, close to the thermal conductivity of some metallic alloys. However, metals usually have a much higher thermal conductivity than insulators, since electrons can carry a large amount of the thermal current. The electron contribution can again be estimated from Eq. 4-61. At high temperatures, ℓ is inversely proportional to the absolute temperature T, since the scattering of electrons is essentially an electron-phonon interaction process, and the phonon concentration is proportional to T at high temperatures. The electron velocity is the Fermi velocity, $v_{el.} = (2E_F/m)^{-1/2}$, which is in first approximation constant. For most metals $c = c_{el}$ is proportional to T even at elevated temperatures. Therefore, κ_{el} is independent of temperature for $T > T_D$. The electronic heat transfer can be described in the following way. The energy distribution function for electrons at the hot end of the sample has more high energy electrons than the low temperature side. Heat transfer should be due to a

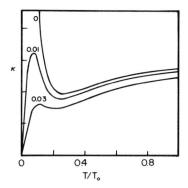

FIG. 4-40. Theoretical curves for thermal conductivity with different amounts of impurity resistance, calculated from the Bloch model. After Ref. 245.

flow of hot electrons to the cold side, losing their excess energy. The opposite process takes place when "cold" electrons travel to the hot side. However, the energy transfer in this diffusion process is much more complex than the scattering process which affects the electrical current in an applied field.   There, small angle scattering will have only a minor effect on the electron transport process. However, it can have a major effect on the heat transfer since the energy exchange in an electron-electron, electron-phonon, or electron-impurity collision is of the order of $k_B T$.   Therefore, an electron can change in any of these collisions from a "hot" state to a "cold" state and vice versa with only a small change in the direction and the absolute value of the velocity. Results of calculations of the thermal conductivity due to electrons by Sondheimer [245] are given in Figure 4-40. It shows a conduction minimum at $T/T_D \simeq 0.2$ both for the pure element, and for a sample with impurities. Such a minimum also shows up in the Bloch equation (Fig. 4-41). Only a Bardeen-type equation (see Fig. 4-41) gives such a small minimum that it can be neglected. No experimental evidence supports this minimum. The rapid increase of κ with decreasing temperature as predicted by these calculations is found experimentally, as data on sodium show (see Fig. 4-41). Sodium has a very simple electronic structure. It should be a good example of an element with a nearly spherical Fermi surface and is therefore the object of numerous calculations. Unfortunately, this element undergoes a diffusionless crystallographic transformation at low temperatures, and then consists of a mixture of two phases. This makes it impossible to obtain good thermal conductivity data at very low temperatures. Not only is κ an average of the thermal conductivity of two phases, it is also affected by crystal defects, since transformations usually introduce large amounts of dislocations and vacancies. Experiments on gold show that these defects change κ drastically (see Fig. 4-42). Cold work in metals increases both the dislocation density and the vacancy concentration. Most of these vacancies will "anneal out" at higher temperatures. Changes in κ during annealing should be due to a rearrangement and annihilation of dislocations and changes in vacancy concentration.

These large changes in the thermal conductivity of pure metals explain why only a few thermal conductivity studies on alloy systems have been published. It is very difficult to control the dislocation density and vacancy concentration, since both may depend on the concentration of the second constituent of the binary alloys.

C. Wiedemann-Franz Law

Thermal conductivity measurements are frequently compared with electrical conductivity measurements. This is due to the fact that the thermal conductivity is theoretically proportional to the product of electrical conductivity times electron energy, provided the thermal conductivity is essentially an electron transport process. One obtains for the Drude free electron model:

$$\kappa = \pi^2 \, nk_B^2 \, T \, . \ell/3mv_F = \sigma T \, \pi^2 k_B^2/3e^2 . \tag{4-65}$$

This gives the Wiedemann-Franz law:

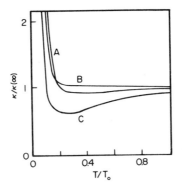

FIG. 4-41. Thermal conductivity of Na (B) compared with the Bardeen type formula (A) and the Bloch formula (C). After Ref. 245

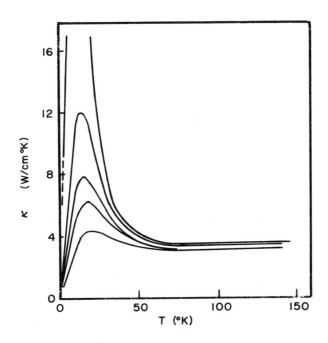

FIG. 4-42. Thermal conductivity of several gold samples tested in different heat treatments and mechanical deformations. Lowest line: 99.9% pure sample, unannealed; second line from bottom: 99.9% pure sample, annealed ; third line from bottom: sample JM freshly drawn; fourth line from bottom: sample JM annealed; top line: sample JM redrawn. After Ref. 247.

$$\kappa/\sigma = LT \tag{4-66}$$

with $L = \pi^2 k_B{}^2/3e^2$ the Lorenz number. This equation and experimental room temperature data agreed reasonably well. This indicated very early in this century that the Drude free electron model should be appropriate to describe the electronic structure of metals. Interestingly, Equation (4-66) should give a good description even for those metals and alloys which have a very complicated Fermi surface, and where the free electron model is certainly not applicable. The general validity of the Wiedemann-Franz law is due to the fact that the ratio of the electron current transport, j, and the thermal energy transport by electrons, $\dot{q}$, is evaluated at the Fermi surface with integrals of similar form:

$$j = (e/\hbar 4\pi^3) \int_{FS} (\partial E/\partial k) f^o dk \tag{4-67}$$

$$\dot{q} = (e\hbar 8\pi^3) \int_{FS} E(\partial E/\partial k) f^o dk \tag{4-68}$$

A calculation, starting with these two equations, leads to the result that the Wiedemann-Franz law is applicable to metallic systems with an arbitrary shape of the Fermi surface, if the electron gas is degenerate ($k_B T \ll E_F$), and if the scattering process is isotropic ($T > T_D$, or sufficiently high impurity scattering).

IV. HALL EFFECT MEASUREMENTS

A. Lorentz force and multiband models

Electrons moving in a magnetic field perpendicular to the electron velocity v are subject to the Lorentz force:

$$\vec{f}_L = e \vec{v} \times \vec{B} \quad \text{(SI)} \tag{4-69a}$$

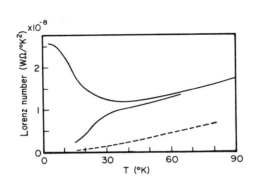

FIG. 4-43. The Lorenz number (Wiedemann-Franz ratio) for copper. Full line (top) gives experimental results. Full line (bottom) is deduced from experiments for ideal pure metal. Dashed line is calculated from Bloch theory. After Ref. 2.

$$\vec{f}_L = (e/c) \; \vec{v} \times \vec{H} \qquad \text{(cgs)} \qquad\qquad (4\text{-}69b)$$

which is perpendicular to both v and the magnetic field H. This force pushes the electrons to the side of the current carrier and leads to an excess or deficit of electrons at opposite sides of the conductor. The electric field F_H (Hall field) associated with this induced charge is opposing the Lorentz field $F_L = f_L/e$. In equilibrium, both fields (or forces) cancel:

$$e\vec{F}_H + e \; \vec{v} \times \vec{B} = 0. \qquad \text{(SI)} \qquad\qquad (4\text{-}70)$$

The electric current j for a system with only one type of current carrier is equal to $Nn_o ev$, if N is the number of atoms per unit volume, n_o is the number of free electrons per atom and v is their velocity. This gives:

$$\vec{F}_H = (1/Nn_o \; e) \; \vec{j} \times \vec{B}. \qquad \text{(SI)} \qquad\qquad (4\text{-}71)$$

Using the definition of the Hall coefficient given in Eq. 4-3 yields:

$$R_H = 1/Nn_o e \ .$$

$$(4\text{-}72)$$

This equation is applicable to a conductor with a spherical Fermi
surface or to the Drude free electron model. For such a system, the
Hall effect makes it possible to determine the number and the sign
of the carriers. Negative R_H values indicate electron transport and
a positive R_H a "hole" transport. In a simple one band model, the
conductivity is given by $\sigma = ne\mu = Nn_o e\mu$. Therefore, $R_H \sigma$ is equal
to the Hall mobility μ_H:

$$R_H \sigma = \mu_H \ .$$

$$(4\text{-}73)$$

As mentioned before, accurate calculations require that one distinguishes between the Hall mobility μ_H and the drift mobility μ_{dr}. The
correlation is discussed in detail by Blatt [198]. We will usually
neglect the difference between μ_H and μ_{dr}. R_H in most metals is temperature dependent. This cannot be explained with changes in n_o,
since the number of charge carriers in metals is essentially independent of temperature. It means that the correlation $n_o = 1/R_H Ne$ is
only an approximate description of the electron transport process in
metals under an applied magnetic field. For most metals it is necessary to use a multi-band model to explain the Hall effect data. For
a two band model, one obtains for low magnetic fields:

$$R_H = (\sigma_1/\sigma)^2 R_1 + (\sigma_2/\sigma)^2 R_2$$

$$(4\text{-}74a)$$

where R_i with $i = 1$ or $i = 2$ would be the Hall coefficient for band
"i". $R_i = 1/Nn_i e$, if these subbands are of standard parabolic form.
σ_i is the conductivity in band 'i', and $\sigma = \sigma_1 + \sigma_2$. These equations
can be easily extended to a system with more than two bands:

$$R_H = \Sigma (\sigma_i/\sigma)^2 R_i \ .$$

$$(4\text{-}74b)$$

By introducing the mobility μ_i of carriers in band "i", which is defined by $\mu_i = |v_i/F|$, one obtains for a multiband system:

$$R_H = (eN/\sigma^2) \sum_{i=1}^{n} n_i \mu_i^2 \qquad (4-74c)$$

R_H is essentially an average of R_1 and R_2 in a two band system weighted by $(\sigma_i/\sigma)^2$. This means that a minority carrier, which is responsible for 20 percent of the current, will usually not affect R_H noticeably, since the squared ratio $(0.2/0.8)^2$ is less than ten percent. It is therefore frequently justified to use the value of:

$$n^* = 1/R_H Ne \qquad (4-74d)$$

as the "effective number of carriers per atom" in the analysis of data. n^* cannot be used in a two band system with electron and hole conduction, if $n_{el.} \mu_{el.}^2$ is close to $n_{ho.} \mu_{ho.}^2$. In this case, n^* can become very large even for a small carrier concentration in both bands, because $n^* = (\sigma/Ne)^2/(|n_{el.} \mu_{el.}^2| - |n_{ho.} \mu_{ho.}^2|)$.

B.   Non-transition Elements

One would expect that a one band model should give reasonable results for metals with simple, nearly spherical Fermi surfaces. Table 4-7 gives a survey of experimental results in the form of $R_H Ne = 1/n^*$. A positive sign indicates that the effective carriers have "hole character", a negative sign "electron character". Copper, silver, and gold have n^* values close to one. This is not surprising since the Fermi surface of these elements is nearly spherical. The few necks in the Fermi surface of these elements will give a 'hole' contribution since the effective mass defined by the equation $m^* = \hbar^2/(\partial^2 E/\partial k^2)$ is negative. This has the same effect as a positive charge in an electric current. Electrons in these necks will therefore reduce $|n^*|$. Flat sections on the Fermi surface with a radius larger than the radius of the free electron sphere may have electrons with an even higher mobility than the mobility of free electrons. This

TABLE 4-7
Hall constants of Metals per Atomic
Unit Volume (RN|e|).   [2]

Li −1.3	Hg −0.6	Sn −0.025
Na −0.9	Al −0.4	Pb +0.05
K −0.9	Ga −0.5	Mo +1.3
Rb −1.0	In −0.04	Ru +2.5
Cs −1.1	Tl +0.11	Pd −0.75
Cu −0.8	Ti −0.2	As + 35
Ag −0.8	V +0.9	Sb + 65
Au −0.7	Cr +5	Bi − 5000
Be +5	Mn −1.2	Ta +0.9
Mg −0.7	Fe + 14	W +1.2
Ca −0.7	Co +4	Re +3.5
Zn +0.4	Ni −4	Ir +0.35
Cd +0.5	Rh +0.6	Pt +0.2

could increase $|n^*|$. A more detailed analysis of the Hall coefficient
of copper is given in Section F   where R_H of copper alloys is discussed. The Fermi surfaces of lithium, sodium and potassium should
be close to spherical in reasonable agreement with measured n^* values.
The R_H value of lithium yields an effective electron concentration of
less than one. This cannot be reconciled with presently accepted models of its Fermi surface.

The experimental Hall effect data for elements in groups IIA
and IIB can be explained with a two band model. These elements are
compensated. They have two conduction electrons per atom, which would
fill the first Brillouin zone completely if the energy gap between
first and second Brillouin zones was sufficiently high. For metals
these zones overlap. Some electrons move into the second or higher
Brillouin zones, leaving the same number of holes back in the first
Brillouin zone. This means that these elements have the same number
of electrons and holes, $n_{el.} = n_{ho.}$. The positive or negative sign
of R_H depends on the relative mobility of these carriers. In zinc

and cadmium the mobility of holes must be larger than the mobility of electrons since R_H is positive; the opposite conclusion has to be reached in magnesium and calcium. If R_H is field dependent, one frequently rewrites Eq. 4-74c for the case of a two band system with holes and obtains:

$$R_H = (B/Ne)[1 +$$
$$(A/B)(H\sigma_o/Ne)^2]/[1 + (n_{el.} - n_{ho.})A(H\sigma_o/Ne)^2] \qquad (4\text{-}75a)$$

with:

$$A = (\sigma_{el.}{}^2\sigma_{ho.}{}^2/\sigma^4)(n_{el.} - n_{ho.})n_{el.}{}^2 n_{ho.}{}^2$$
$$B = (\sigma_{el.}/\sigma)^2/n_{el.} - (\sigma_{ho.}/\sigma)^2/n_{ho.} \qquad (4\text{-}75b)$$

The subscript "o" indicates zero field values. This equation also explains a temperature dependent R_H function as in the regular two-band model, since (σ_i/σ) or μ_i can be temperature dependent.

C. Transition Elements

The Hall coefficient of only a few selected transition elements, which show special effects due to spontaneous magnetic ordering will be discussed. Elements of the chromium group are examples of compensated metals. They have six outer electrons in unfilled bands. These electrons do not completely fill the two electron states in the s-band and the four electron states in the two lower d-subbands, but rather spill partly into the three higher d-subbands, leaving the same number of "holes" behind. R_H of chromium at room temperature yields an n* value of 0.23 holes per atom. Liquid He temperature measurements in high fields indicate that a two band model, with $n_{el.}$ close to $n_{ho.}$, is more appropriate (see Fig. 4-44), with a "magnetic breakdown" close to 8×10^4 Oe.

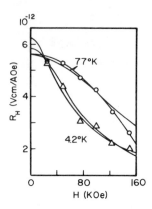

FIG. 4-44. Hall coefficient of chromium as a function of field. Full lines are calculated values. See Ref. 248 for details.

Equation 4-3 cannot be used for the analysis of the Hall effect of ferromagnetic materials, because F_H is affected not only by the applied field H, but also by the magnetization of the sample, M. Therefore, $R_H \cdot H = F_H/j$ should not be a linear function of the applied field H (see Fig. 4-45). The effect of the magnetization on the Hall field is frequently much more pronounced than the applied field. Experiments are usually described with the equation:

$$F_H/j = R_o H + R_1 M \qquad\qquad (4\text{-}76a)$$

R_o is the ordinary Hall coefficient and R_1 is the extraordinary Hall coefficient. Figures 4-46 and 4-47 give R_o and R_1 as functions of temperature for nickel. R_o is similar to the Hall coefficient of non-ferromagnetic metals. An understanding of the extraordinary Hall coefficient is difficult since the magnetization acts only on the spin of the electron. The magnetization does not lead directly to a force like the Lorentz force which is responsible for the ordinary Hall coefficient. In this sense, the molecular field is not a "real" field for the movement of electrons. The interpretation of the extraordinary Hall coefficient is still controversial. For instance, one attempts to claim a correlation between spin orientation and the movement of electrons. This means that the molecular field or

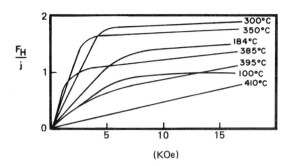

FIG. 4-45.   The Hall field per unit current density in nickel as a function of magnetic induction. After Ref. 249.

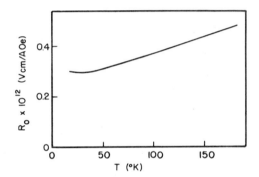

FIG. 4-46. The ordinary Hall coefficient of nickel as a function of temperature. After Ref. 250.

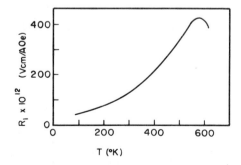

FIG. 4-47. The extraordinary Hall coefficient of nickel as a function of temperature. After Ref. 251.

exchange forces act on the spin, and the force on the spin leads to
an effect on the movement of the electron. This is called "spin-
orbit coupling". The interpretation of the individual features of
the Hall coefficient of ferromagnetic elements requires a multiband model. Hall coefficients can be more easily analyzed in alloy
systems than for pure elements. Examples are given in Section I.

D. Hall Coefficient of Semiconductor Elements

The current of germanium and silicon is carried by electrons or
holes in three bands. Therefore, Eq. (4-74c) with n = 3 may be
appropriate. Even this multiparameter equation is a simplification.
Some of these parameters of the multiband models are known from
theory or experiments, and it is possible to find a self-consistent
model to explain the results not only of the Hall coefficient and
its magnetic dependence but also of the electrical resistivity with
and without a magnetic field. Figure 4-48 gives experimental results
and calculations for R_H as a function of H. The essential features
of the curve were obtained by using a model with three carriers,
assuming that the ratio of the light to heavy masses of the holes
(subscripts 2 and 3) was 0.02, and that the ratio of the mobilities
was 7.5. The positive value of the Hall coefficient indicates that
the essential contribution to the current is from mobile holes.

The Hall coefficient for semiconductors should, as mentioned
above, be described with a multiband equation (Eq. 4-74c). Frequently
the two band equation gives the essential features if one has a hole
and an electron band. This equation is rewritten for non-degenerate
semiconductor applications in small fields in the form [107]:

$$R_H = -(r_R/e)(b^2 n_{el.} - n_{ho.})/(bn_{el.} + n_{ho.})^2 \qquad (4\text{-}76b)$$

with $b = \mu_{ho.}/\mu_{el.}$. r_R is determined by the scattering mechanism,
$r_R \simeq 1$.

The low field Hall coefficient for germanium is given in

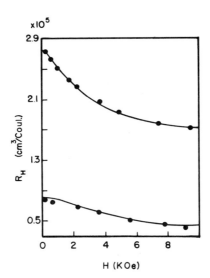

FIG. 4-48. Variation of Hall constant with magnetic field for p-type germanium. The solid curve is the theoretical curve computed with $n_3/n_2 = 0.02$ and $\mu_3/\mu_2 = 7.5$, assuming the mean free path to be independent of energy. Top line: T = 205°K, bottom line : 297.7°K. After Ref. 252.

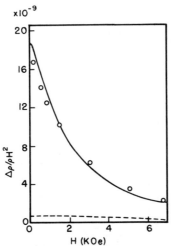

FIG. 4-49 Variation of magnetoresistance coefficient $\Delta\rho/\rho H^2$ with magnetic field in p-type germanium. The solid curve is a theoretical curve, the dotted curve is the result which would be obtained on a one band model. T = 205°K. After Ref. 252, with permission.

Fig. 4-50 in which log R_H is plotted as a function of the reciprocal
temperature. In the intrinsic range, this curve is slightly bent.
The full line is calculated from conductivity mobility values and
electron and hole concentration. The correction for the change
from μ_{cond} to μ_{Hall} has been taken into consideration. In the extrinsic conduction region, the Hall coefficient should at first become constant with decreasing temperature. This indicates the carrier
concentration is constant (if one has one dominant type of carrier).
At lower temperatures, the carrier concentration again becomes an
exponential function of temperature. However, it is difficult in
this range to describe the Hall coefficient with simple models, and
the concept of impurity electron bands has to be used. As mentioned
above, the first Bohr radius of outer impurity electrons may be of
the order of 10 to 100 $\overset{o}{A}$. Therefore, even impurity concentrations
where the distance of impurities is of this order of magnitude give
rise to impurity bands.

Since the mass of electrons is much smaller than the mass of
holes, one would expect that the Hall coefficient in the intrinsic
range (that is at high temperature) should always be negative. Even
in some sections of the extrinsic range where the hole concentration

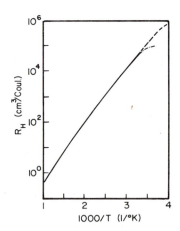

FIG. 4-50. Hall coefficient of germanium versus the reciprocal
temperature. After Ref. 253.

may be larger than the electron concentration, the Hall coefficient
will be negative. Only in those samples where the hole concentration is noticeably larger than the electron concentration, R_H will
change sign and become positive at low temperatures. Increasing
impurity concentrations will decrease the value of R_H, since the Hall
coefficient with one type of dominant carrier is inversely proportional to the carrier concentration. Pearson and Bardeen found in
their study of p-type silicon that R_H at low temperature was in the
extrinsic exhaustion range approximately inversely proportional to
the impurity concentration for their samples with relatively high
dopant levels [210]. This is the same range where the electrical
resistivity is nearly temperature independent.

The Hall coefficient should be an exponential function of the
temperature in the intrinsic range, since there the carrier concentration follows this temperature dependence. The temperature dependence of the energy gap is usually taken into account in the form
$E = E^o + const. \cdot T$.

E. Hall Coefficient of Semiconducting Compounds

Results for the Hall coefficient of InAs are qualitatively
similar to those found for Ge and Si (see Fig. 4-51). Again, log R_H
is a nearly linear function of 1/T for high temperature in the intrinsic conduction range. R_H is negative since the conductivity is
due to the very mobile electrons with exceptionally low effective
masses. R_H is nearly constant for low temperatures in the extrinsic
range, indicating that the carrier concentration is constant. This
is the same range where σ is constant (Fig. 4-18). The sign of the
Hall coefficient changes at low temperatures for the p-type samples
where it is then positive. An analysis by Folbert, Madelung, and
Weiss [215] gives for this system a forbidden energy gap of $(0.47 - 4.5 \cdot 10^{-4} \, T)$ eV. The Hall mobility for electrons in the purest sample
is 27,000 cm^2/V sec. at room temperature and the temperature

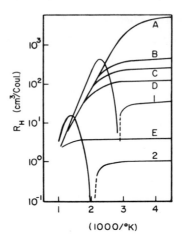

(1000/°K)

FIG. 4-51. Hall coefficient vs. 1/T of InAs samples. A to E
n-type, 1 and 2 p-type samples. The conductivity of these samples
is shown in Figure 4-18 . After Ref. 215.

dependence follows a $T^{-1.5}$ law. The hole mobility is about two
orders of magnitude smaller ($\mu_{ho.}$ = 200 cm^2/V sec. at room temperature)
and is proportional to $T^{-\alpha}$ with $\alpha \geq 1.66$.

The very low effective mass of the conduction electron in InSb
leads to even larger values for the room temperature electron mobility:
$\mu_{el.}$ = 77000 cm^2/V sec. The temperature dependence is given by $T^{-1.66}$,
as long as ionized - impurity scattering does not limit the mobility.
E_g follows E_g = (0.27 - $3 \cdot 10^{-4}T$) eV[254]. Since the mobility of holes
is much smaller than the mobility of electrons, one can neglect their
contribution to σ and calculate σ with a one band model. The intrinsic concentration of electrons or holes between 200 and 600 °K has a
value given by n_i = 6.0 x 10^{14} $T^{3/2}$ exp $(-0.26/2k_BT)cm^{-3}$.

The mathematical effort to analyze Hall coefficient data of semiconducting compounds is substantial. The following section gives
only a broad view of the parameters used in the analysis of R_H
to indicate the complexity of the analysis. The original references
have to be consulted for the definition of some terms. A comparison
of experiments with the theory for the mobility of the InSb has been
made by Ehrenreich [255,256]. The mobility equation given by Bardeen

and Schockley for acoustic scattering was used. This equation depends on the value of the deformation potential which is only approximately known. An estimate from the pressure dependence of the conductivity indicates a value of about -7 eV for this. Using this
value, one obtains for the mobility a value of 10^7 cm^2/V sec. This
is much too large. Therefore, other scattering mechanisms are important. It seems likely that scattering due to polar and to electron–hole interaction has to be taken into account. This gives good
agreement between experiments and theory above 200°K.

In addition to acoustic mode scattering through the deformation
potential and polar optical mode scattering, as used by Ehrenreich,
Rode [257] evaluated piezoelectric scattering effects in direct gap
polar semiconductors. These calculations on III-V semiconductors
agree better with low temperature data than Ehrenreich's theory.
Typical examples of the III-V group compounds are the GaAs, GaSb,
InP, InAs, and InSb compounds, which exhibit some of the highest electron mobilities found for semiconductors in the liquid N$_2$ - room
temperature range. This makes them interesting for Hall effect
detector devices. It turns out that piezoelectric coupling of
electrons to longitudinal modes is the dominant type of scattering
in GaAs for temperatures below 60°K and is noticeable at 100 to
150°K. These scattering mechanisms are combined before the effect
on the electron distribution is calculated.

Rode then determined the structure of the electronic wave functions, taking into account both conduction band–valency band interactions and the atomic spin-orbit interaction. Electrons near the
conduction band edge have wavefunctions of s-type character. They
are non-localized and are similar to plane waves. Above the band
edge electrons become more localized. They are able to sense the p-
states of lattice electrons for a short part of the time. The electronic wavefunction is then given by a linear combination of s- and
p-wavefunctions. The electron energy under such conditions is no
longer a parabolic function of $\vec{k}$. Rode then calculates in a self-
consistant way the scattering process in these III-V semiconductors.

taking both the non-parabolic band structure and the functional dependence of the wavefunction on crystal momentum into account.

With this model of the energy band and the electronic wavefunctions, Rode calculates the electron mobility with a variational approach. Typical results of Rode's calculations are shown in Fig. 4-52
for the case of GaAs, where experimentally obtained mobility values
are also given. These data were obtained from Hall effect measurements and the drift mobilities were then calculated from the Hall
mobilities. Agreement between theory and data is surprisingly good.
The physical parameters used in this calculation are given in Table 4-
8. Deviations between data and calculations for temperatures below
80°K may be due to impurity scattering which was not considered in
the calculation. However, agreement between theory and data at higher temperatures is excellent [257, 258].

F. Noble Metal Base Alloys

It is not possible to explain the composition dependence of R_H
in noble metal base alloys with the free electron model. It is necessary to consider both the anisotrophy of the Fermi surface and the
anisotrophy of scattering. There are several ways to take the anisotrophy of the Fermi surface in the calculation of R_H in metals and
alloys into account. Figure 4-53 shows a schematic cross-section of
the k-space of noble metals. The Fermi surface is not spherical, but
has cones which touch the L point (coordinates (π/a) <111> in reciprocal space). The E(k) curve has an energy gap for $\vec{k} = \Gamma L$. Figure
4-54 gives schematically the correlation between some parameters which
have been used to characterize the Fermi surface, and their correlation with the energy of electrons near the Brillouin zone boundary.

The Hall coefficient of silver-gold and silver base silver-palladium alloys is given for several temperatures in Fig. 4-55. Full
lines give predicted results from the free electron model. The effective carrier concentration $n^* = 1/NeR_H$ obtained from experiments

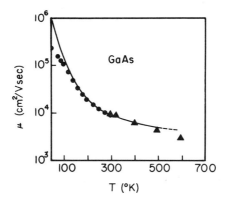

FIG. 4-52. Comparison between electron drift mobility in GaAs
as calculated from data given in Table 4-8 (solid curve) and drift
mobility derived from experimental Hall mobility data (circles and
triangles). After Ref. 257, with permission.

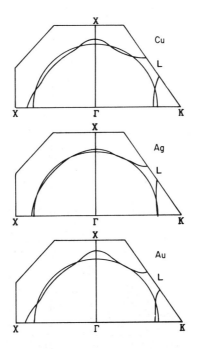

FIG. 4-53. Schematic cross sections of Fermi surfaces for copper,
silver, and gold. The circles represent the free electron Fermi surfaces. After Ref. 259, with permission.

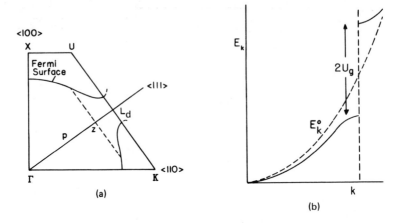

FIG. 4-54. Characteristic parameters of the Fermi surface (a)
and the E(k)-diagram (b, see text).

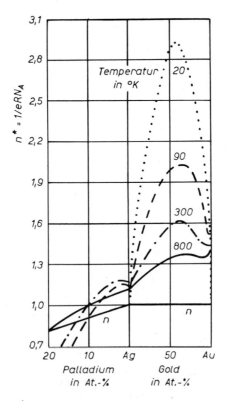

FIG. 4-55. n^* of Ag-Au and Ag-Pd alloys. n: free electron model.
After Ref. 260.

TABLE 4-8

Material Parameter of Semiconducting Compounds

Quantity	GaAs	GaSb	InP	InAs	InSb [257]
Effective mass $m*/m$	0.0655	0.047	0.067	0.022	0.013
Zero-temperature energy gap $E_g(0)$ (eV)	1.58	0.80	1.48	0.46	0.265
Energy-gap temperature coefficient $3\ell(\partial E_g/\partial P)_t/K(10^{-4}eV/{}^{o}K)$	1.20	1.16	0.50	0.79	1.12
Low-frequency dielectric constant ε_o	12.90	15.00	12.30	14.55	18.70
Polar-phonon Debye temperature T_{po} (${}^{o}K$)	420	346	501	350	278
Material density ρ (g/cm^3)	5.36	5.66	4.83	5.71	5.82
Sound speed v_a (km/sec)	5.24	4.36	5.16	4.28	3.80
Acoustic deformation potential E_1 (eV)	7.0	6.7	6.5	4.9	7.2
Piezoelectric constant e_{14} (C/m^2)	0.160	0.126	0.035	0.045	0.071
Sound speed v_p (km/sec)	4.03	3.35	3.85	3.17	2.78
High-frequency dielectric constant ε_∞	10.92	13.80	9.56	11.78	16.76

is larger than n obtained from the free electron model for all Ag-Au
alloys. The calculation of n* by Ziman [159] for a nearly spherical
Fermi surface (Fig. 4-54) with cones leads to increases of n* above n.
Ziman characterizes the anisotrophy by $\delta = d/p$ where d is the diameter
of the contact area of the Fermi surface with the first Brillouin
zone and p is the distance from Γ to L. $U = U_g/E_k^o$ is the ratio
of the energy gap at L divided by one half the Fermi energy at the
first contact between the Fermi surface and the Brillouin zone boundary. Figure 4-56 shows that an increase in δ and u, which corres-

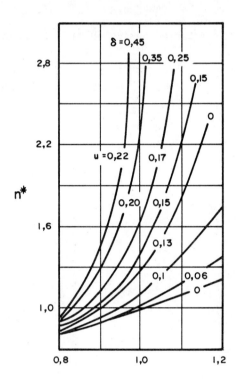

FIG. 4-56. Calculated n* values as functions of the el./at.
ratio. After Ref. 260.

sponds to an increase in anisotrophy, always leads to an increase in
n*. Another calculation by Cooper and Raimes [261] makes both an
increase or a decrease of R_H with respect to the free electron model
possible. The anisotropy of both the Fermi surface and the scattering process was taken into account.

Figure 4-55 shows that n* of both silver and gold is well above
one, indicating that the Fermi surface of both elements is distorted.
The largest deviation from the free electron model is found for
alloys with about 60 at.% Au, with n* close to 3 at 20°K instead of
the ideal value of 1 for all these alloys. The alloy system comes
closer to the ideal value with increasing temperature. Using the

appropriate band parameters of Fig. 4-56 at n = 1, it should be
possible to explain the Au-Ag data with the anisotropy of the Fermi
surface. However, the experiments indicate that (n* - n) is approximately proportional to c(1 - c). This indicates that the deviation
from the ideal system, (n* - n), is essentially proportional to the
residual resistivity, suggesting that the scattering process may be
at least partially responsible for the deviation from the ideal
system.

n* of noble metals with higher valency B-group elements is given
in Figure 4-57. The free electron model would give n* = n = el./at.
The experimental values deviate markedly from this simple model for
el./at. near one. n* is a nearly linear function of el./at. only for
larger impurity concentrations. This concentration dependence may
be explained qualitatively with models similar to those used to
analyze the electronic specific heat coefficients. The addition of
impurity atoms may make the Fermi surface more spherical. It could
approach the free electron model for larger impurity concentrations
in noble metal alloys. Further, extra electrons of impurity atoms
are kept in clouds near the positively charged impurity atom due to
screening. The Fermi energy initially increases less rapidly with
el./at. than the free electron model predicts. This should lead
to a notably slower increase of n* compared with n. Finally, increased impurity scattering could also reduce the anisotropy of the
transport process.

The addition of palladium and platinum to noble metals should
reduce the number of outer electrons. This agrees with data given
in Figure 4-55 in silver-rich alloys. A further increase in palladium concentration leads to unfilled d-states.

G. Transition Element Alloys

The density of states curve of transition element alloys of the
first long period has a sharp minimum for el./at. near 6. This minimum is found at chromium for Cr̲Fe and Cr̲V alloys and for Fe-V

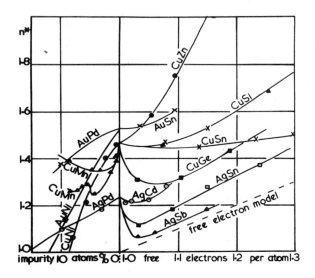

FIG. 4-57. n* of noble metal base alloys. After Ref. 262.

alloys near $Fe_{30}V_{70}$. The minimum is not as pronounced in the second
long period of the periodic system. It would seem reasonable to
assume that a three band model would describe the electronic transport mechanism in these alloys in first approximation if magnetic
ordering can be neglected. One would have an s-band (conduction
band) responsible for most of the current, and two separate d-bands.
The d-subbands for el./at. > 6 in the case of chromium alloys contain localized magnetic moments. The mobility of its (e_g-) electrons
should therefore be small. The d-band below el./at. = 6 has electrons
which are relatively mobile (t_{2g}-electrons). Whereas the s-electrons
and e_g-electrons are expected to give a negative contribution to R_H
(the e_g-band is nearly empty near chromium), the t_{2g}-electrons should
give a positive contribution because the t_{2g}-electron band is nearly
filled. The Hall coefficient for this three-band system would be:

$$R_H = (eN/\sigma^2) \ [-\mu^2(s)n(s) + \mu^2(t_{2g}) \ n(t_{2g}) - u^2(e_g)n(e_g)] \quad (4-77)$$

One would expect that $\mu(s) > \mu(t_{2g}) > \mu(e_g)$, $n(e_g) \lesssim n(s)$ and $n(t_{2g})$
$\lesssim n(s)$ near the minimum of $N(E)$, because $n(s)$ should be of the
order of one, and the d-subbands are either nearly filled or nearly
empty. Therefore, the first term in Eq. 4-77 dominates, and R_H should
be negative and nearly independent of alloy concentration. This dis-
agrees with the experimental evidence. Figure 4-58 shows that R_H
is positive and has a sharp peak between el./at. = 6 to 6.2. The
following interpretation of the data may have been proposed. The t_{2g}
electrons have wavefunctions similar to s-electrons. These two types
of electrons therefore "hybridize". This means they combine and
form one common band. Below el./at. = 6 for Cr alloys, electrons in
this band behave as if they are in a nearly filled band. Their
mobility is much larger than that of the e_g-electron. Therefore, the
hybridized band gives a large positive contribution to R_H and makes
it positive [242].

I. Ferromagnetic Alloys

The analysis of ferromagnetic alloys requires multiband models.
Magnetization measurements in nickel-rich alloys indicate that the
number of s-electrons per atom is $n_s \approx 0.6$. This result seems to
disagree with Hall effect measurements (Fig. 4-59), which give $n_s \leq$
0.3 for an alloy with 80 at.% Ni if one assumes a two band model with
s- and d-electrons, and assumes that the mobility in the d-band is
not more than 4 or 5 times smaller than the mobility in the 4 s-band
[264]. However, one obtains much better correlation between theory
and measured R_o values if one assumes that the s-band is split into
subbands with spins up and down, each subband containing about 0.28
electrons per atom at saturation magnetization. Only conduction
electrons in the half band with spins parallel to the magnetization
have high mobility, since electrons in the other half band can be more

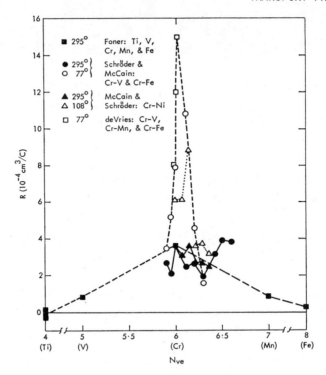

FIG. 4-58. Hall coefficient R of selected transition elements and their alloys. After Ref. 263, with permission.

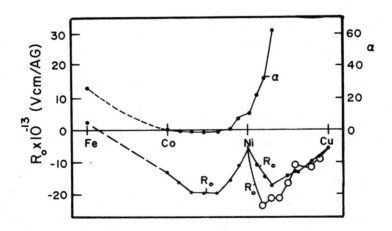

FIG. 4-59. Ordinary Hall coefficients for Fe, Co-Ni and Cu-Ni alloys. Open circles: 4.2°K; solid circles: room temperature. After Ref. 264, with permission.

easily scattered into the partially filled d-band. Therefore, n* of
nickel-rich alloys can be as small as 0.28. This gives the right
value for R_o of the 80 at.% alloy. The Hall coefficient of this sample should decrease with increasing temperature because the magnetization decreases. Above the Curie temperature, $n_s \simeq 0.6$ should be
reached.

Figure 4-59 shows the ordinary Hall coefficient of Ni-Cu alloys
as a function of composition. $|R_o|$ has a maximum between 70 and 80
at.% Ni. The data indicate that n_s decreases from about 1 for pure
Cu to 0.6 for 40 at.% Ni which is the onset ferromagnetism. A further
increase in Ni concentration leads to open d-states in one half of a
d-band. The current is now carried predominantly in one half section
of the s-band, so that n* is further reduced. n* = 1/ReN
reaches a minimum. The increase of n* = $1/R_o eN$ with increasing
nickel concentration for Ni-rich alloys with more than 80% Ni may be
due to conduction in the d-band.

The addition of cobalt to nickel again leads first to an increase
of $|R_o|$, then to a decrease. The maximum of $|R_o|$ occurs near 50 at.%
Co. n* is again rather small near this composition and corresponds
to n* $\sim$ 0.4 electrons/atom.

J. Hall Coefficient of Semimetal and Semiconducting Alloys

The Hall coefficients of several Bi-Sb alloys are given in Figs.
4-60a and b. The samples with up to 11 at.% Sb show a nearly temperature-independent R_H value below $10°K$. R_H decreases with increasing temperature above $10°K$. The low temperature data indicate a
temperature-independent carrier concentration. The decrease of R_H
with temperature above $10°K$ could be explained with a rapid increase
in the carrier concentration with temperature. Alloys with 12 to 19
at.% Sb show the R_H temperature dependence typically for semiconductors, with the change of sign near T $\simeq$ 25 to $30°K$. The semiconductor-
semimetal transition occurs near the 11 at.% Sb concentration, not at

the 5 at.% Sb concentration given in Fig. 4-33c. Magnetic susceptibility data given in Fig. 2-26 suggest that the semimetal-semiconductor transition occurs near 7 at.% Sb. The interpretation of the
Hall effect data has to take the anisotropy of the crystal into
account since experiments show that R_H is markedly orientation dependent. Figure 4-60c gives R_H obtained in experiments in which the
current moves along the binary axis , and where H is applied parallel
to the bisectrix ($R_H = R_S$), and for the case where the current flows
parallel to the binary axis and H is applied parallel to the trigonal
axis ($R_H = R_p$). This marked anisotropy makes it difficult to analyze
the data with a simple model. Jain tried to calculate the carrier
concentration under the assumption that there exist six ellipsoids for
electrons and two ellipsoids for holes. The total number of carriers
is calculated as a function of temperature under the assumption that
the distance between the valency band and the conduction band at L,
$C_L - V_L$, is constant and equal to 0.0072 eV, whereas $C_L - V_H$ varies
from 0.0185 eV for pure bismuth (curve A in Fig. 4-60d) to such high
values that a further increase in V_H has no effect on the carrier
concentration. V_H is the top of the valency band at H.

Jain also determined the mobility of electrons $\mu_{el.} \propto \sigma/N_e e$.
$\mu_{el.}$ is given for several semimetallic alloys in Fig. 4-60e as a
function of temperature. The temperature dependence of the mobility
follows power laws of the form $\mu_{el.} \propto T^{-\alpha}$ with $1.4 \leq \alpha \leq 1.8$. α
is close to 3/2, the exponent predicted for lattice scattering of
electrons in semiconductors.

FIG. 4-60 (opposite page). Hall coefficient R_p versus 1/T of
Bi-Sb alloys, with current along the binary axis and H parallel to
trigonal axis for 0-11 at.%Sb (a). R_p versus 1/T for current along
binary axis, H parallel to trigonal axis for 12.5 to 20 at.% Sb (b).
Behavior of R_p and R_s at 4.2°K as a function of Sb-concentration (c).
Carrier concentration N versus 1/T between 200K and 1000K (d);
$C_L - V_H$ = 0.018 (A), 0 (C), -0.0044 (E), -0.0144 (G), -∞(J) (units in
eV). Log($\sigma/N_e e$) versus log T for samples with 0%Sb (A), 4.8%Sb
(B), 7.2%Sb (C), and 5.8%Sb (D). After Ref. 235.

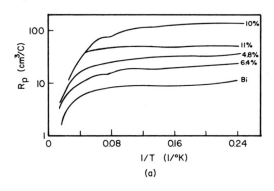

(a)

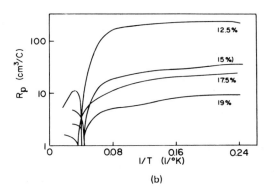

(b)

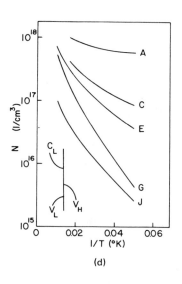

(d)

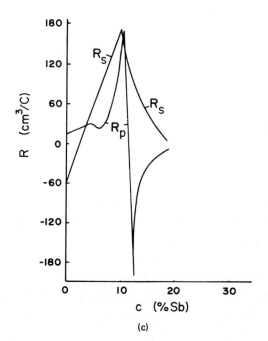

(c)

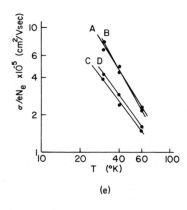

(e)

V. THERMOPOWER

A. Phonon-drag, Magnon-drag, and Diffusion Terms

The electrical conductivity and the Hall coefficient can be
easily visualized with models of the electron movement in metals.
This is not so easy   with the interpretation of the thermopower.
A simple model can explain qualitatively only the thermopower due to
the "phonon-drag" effect. A temperature gradient in a sample leads
to a heat flow. This is partially due to a flow of phonons from
one end of the sample to the other end. These phonons have a momentum. In scattering, some of the momentum is transferred to electrons, which will be "dragged along". This leads to an excess of
electrons on one end of the sample and a deficit at the other end,
and is associated with an electric field gradient $dV/d\ell$ in the sample, which should in first approximation be proportional to the
temperature gradient, $dT/d\ell$. The ratio of these two terms is the
Seebeck coefficient associated with the phonon drag effect. This
phonon drag Seebeck coefficient, $S_{ph.dr.}$, is important only in those
systems in which scattering is anisotropic. This means that
$S_{ph.dr.}$ is only important for pure elements at temperatures well
below the Debye temperature, and for dilute alloys at low temperatures
with low residual resistivities.

The phonon drag thermopower Seebeck coefficient $S_{ph.dr.}$ is proportional to T^3 at very low temperatures. For pure elements it
reaches a peak well below the Debye temperature and then it decreases
with $S_{ph.dr.} \propto 1/T$.

Since a large peak has been found also in the Seebeck coefficient
of pure iron at $200^\circ K$, it has been proposed that not only lattice
waves or phonons, but also spin waves or magnons, contribute to
the Seebeck coefficient, which is then defined as the
"magnon-drag" effect [265]. Another possible explanation for this
effect has been given in terms of the spin-orbit coupling, as discussed in the section on the Hall effect. Molecular fields, which

line up the spin of electrons in magnetically ordered systems, should
not lead to a Lorentz force, and should therefore not affect the
orbital movement of traveling electrons. However, the extraordinary
Hall coefficient can be explained if one assumes that the molecular
field (or Weiss field) affects the traveling electron in a manner
similar to an applied field. This is called the spin-orbit coupling.
Sondheimer and Wilson [266,267] showed that applied fields lead to
a maximum in the Seebeck coefficient if plotted as a function of
temperature. MacInnes and Schroder [268] showed that the Sondheimer
equation and the concept of spin-orbit coupling leads to a maximum
in the Seebeck coefficient, and they obtained good agreement between
various physical parameters, such as the molecular field, Hall coefficient, thermal conductivity et cetera and the measured Seebeck
coefficient of iron, using Sondheimer's correlations.

The heat flow in metals due to a temperature gradient for systems with isotropic scattering ($T > T_D$, or sufficiently high residual
resistivities, typically $\rho_{re} > 1$ $\mu\Omega$ cm) is predominantly due to a
diffusion of electrons. If "hot" electrons diffuse with a different
speed through the lattice than "cold" electrons, an electric field
gradient will be produced. The electron current and the energy flow
(or heat flow) can be described with two equations of very similar
form. A calculation based on Eq. 4-67 and 68 leads to the following equation for the "diffusion Seebeck coefficient":

$$S_{diff.} = -(\pi^2 k_B^2 T/3|e|)[\partial \ln \sigma (E)/\partial E]_{E_F} \quad . \tag{4-78}$$

$S_{diff.}$ is independent of the detailed shape of the Fermi surface,
provided that $T > T_D$, or that the residual resistivity is not too
small. Naturally, T has to be small compared with the degeneracy
temperature E_F/k_B.

The electrical conductivity can be measured easily, and also the
Seebeck coefficient. Therefore, one may suspect that Eq. 4-78 could
be used to obtain information on the energy dependence of the electron transport process. One needs a specific model for the energy

dependence of the electron transport process. The electrical conductivity of metals is given by Eq. 4-20d. The energy dependence of
σ depends predominantly on the electron mean free path ℓ and the
area of the Fermi surface A_F. One can write:

$$\sigma \propto \ell A_F. \qquad (4\text{-}79)$$

This gives:

$$S = -(\pi^2 k_B^2 T/3|e|)(\partial \ln A_F/\partial E + \partial \ln \ell/\partial E). \qquad (4\text{-}80)$$

The mean free path for fast high energy electrons is expected to be
larger than that of slow electrons. Therefore, $\partial \ln \ell/\partial E$ should be
positive. Assuming an energy dependence of:

$$\ell \propto E^s \qquad (4\text{-}81a)$$

gives:

$$\partial \ln \ell/\partial E = s/E. \qquad (4\text{-}81b)$$

The surface area of a spherical Fermi surface is:

$$A_F = 4\pi k_F^2 \qquad (4\text{-}81c)$$

where k_F is the electron wave number at the Fermi energy level. This
gives $\partial \ln A_F/\partial E = 1/E$. One obtains therefore for the Seebeck coefficient for an electron system with a spherical Fermi surface:

$$S = -(\pi^2 k_B^2 T/3|e|)(s + 1)/E_F. \qquad (4\text{-}82)$$

If the energy dependence is predominantly given by the energy dependence of the Fermi surface rather than the mean free path, this
equation simplifies to:

$$S = - (\pi^2 k_B^2 T/3|e|)/E_F. \tag{4-83}$$

For a parabolic energy band with the origin at E_o, one obtains in the same way:

$$S = -(\pi^2 k_B^2 T/3|e|)/(E_F - E_o). \tag{4-84}$$

For "hole conduction" one obtains with $E_F < E_o$ positive $S_{diff.}$ values with this equation, whereas a spherical electron band (Eq. 4-83) gives negative $S_{diff.}$ values. The diffusion thermopower for a multi-band system is given by:

$$S = -(\pi^2 k_B^2 T/3|e|) \ \partial[\ln \sum_i \sigma_i]/\partial E$$

$$= -(\pi^2 k_B^2 T/3|e|) \ \sum_i [(\sigma_i/\sigma) \ \partial \ln \sigma_i/\delta E] \tag{4-85}$$

$$= \sum_i (\sigma_i/\sigma) \ S_i$$

where:

$$S_i = -(\pi^2 k_B^2 T/3|e|) \ \partial \ln \sigma_i/\partial E. \tag{4-86}$$

In the frequently used two-band model, one obtains:

$$S = (\sigma_1/\sigma) \ S_1 + (\sigma_2/\sigma) S_2. \tag{4-87}$$

Assuming for each band a parabolic energy dependence, one obtains:

$$S = -(\pi^2 k_B^2 \ T/3|e|) \ [1/(E_F - E_{o_1}) + 1/(E_F - E_{o_2})]. \tag{4-88}$$

provided the energy dependence of the mean free path can be neglected. This equation shows that one cannot neglect the contribution to the diffusion thermopower from a subband with a small number of carriers, since $E_F - E_{o_1}$ can become quite small. Comparing this equation with the corresponding equation for the Hall coefficient shows (Eq. 4-74a)

that the contribution of each band in the case of the Seebeck coefficient is proportional to the fraction of the current carried in
each band, whereas in the case of the Hall coefficient, it is proportional to the square of this term. It is therefore possible
that the hole contribution to the current in copper due to the necks
of the Fermi surface near contact points with the first Brillouin
zone affects the thermopower of copper markedly, even if it has
only a small effect on the Hall coefficient.

In transition elements, where one frequently neglects the conduction in the d-band since the d-electrons have a very high effective mass, the energy dependence of the conduction process should be
affected markedly by the mean free path energy dependence since one
frequently assumes:

$$\ell \propto 1/N(E).$$
(4-89)

This gives:

$$S = -(\pi^2 k_B^2 T/3|e|) \, [\partial \ln A_F/\partial E - \partial \ln N(E)/\partial E]$$
(4-90)

It is difficult to predict which of the two terms inside the bracket
will dominate. One usually assumes a wide conduction band and a
narrow d-band. Then $|\partial \ln N(E)/\partial E|$ is expected to be larger than
$|\partial \ln A_F/\partial E|$.

The thermopower of semiconductors shows a different temperature
dependence than that found in metals because one has to replace the
Fermi-Dirac statistic by the Boltzmann statistic. If one assumes
for this case that the energy dependence of the relaxation time is
given by $\tau(E) \propto E^\alpha$, then one obtains for a one band system:

$$S = (k_B/e)\{(\alpha + 5/2) + E/k_B T\} \, .$$
(4-91)

The first term of Eq. 4-91 gives a thermovoltage of the order of
millivolts, not microvolts as found for metals. The effect of the

second term increases with decreasing temperature. For a two band
system with holes and electrons, this equation changes for a non-
degenerate semiconductor to:

$$S = k_B[(\frac{5}{2} + r + \ln\frac{p}{p_o})p\mu_{ho.} - (\frac{5}{2} + r + \ln\frac{n}{n_o})n\mu_{el.}]/\sigma \qquad (4-92)$$

where n_o is given by $2(2\pi m_{el.}kT/h^2)^{3/2}$, and p_o is $2(2\pi m_{ho.}kT/h^2)^{3/2}$.
This equation shows that the Seebeck coefficient should be positive
if the current is carried predominantly by holes and negative if
carried predominantly by electrons.

B. Metallic Elements

The absolute Seebeck coefficient of noble metals and some transition elements is plotted in Fig. 4-61 as a function of temperature.
Noble metals have positive S values. At high temperatures their
Seebeck coefficients are frequently proportional to the absolute
temperature. This linear dependence is expected from Eq. 4-78 for
the diffusion thermopower. At low temperatures the S(T) functions
have maxima, associated with the phonon-drag effect. The Seebeck coefficient in this temperature range can be markedly affected by

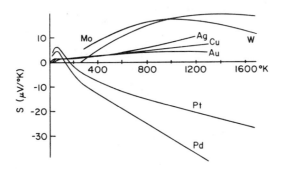

FIG. 4-61. Seebeck coefficient of noble metals and of some transition elements. After Ref. 269.

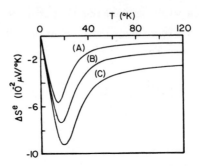

FIG. 4-62. Change of the Seebeck coefficient ΔS^e due to lattice vacancies in gold versus temperature at the following vacancy concentrations: (A): 0.0004 at.%; (B): 0.0087 at.%; (C): 0.0142 at.%. After Ref. 270, with permission.

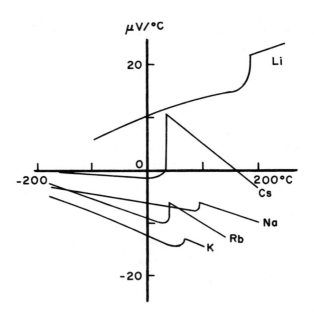

FIG. 4-63. Absolute Seebeck coefficient of alkali metals. After Ref. 271.

lattice defects. Figure 4-62 shows the influence of vacancies on the Seebeck coefficient of gold.

Data for group AI elements are given in Fig. 4-63. Again, the approximately linear temperature dependence of S(T) is noticeable above the liquid nitrogen temperature range. Peaks typical for phonon-drag effects are found at lower temperatures (Fig. 4-64). The abrupt change of S above room temperature is due to a first order phase transition (solid-liquid).

One knows that the Fermi surfaces of IA elements and IB elements are nearly spherical. One would therefore expect that,if simple models are sufficient to give information on the Seebeck coefficient, (Eq. 4-83) should give the correct order of magnitude for S of these elements. This is found for sodium and potassium, but not for lithium, copper, silver, or gold. The predicted approximately linear temperature dependence of S is frequently found. It was possible to

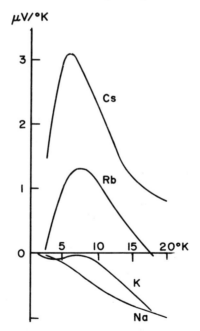

FIG. 4-64. Thermoelectric power of alkali metals at low temperatures. After Ref. 272.

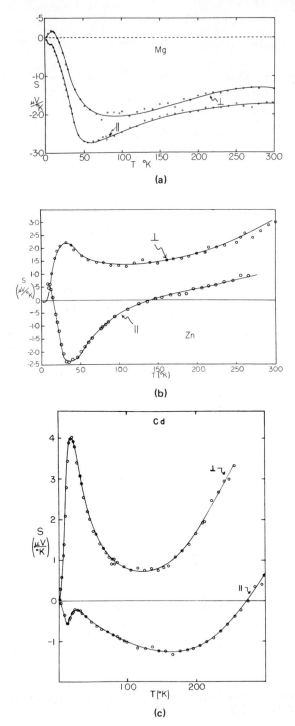

FIG. 4-65. Thermoelectric power of Mg, Zn, and Cd. ∥ and ⊥ give the component parallel and perpendicular to the principle axis, respectively. Reproduced with permission from Ref. 273.

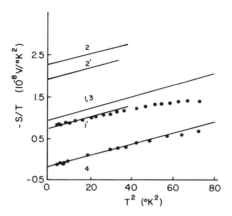

FIG. 4-66. The phonon drag contribution to the thermopower of aluminum. Data points are given only for two samples. S/T is plotted versus T^2. After Ref. 274.

show with the pseudopotential theory that the mean free path of electrons can decrease with increasing energy. This can give positive S-values for noble metals.

The thermopower of AII and BII elements like magnesium, zinc, and cadmium is orientation dependent. Figure 4-65 gives results. These elements are compensated. The number of electrons is equal to the number of holes. The Seebeck coefficient should be the difference of the Seebeck coefficients of each band times the fraction of the current in electron and hole subbands. Small changes in the mobility of electrons and holes with temperature and crystal orientation can lead to changes in the sign of the Seebeck coefficient, as found for all three elements.

Figure 4-66 gives S/T of several aluminum samples. S/T is plotted as a function of T^2. Since $S = S_{ph.dr.} + S_{diff.}$, $S_{ph.dr.} = C'T^3$ and $S_{diff.} = C''T$, a plot of S/T versus T^2 should give a straight line. The slope should be equal to C', and the intercept with the S/T-axis should be equal to C''. The S/T values at very low temperatures follow parallel straight lines, indicating that the phonon drag effect is similar in all samples, but that the diffusion thermopower changes from sample to sample. This seems to imply that the

phonon drag effect is less influenced by impurities than the diffusion thermopower.

Equation 4-90 may be used for simple transition elements in
which the conduction electrons are found only in one band and where
the conduction electrons can be scattered into another band, usually
the d-band with states for d-electrons of very high mass. These d-
electrons do not contribute noticeably to the current. One would
expect in typical transition elements and their alloys that the
energy dependence of N(E) is much larger than the energy dependence
of the Fermi surface area, since the d-band is much narrower than
the conduction band. One would therefore try to analyze the thermopower of transition elements with the equation:

$$S = (\pi^2 k_B^2 T/3|e|) \, \partial \ln N(E)/\partial E. \tag{4-93}$$

$\partial \ln N(E)/\partial E$ for nickel, palladium and platinum should be negative
because the energy band for the d-electrons is nearly filled. This
would give negative Seebeck coefficients above the Debye temperature
as found in experiments (see Figs. 4-61 and 4-67).

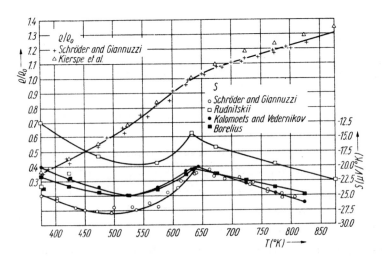

FIG. 4-67. Resistivity and thermoelectric power of nickel between 375 and 825°K. See Ref. 275 for details.

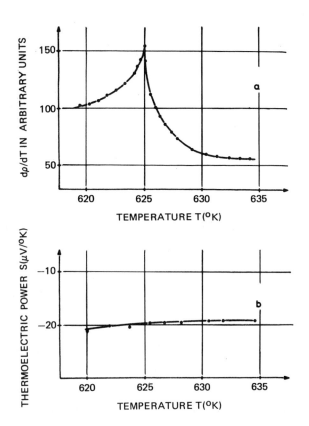

FIG. 4-68. Variation of the absolute Seebeck coefficient (b) and dρ/dT (a) with temperature for a high purity nickel sample. A second sample showed near the critical temperature a sharp minimum and maximum, with changes in S from -24 μV/°K to -17μV/°K within 1°K. After Ref. 276, with permission.

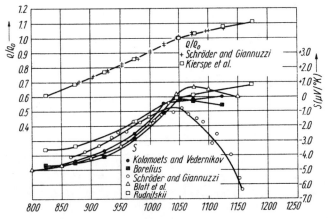

FIG. 4-69. Resistivity and Seebeck coefficient of iron. After Ref. 275.

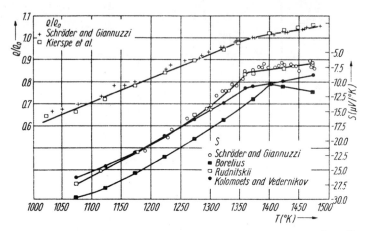

FIG. 4-70. Resistivity and thermopower of cobalt [275].

Magnetic second order transformation leads to a change in the
slope of S(T) as Figs. 4-67, 4-69 and 4-70 with the data on iron,
cobalt, and nickel show. In all these cases, S decreases with ordering with respect to the same temperature-extrapolated
paramagnetic S values. The behavior of S near the critical temperature depends markedly on the state of the sample. Figure 4-68
gives S of one high purity nickel sample. This sample shows a
smooth S vs composition curve. A second sample showed a sharp minimum and maximum with changes of S from -24 $\mu V/^{o}K$ to -17 $\mu v^{o}K$ within
$1^{o}K$. The critical exponent λ for the resistivity was in the first
case zero for both T > T_{c} and T < T_{c}. However λ = 0.95 for T > T_{c}
for the second sample and λ = 0 for T < T_{c}.

Antiferromagnetic chromium is a compensated element with an equal
number of electrons and holes. It seems appropriate to explain its
Seebeck coefficient in first approximation with a two band model
where most of the current is carried by holes of higher mobility
than the electrons, since the Hall coefficient S is positive. One
would therefore expect from Eq. 4-87 that the Seebeck coefficient
of chromium is positive, in good agreement with experimental results.

It is disappointing that Eq. 4-78 together with simple models
fails so frequently to predict the correct sign of the Seebeck

coefficient. One may therefore wonder if one cannot obtain some
direct information on possible energy changes of the electronic
transport process from Seebeck coefficient data and the resistivity
without using special models. Experiments on ferromagnetic and antiferromagnetic materials show that there is a systematic correlation
between changes in the electrical resistivity, changes in the Seebeck
coefficient, and changes in the magnetic structure. The onset of
ferromagnetic ordering leads in iron, cobalt and nickel to a decrease in both resistivity and the Seebeck coefficient, and antiferromagnetic ordering in chromium leads to an increase of both
these properties.

It has been proposed that it may be possible for second order
type ordering phenomena to develop the Seebeck coefficient in a
power series near the critical temperature. These plausibility argu-
ments lead to the correlation $\Delta S = (\pi^2 k_B^2 T/3|e|)(\Delta\rho/\rho)^2/\Delta E_F$. Both
ΔS and $\Delta\rho$ are determined by extrapolating the Seebeck coefficient
and the resistivity from the paramagnetic region into the magnetically ordered region. ΔS and $\Delta\rho$ are the difference between these
two properties at a given temperature in two states, which differ
by ΔE_F in their Fermi energy levels. One may wonder if it is
justifiable to develop S in a power series near the critical temperature. However, the equation for ΔS gives reasonable values
for shifts in the Fermi energy level [275]. The equation for ΔS
even gives a good value for ΔE_F for Constantan (a copper-nickel
alloy) where both ΔS, ΔE_F and $\Delta\rho$ can be obtained, if one assumes
that the Fermi energy level shifts due to an applied magnetic field
by about $\Delta E_F \simeq \mu_B H$ [277,278].

C. Semiconductor Elements

It is more difficult to obtain reproducible Seebeck coefficient
data in semiconductors than in metals (see Fig. 4-71). It was found
for instance, that the thermopower of a germanium sample tested in

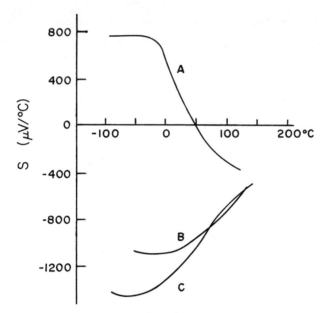

FIG. 4-71. The effect of heat treatment on S of germanium sample
21V. B: initial state; A: after 800°C in vacuum for four hours; C:
after holding the sample an additional 18 hours in vacuum at 500°C.
After Ref. 279, with permission.

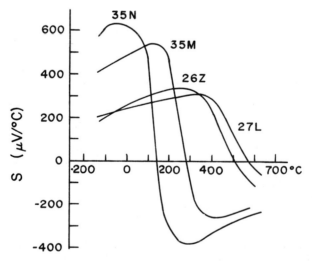

FIG. 4-72. S of four polycrystalline, Al-doped germanium samples.
The number of hole/cm^3 at exhaustion is $5.6 \cdot 10^{15}$ for sample 35N,
$1.5 \cdot 10^{17}$ for 35M, $4.1 \cdot 10^{18}$ for 26Z, and $7.0 \cdot 10^{18}$ for 27L. After Ref.
279, with permission.

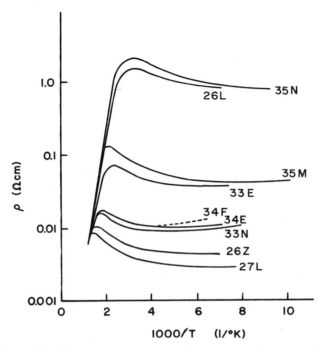

FIG. 4-73. Electrical resistivity curves of germanium samples for which S-curves are given in Figs. 4-72 + 75. After Ref. 279, with permission.

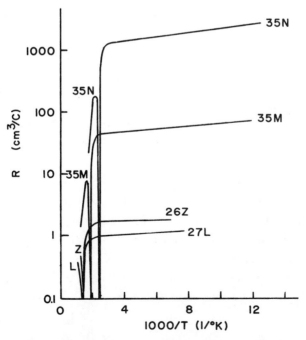

FIG. 4-74. Hall coefficients of p-type germanium samples. S-curves are given in Fig. 4-72. $R < 0$ for $T \to \infty$, $R > 0$ for $T \to 0$[279].

the as received condition at 0°C was −1 mV/°C. After holding the
sample in a vacuum for four hours at 800°C it was +0.8 mV/°C, and
after holding the sample an additional 18 hours at 500°C, again in
a vacuum, the thermopower at room temperature had a value of 1.3
mV/°C. This indicates that a high temperature heat treatment can
change the energy states of the defect structure.

     Figure 4-72 gives the thermopower of several germanium samples
doped with aluminum. Figure 4-73 gives their resistivities, and
Fig. 4-74 their Hall coefficients. Figure 4-75 gives the thermopower for antimony doped samples. There is a close correlation between Hall effect (and conductivity) measurements and the Seebeck
coefficient measurements in aluminum doped germanium samples. The
sign of the thermopower is the same as the sign of the Hall coefficient in most temperature ranges. The positive Seebeck coefficient
of the sample with $5.6 \cdot 10^{15}$ holes at exhaustion (sample 35 N)
becomes negative at 410°K. The Hall coefficient of this sample
changes its sign close to 400°K, as Fig. 4-74 shows. The minimum
in the Seebeck coefficient curves is again always close to the onset of the exhaustion range for the Al-doped samples. The Seebeck
coefficient of all antimony-doped samples is negative over the complete temperature range, just as the Hall coefficient is (see Fig.
4-76). Even the heat treatment of the germanium sample has a similar effect on the thermopower and the Hall coefficient (Figs. 4-71
and 4-77) where the change in sign of the Seebeck effect at about
320°K has a corresponding change in sign of the Hall coefficient at
270°K. Naturally, one would not expect perfect coincidence in temperature because the Seebeck coefficient of a multiband system is a
differently weighted sum of the Seebeck coefficients of the individual bands than the weighted sum of the Hall coefficients. However,
the general features follow the expected pattern of a simple model:
the dominant charge transport by electrons leads to negative Hall
and Seebeck coefficients and dominant hole transport leads to positive S and R_H values. Similar results are found for silicon
samples [279].

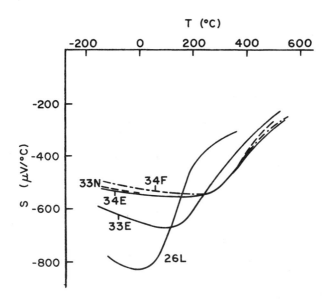

FIG. 4-75. S-curves of polycrystalline, Sb-doped germanium samples. The number of conduction electrons/cm^3 at exhaustion is: $3.4 \cdot 10^{15}$ for sample 26L; $9.6 \cdot 10^{16}$ for 33E; $5.8 \cdot 10^{17}$ for 34E; $7.1 \cdot 10^{17}$ for 34F; and $7.4 \cdot 10^{17}$ for 33N. After Ref. 279, with permission.

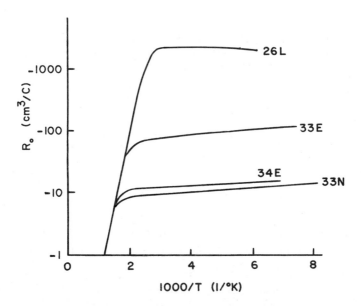

FIG. 4-76. Hall coefficient curves for n-type germanium samples for which the thermoelectric power curves are given in Fig. 4-75. Sample 34F, not given in this diagram, is close to 33N. All Hall coefficients are negative in this diagram. After Ref. 279, with permission.

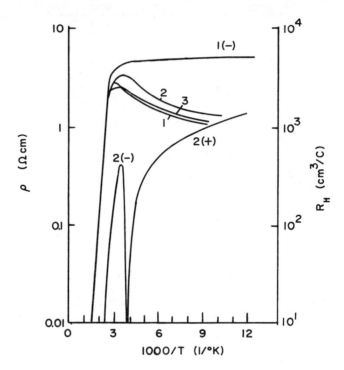

FIG. 4-77. The resistivity (1,2,3) and Hall curves [1(-), 2(-) and 2(+)] of heat-treated germanium sample 21V$_a$. Curves 1, 2 and 3 correspond to the conditions B,A,C, respectively, given in Fig. 4-71. After Ref. 279.

D.   Semiconducting Compounds

Figures 4-78 and 4-79 show the Seebeck coefficient of InSb and InAs compounds. The Hall coefficient of such compounds shows, if doped for p-type characteristics, the typical change in the sign of the Hall coefficient. This gives for samples in Fig. 4-51 a transition from predominantly electron to predominantly hole transport between room temperature and 800°K. The same results are found qualitatively for the InAs compound samples.   The n-type samples always have negative Seebeck coefficients while the p-type samples have negative Seebeck coefficients only at high temperatures.   The n-type InSb sample in Fig. 4-78 has again only negative S values.

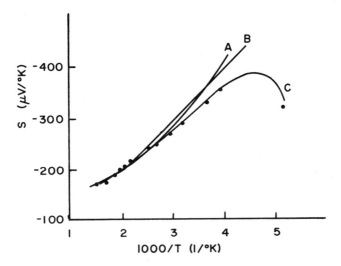

FIG. 4-78. Thermopower of n-InSb. A; Theory. B and C experimental results. After Ref. 107.

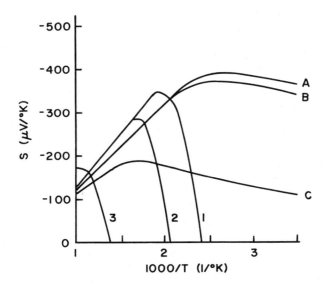

FIG. 4-79. Temperature dependence of the absolute thermopower of InAs samples. A to C give experimental results for three n-type samples, and 1 to 3 give experimental results for p-type samples. After Ref. 107.

A calculation of the Seebeck coefficient by Ehrenreich again gives
good agreement with experimental data.   He took both polar and elec-
tron-hole scattering into account.

E. Non-transition Element Alloys

As shown in section B, it is difficult to analyze the thermopower of pure elements with simple models.   One may therefore suspect that the analysis of the Seebeck coefficient of alloys is even
more complicated.   There one may expect additional composition effects.   Surprisingly, the interpretation of the thermopower of
alloys can frequently be given with simple rules, indicating that an
alloy is a more "ideal" substance for thermopower effect studies
than pure elements.   This may be due to the fact that scattering of
electrons in alloys is more isotropic than in pure elements.   It is
also possible that, for example, the Fermi surface of pure Au-Ag
alloys is more spherical than the Fermi surface of the pure elements.
Small amounts of impurities change the resistivity of a metal in
a systematic way.   This is given by Matthiessen's rule $\rho_{alloy} =$
$\rho_{la.} + c\, \partial\rho/\partial c = \rho_{la.} + \rho_{re.}$. $\partial\rho/\partial c$ is in first approximation both
composition and temperature independent.   One can derive with this
equation the following relation to describe the effect of small
amounts of impurities on the thermopower.

$$S = (\pi^2 k_B^2 T/3|e|)\partial \ln(\rho_{la.} + \rho_{re.})/\partial E$$

$$= S_F(\rho_{la.}\zeta + \rho_{re.}\Delta\zeta)/(\rho_{la.} + \rho_{re.})$$

(4-94)

with

$$\zeta = -E_F(\partial \ln \rho_{la.}/\partial E)_{E_F} \quad ,$$

$$\Delta\zeta = -E_F\, \partial \ln \rho_{re.}/\partial E \quad ,$$

$$S_F = \pi^2 k_B^2 T/3|e|E_F \quad .$$

These equations are equivalent to the Gorter-Nordheim relation:

$$S = S_{im.} + (\rho_{la.}/\rho)(S_{la.} - S_{im.}) \tag{4-95}$$

if one sets:

$$S_{la.} = (\pi^2 k_B^2 T/3|e|)\zeta/E_F$$

$$S_{im.} = (\pi^2 k_B^2 T/3|e|)\Delta\zeta/E_F . \tag{4-96}$$

In these equations, S is the measured Seebeck coefficient of the alloy, ρ is the measured resistivity of the alloy, $S_{la.}$ and $\rho_{la.}$ are the corresponding properties of the host metal in its pure state (the subscript "la." stands for lattice), and $S_{im.}$ and $\rho_{im.} = \rho_{re.}$ would be values associated with a given impurity atom. $S_{im.}$ and $\rho_{im.}$ are not the properties of the impurity atom in its bulk state, but in its state in the alloy.

Equation 4-94 predicts for non-dilute alloys, if $\rho_{im.}$ is proportional to c and large compared with $\rho_{la.}$ (more accurately $\rho_{la.}\rho \ll \rho_{re.}\Delta\rho$), that S of alloys will be a rather flat function of composition. This is found, for instance, in silver-gold alloys, as shown in Fig. 4-80. The Seebeck coefficient is nearly constant for alloys with 30 to 80 at.% silver. In these alloys, where S has values between -1 $\mu V/°K$ and -2 $\mu V/°K$, the Seebeck coefficient is very close to the value predicted for a system with a nearly spherical Fermi surface and one electron per atom.

One would expect that ordering influences the thermoelectric power, because S depends sensitively on the Fermi surface and the scattering process of conduction electrons. New Brillouin zones are formed during ordering. Experimentally measured S values between 180°K and 1000°K of ordered and disordered Cu_3Au are given in Fig. 4-81. It shows that the disordered alloy has the same sign of S as Au and Cu; however, S is smaller in the alloy. Ordering changes the sign of S, just as it changes the sign of the Hall coefficient. The positive sign of S for the disordered alloy

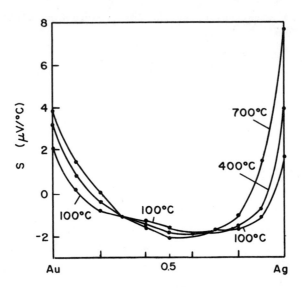

FIG. 4-80. Isotherms of the Seebeck coefficient of Au-Ag alloys.
After Ref. 271.

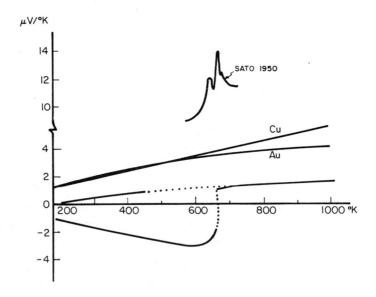

FIG. 4-81. Absolute Seebeck coefficients of Cu, Au, and AuCu$_3$.
The data of AuCu$_3$ are given by the combination of full lines and dots,
and are given on a sample in the quenched state (upper curve) or in
the equilibrium state (lower curve). After Ref. 280, with permission.

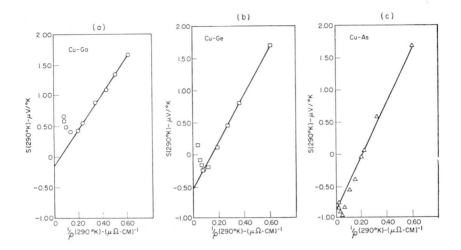

FIG. 4-82. Gorter-Nordheim plots at 290°K for solid solutions: gallium (a), germanium (b) and arsenic (c) in copper. After Ref. 281.

indicates that disordered Cu$_3$Au, copper, and gold probably all have similar Fermi surfaces. The Fermi surfaces of copper and gold are in most sections spherical but have necks which contact the hexagonal faces of the first Brillouin zone. Ordering in Cu$_3$Au produces a new Brillouin zone. The Fermi surface containing one electron per atom should intersect these new Brillouin boundaries of the ordered lattice. There should be both electron and hole conduction (see Fig. 1-61). The current carried by holes should be larger than for the disordered alloys where only the "necks" give this type of current contribution. However, the complex Fermi surface of ordered Cu$_3$Au makes it very difficult to predict the value of the Seebeck coefficient, since electrons and holes in four bands have to be considered.

The Nordheim-Gorter relationship (Eq. 4-95) predicts that S is a linear function of 1/ρ. This relationship has been tested on alloys of noble metals with impurities of higher valency. Figure 4-82 shows that S of copper alloys with germanium and gallium usually follows the Nordheim-Gorter rule. S is larger than expected from a straight line fit only for low 1/ρ values. This result seems to be different for CuAs. However, if one neglects the highest S-value, one may draw a straight line through most experimental data excluding again S-values associated with low $1/\rho$ values.

It is surprising that the Gorter-Nordheim plot can give a linear correlation between S and $1/\rho$ over a larger composition range than found for Matthiessen's rule in the resistivity of dilute alloys, in spite of the fact that the Gorter-Nordheim equation is derived with Matthiessen's rule.

It has been proposed that the impurity resistance $\rho_{im.}$ is proportional to [282]:

$$\rho_{im.}(E) = [(2\ m)^{1/2}\ n_{im.}/n_o\ e^2]E^{1/2}\ Q_{im.}(E) \qquad (4-97a)$$

where $Q_{im.}(E)$ is the scattering cross section of an impurity, and $n_{im.}/n_o$ is the ratio of impurity atoms to the number of free electrons. $Q_{im.}(E)$ is given by:

$$Q_{im.}(E) = 2\pi(\Delta z\ e^2/m\ v_F^2)^2 \ln(1 + 1/Y) - 1/(1 + Y) \qquad (4-97b)$$

Δz is the number of outer electrons per impurity atom, v_F the Fermi velocity, and $Y = \hbar^2/4\ m^2 v_F^2$. r_o is the screening radius of the screening potential. Developing $Q_{im.}$ in a power series of E gives [282]:

$$Q_{im.} = Q_o \times [1 + \alpha_1(E - E_o) + \alpha_2(E - E_o)^2 +] \qquad (4-97c)$$

S of a hypothetical copper alloy depends on α_1, (α_2 is assumed to be zero to reduce the number of arbitrary constants). The fit of a theoretical curve to experimental data of S is quite good for copper alloys with silicon, tin and zinc (4-83a and b). In all these alloys α_1 is negative, indicating that the mean free path of conduction electrons increases with increasing energy. This result is explained by Domenicali and Otter [282] with scattering on screened charge clouds round impurity atoms. A calculation shows that α_1 is a function of the screening radius and Fermi energy (4-83c).

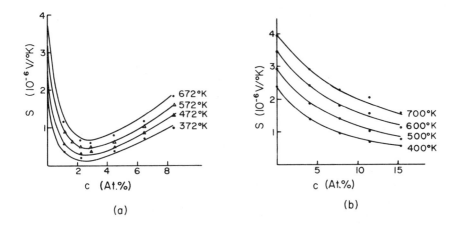

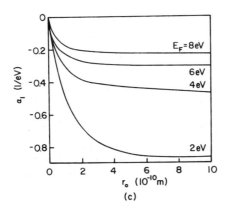

FIG. 4-83. Thermopower as a function of Si-concentration for Cu-Si alloys (a). Solid curves are theoretical, triangles and circles give experimental results. Thermopower as a function of Zn-concentration for Cu-Zn alloys (b). Solid curves are theoretical, dots are experimental results. First scattering parameter α_1 versus effective radius r_0 for various Fermi energy levels E_F (c). After Ref. 282.

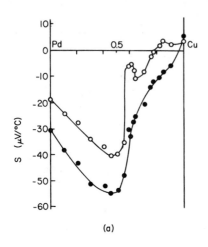

(a)

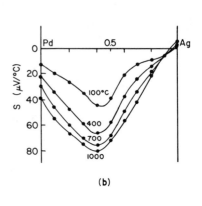

(b)

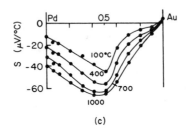

(c)

FIG. 4-84. Isotherms of the absolute thermoelectric power of palladium-gold alloys (a), [open circles at 300°C, full circles at 700°C]; palladium-silver alloys (b); and palladium-gold alloys (c). After Ref. 271.

F. Noble Metal and Group VIII Element Alloys

The thermoelectric power of the noble metals copper, silver, and gold with the transition elements palladium and platinum is given in Fig. 4-84 as a function of composition. Palladium forms a complete

series of solid solutions with noble metals at elevated temperatures.
Palladium-copper alloys show superlattices for Pd_3Cu_5 and $PdCu_5$.
The disordered alloys are metastable at low temperatures. The Seebeck
coefficient of these disordered alloys all show a similar el./at.
dependence. This indicates again that a rigid band model may be
appropriate to describe their properties. The pronounced minimum
in S occurs between 50 and 60 at.% palladium in these alloys.
This is close to the concentration where the d-band is filled, as
shown by magnetic susceptibility and low temperature electronic
specific heat measurements. These measurements indicate that the
d-band is filled for el./at. $\gtrsim$ 10.5. Open d-electron states are found
for lower el./at. values. One would expect in a simple model that the
Seebeck coefficient S would be determined essentially by the energy dependence of the conduction electron band Fermi surface
in palladium-noble metal alloys with less than about 50 at.% Pd, and
not by the energy dependence of the scattering process, since the
d-band is filled. If one assumes that this Fermi surface is spheri-
cal, then $\partial \ln A_F / \partial E$ would be positive and would be the dominant
term in Eq. 4-90. This would yield a negative Seebeck coefficient
for noble metal-rich alloys, about -1 $\mu V/°K$. In palladium-rich
alloys, the term $\partial \ln N(E)/\partial E$ would dominate over the $\partial \ln A_F/\partial E$
term. Since, in a rigid model, $\partial \ln N(E)/\partial E \sim \partial N(E)/\partial c$, one would
expect that $\partial \ln N(E)/\partial E$ is negative. This gives a negative contribution to S in Eq. 4-90. If the density of states N(E) increases linearly with palladium concentration in palladium rich
alloys, $\partial \ln N(E)/\partial E$ would be constant. A parabolic increase of
N(E), as suggested from low temperature specific heat measurements,
would give a maximum in $|\partial \ln N(E)/\partial E|$ at el./at. $\approx$ 10.5. This simple
model gives, as Fig. 4-84 shows, qualitatively the correct composi-
tion dependence of the Seebeck coefficient of palladium-noble metal
alloys at 100°C, provided one excludes noble metal alloys with only a
few percent palladium. It predicts negative S values for all
alloys, a minimum of S at the equiatomic composition and very low
S values for alloys with more than 50 at.% palladium. Nearly

constant S values would be expected from this model for alloys
with 10 to 40 at.% palladium (the noble metal alloys with less 10
at.% palladium have been excluded from our discussion). However, S
decreases for the noble metal-rich alloys smoothly with increasing
palladium concentration. This decrease is more pronounced with
high than with low temperatures.

A more quantitative approach to determine the Seebeck coefficient of palladium rich alloys has been given by Kimura and Shimizu
[283]. They started with the model used by Coles and Taylor [226],
who calculated the electrical resistivity as a function of composition and temperature. It was assumed in the calculation by Coles
and Taylor that only s-electrons are responsible for the current
but that the scattering mechanism involves both s-s and s-d transitions. This model leads to an equation which describes the experimentally found ρ and $\partial\rho/\partial T$ results adequately, if one uses N(E)
values as determined from low temperature specific heat values.

The same model is used to calculate the Seebeck coefficient of
these alloys (see Fig. 4-85). Kimura and Shimizu [283] assume that
the conduction electrons are similar to free electrons with the
effective mass of silver. The Mott model is used to determine electron scattering into s- and d-states. Equation 4-78 shows that the
Seebeck effect can be determined if the energy dependence of the resistivity is known. However, S as given in Fig. 4-84 as full lines
is not calculated with this equation because the development of S
in a power series of $k_B T/E_F$ together with the low temperature
approximation of the Fermi-Dirac statistic may not be justified for
these transition elements; instead S is described by integrals.
The results of the calculation agree qualitatively with experimental data at high temperatures. It may be possible to improve the
correlation between theory and experiment, if one takes conduction
by holes into account. This was neglected in the resistivity calculation. As the two band Eq. 4-87 shows, the measured S value
should be an average of $S_{el.}$ and $S_{ho.}$, where "el." stands for
electron band and and "ho." for the hole band. The contribution of
the hole band should be of opposite sign to that of the electron

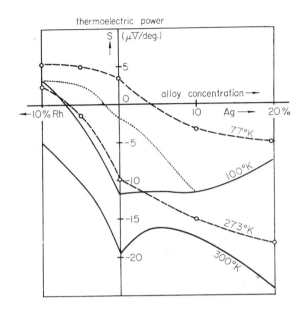

FIG. 4-85. Thermoelectric powers of Pd-Rh and Pd-Ag alloys as functions of alloy composition. The full curves are the calculated results for T = 100°K and 300°K. The broken curves are drawn through the experimental results by Taylor and Coles (which are indicated by circles). The dotted line indicates the hypothetical contribution of the phonon drag effect. After Ref. 283.

band. $S_{ho.}$ is of comparable magnitude to $S_{el.}$, but $(\sigma_{ho.}/\sigma)$ is, according to an estimate of Sondheimer [284], about 0.2 to 0.3 at room temperature. Therefore, the total thermopower S is about 0.4 to 0.6 of $S_{el.}$. This gives reasonable agreement between theory and experiments. A more recent calculation by Dugdale and Guenault [285] shows good agreement between calculated and measured S values at 1°K (4-86). These authors followed closely the Mott model as used by Coles and Taylor.

Deviations between experiments at liquid N_2 temperatures and calculations by Kimura and Shimizu may be partly due to the phonon-drag effect for pure metals and slightly contaminated alloyed

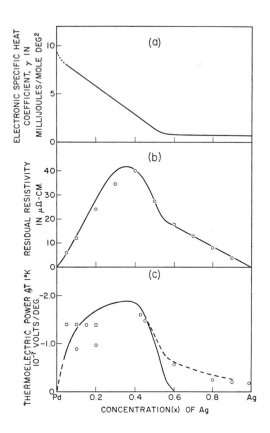

FIG. 4-86. Summary of variation of some electronic properties
with alloy concentration. Electronic specific heat coefficient (a).
Residual resistivity: circles give experimental values from Coles
and Taylor [226] (b). Thermoelectric power at 1°K (c). Experimental values by other authors are given by circles and squares. The
full line gives a theoretical estimate based on the model of Ref. 285.
The dashed line is an estimate based on the assumption that S_d remains constant even in the silver-rich alloys. After Ref. 285.

samples. The dotted line in Fig. 4-85 gives an estimate of the
phonon drag $S_{pd.}$ contribution. $S_{pd.}$ is about 10 μV/°K at 100°K
for pure palladium. It decreases nearly linearly with composition
if one adds silver or rhodium and is approximately zero at 10 at.%
Ag or 10 at.% Rh. For palladium, the largest hypothetical $S_{pd.}$ value
of 20 μV/°K should be found near 50°K.

It is possible for non-dilute alloys to avoid both the mathematical problem associated with the analysis of S in the electron
gas with a low degeneracy temperature (which makes it difficult to
use quantitatively Eq. 4-79 at higher temperatures for palladium and
platinum alloys) and the phonon drag effect by measuring S of
alloys at very low temperatures. $S_{pd.}$ is at low temperatures proportional to T^3, whereas $S_{df.}$ is a linear function in T. Therefore for T → 0, $S_{df.}$ should finally dominate over $S_{pd.}$. Naturally,
one could separate both terms by plotting S/T versus T^2, similar
to the way in which lattice and electronic specific heats are separated. Measurements on palladium and platinum alloys between 2°K
and 120°K (4-87) show that the Seebeck coefficient S has a marked
peak between 60°K and 120°K. This should be due to $S_{pd.}$. Below 5°K,
$S_{pd.}$ is usually of less importance than S_{dif}.

The measurements on Pd-alloys in the liquid He-temperature
range by Fletcher and Graig [286] are described with similar models
as the previously discussed measurements at higher temperatures.
The analysis of S of platinum alloys follows the same line. The
density of states curve for PtIr and PtAu alloys has a steep slope
for PtAu alloys, a rather flat maximum from pure Pt to $Pt_{98}Ir_2$ (the
highest $\gamma \propto N(E)$ value is found for the $Pt_{98}Ir_2$ alloy), and a shallow
decline for alloys with increasing Ir-concentration for
$0.02 \leq c(Ir) \leq 0.1$. If one assumes that S is given by Eq. 4-93,
then one would expect that S would be negative for platinum and
PtAu alloys, and positive for PtIr alloys with more than 2 at.% of
iridium. S = 0 should be expected for the alloy close to 2 at.% Ir.
The trend of experimental results agrees reasonably well with this
prediction, as Fig. 4-87 shows.

A quantitative calculation of the scattering process of platinum and palladium-rich alloys with gold impurities seems to require
the assumption that the reciprocal relaxation time is $1/\tau_s \propto c[AN(E)_s$
$+ BcN(E)_d]$, where c is the impurity concentration. The extra c
factor in the last term is introduced to take into account that
noble metal impurities of gold in platinum have tightly bound d-
shells. The d-wave functions of the matrix do not penetrate into the

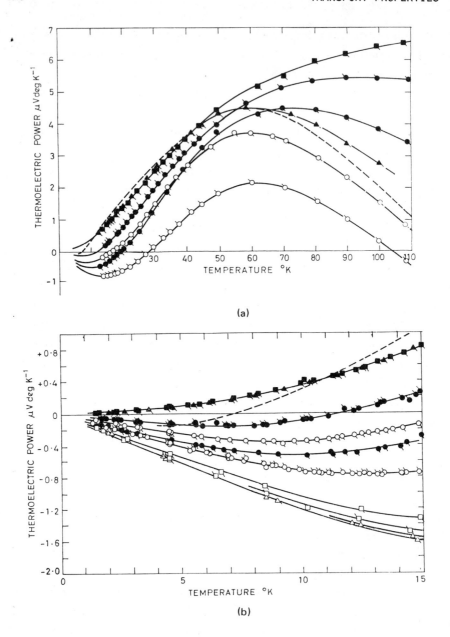

FIG. 4-87. Variation of thermoelectric power with temperature
of platinum and its alloy above 15°K (a) and below 15°K (b). Pt: ----
Q : 2 at.% Au; ð : 4 at.% Au; ⊡: 6 at.% Au; ⊡: 8 at.% Au;
⬤: 2 at.% Ir; ◖: 4 at.% Ir; ▪: 6 at.% Ir; ▪: 8 at.% Ir;
▲ : 10 at.% Ir. After Ref. 286, with permission.

gold core to any appreciable extent. s-electrons will not usually
be scattered by these cores, unless the concentration of noble metal
atoms is sufficiently large.

G. Transition Element Alloys

The multiband structure of transition elements is difficult to
evaluate from first principles. However, it is frequently possible
to correlate results of thermopower measurements with the energy
band models as derived from, e.g., electronic specific heat, magnetization and Hall-coefficient measurements. The large number of
possible alloys makes it impossible to give a systematic review.
The following discussion is therefore limited to alloys with el./at.
values close to 6, such as chromium alloys.

The thermopower of iron-vanadium alloys can be analyzed with
the same energy band model which has been used to explain the Hall
coefficient. As discussed before, this model assumes for alloys
with el./at. values close to 6 that conduction (s and p) electrons
and d-band electrons form a hybridized band for el./at. values below
the minimum in γ_o. The electrons behave as if they are in a "hole"
band which has its top at the energy E_{ho}. close to the minimum of
N(E). Therefore, if one uses as a first approximation for the See-
beck coefficient the equation for a parabolic band, $S \propto -T/(E_F - E_{ho}.)$, S should be positive and increase with el./at. below the
minimum in N(E). For el./at. values above the minimum, the d-band
electrons would be of the antibonding type. They have a high effective mass. Their contribution to the current can be neglected.
Their contribution to the Seebeck coefficient may be noticeable since
the distance between the Fermi level and the bottom of this band at
the minimum of N(E) is small. S of this band would be negative.
The same is expected from the conduction electrons which should not
hybridize with the localized d-electrons above the minimum in N(E).

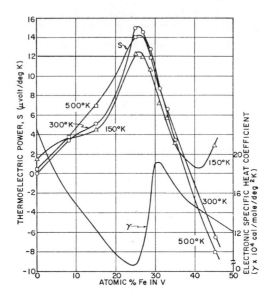

FIG. 4-88. Thermoelectric power vs. composition of vanadium and vanadium-iron alloys at three different temperatures (upper curves). γ is the liquid He temperature electronic specific heat coefficient. After Ref. 287.

One would therefore expect that the total Seebeck coefficient in this electron concentration range is negative. Naturally, one should not expect that S in the region near the minimum of N(E) will have a singularity. The band structure in this range is much too complex for a detailed description. It is reasonable to expect that the increasing positive S values below the minimum in N(E) will increase with increasing el./at. values, reach a maximum close to the minimum in N(E), and then decrease smoothly to the negative S-values above the minimum of N(E). This is found in experiments, as shown for V-Fe alloys in Fig. 4-88. This figure also gives the electronic specific heat coefficient γ. The increase in S for the alloy with about 45 at.% Fe at low temperature is associated with ferromagnetic ordering.

Not all experimental Seebeck coefficient data of transition element alloys can be explained with simple models. For instance, Cr-Fe alloys seem to have a sharp peak in N(E) near $Cr_{81}Fe_{19}$.

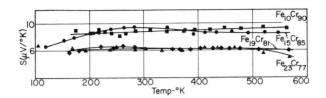

FIG. 4-89. Absolute Seebeck coefficient S of Cr-Fe alloys with 10 to 23 at.% Fe below 600°K. After Ref. 288.

dN(E)/dc, which in the rigid band model should be proportional to dlnN(E)/dE, should be positive for el./at. < 6.36, and negative for el./at. > 6.36. One should therefore expect that S values rapidly change in this composition region, but experiments (4-89) show practically no composition dependence of S for alloys with 15 to 25 at.% Fe. This would indicate that the rigid band model cannot be applied to these alloys. Optical data, as discussed in Chapter 5, confirm this view. Those measurements show that the concept of "virtual states" has to be used to explain properties of chromium-rich chromium-iron alloys.

S of ferromagnetic iron alloys has been studied only on few examples. S of iron alloys (4-90) shows a minimum-maximum near the critical temperature similar to that of pure nickel. Figure 4-91 gives S of several Ni-Co and Ni-Mn alloys. The Ni-Co and Ni-Mn samples show also a rapid change of dS/dT near the Curie temperature.

Antiferromagnetic ordering in chromium base alloys has been investigated by several authors [288]. Antiferromagnetic ordering leads to an increase in the absolute Seebeck coefficient, as shown in Fig. 4-91. This increase can be very pronounced, as data on chromium-iron and chromium-manganese samples show.

Antiferromagnetic ordering leads to the formation of an energy gap $\Delta_g E$ at the Fermi surface. This energy gap is of the order of $k_B T_N$. The first theoretical arguments by Overhauser [125] predicted $\Delta_g E \sim 3.5 k_B T_N$. Antiferromagnetic ordering leads to a decrease in the area of the Fermi surface A_F. Specific heat measurements on Cr-Mo and Cr-W alloys indicate that the total area

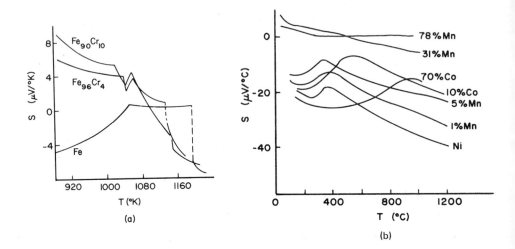

FIG. 4-90. Absolute thermoelectric power of iron and iron-chromium alloys near the Curie temperature (a). After Ref. 185. Absolute thermoelectric power of purified nickel and nickel base alloys (b). After Ref. 289.

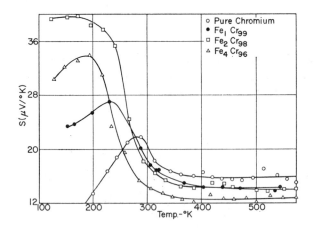

FIG. 4-91. Absolute thermoelectric power of chromium and chromium-iron alloys. After Ref. 288.

of A may decrease by 50%. If one assumes as a first approximation that the associated shift of the Fermi energy may be of the
same order of magnitude as the energy gap, $\Delta E_F \sim k_B T_N$ then one
expects terms of the order of $(\pi^2 k_B^2 T_N/3|e|)[d \ln A_F/dE]$
$\sim (\pi^2 k_B^2 T_N/3|e|)(1/2 \, k_B T_N)$ to show up for the thermopower associated
with antiferromagnetic ordering. This gives the right order of magnitude for the experimentally found increase in the Seebeck coefficient.

VI.   SUPERCONDUCTIVITY

A.   Thermodynamics of Phase Transitions

A few years after 1911, when he was able to liquify helium,
Kamerlingh Onnes found that the resistivity of lead dropped by more
than a factor of 10^4 in a small temperature interval near 4.18°K.
One now says that lead undergoes a phase transition from the normal
to the superconducting state at a critical temperature T_c. One knows
now that this phase transition is of the first order. The resistance below T_c is extremely low. Kamerlingh Onnes [290] found that
the resistance of his sample dropped from 0.08 Ω above T_c to less
than 3×10^{-6} Ω at 3°K. Attempts to determine the current decay in
Nb$_{.75}$Zr$_{.25}$ [291] with nuclear magnetic resonance measurements showed
that the decay time for the superconducting process was larger than
10^5 years.

Just as surprising as the electrical properties are the magnetic properties of superconductors. Meissner and Ochsenfeld found
[292], using high purity tin, that magnetic field lines in their
sample were pushed out of the sample during the transition from the
normal to the superconducting state. This means that this superconductor is a "perfect" diamagnetic material since "normal" diamagnetism pushes out only a small amount of magnetic field lines.
The repulsion of the field lines is called the "Meissner effect",

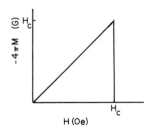

FIG. 4-92. Magnetization of a Type I Superconductor.

and one thinks now that the perfect diamagnetism associated with
superconductivity is the primary effect, and zero resistivity is
the secondary effect. Susceptibilities can be measured very easily,
so that superconductivity and critical temperatures are frequently
obtained from magnetic measurements. One finds that the critical
temperature T_c depends on the applied field and that T_c decreases with increasing field. T_c approaches zero for a critical
field H_c. Since an electric current is also associated with a magnetic field, one expects that T_c is also a function of the current
in the sample, and one defines as "critical current" a current I_c
which destroys superconductivity. T_c values from above 22°K down
to 0.01°K have been found. This lower value is set by experimental
limitations. It is presently not possible to determine superconductivity below this temperature. The question if all materials can
become superconductive at very low temperatures can therefore not be
answered.

Most superconducting elements have properties like those described above from experiments on lead and tin. These are "ideal Type I
superconductors" characterized by one critical temperature T_c (4-92).
However, some pure elements and also alloys show a different behavior. There magnetic fields can penetrate deep into the sample,
which is still superconducting. It consists essentially of two
phases, a "normal" phase with finite resistivity, and a "superconducting" phase with zero resistivity. This is a Type II superconductor. This superconductor is characterized by two critical fields.

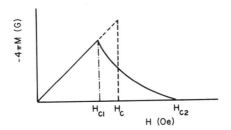

FIG. 4-93. Magnetization curve of a Type II superconductor.
The flux begins to penetrate the specimen at a field H_{c1}. The
sample is in a vortex state between H_{c1} and H_{c2}. The sample is a
normal conductor above H_{c2}.

For fields below H_{c1}, the sample behaves like a Type I superconductor, with no magnetic field lines deep in the interior of the sample
(even in a Type I superconductor field lines will penetrate into a
skin layer of the sample). Above H_{c1} and below H_{c2}, the sample will
consist of two phases, the normal and the superconducting phase, and
above H_{c2} the sample is paramagnetic or diamagnetic (see Fig. 4-93).
Figure 4-94 shows the magnetization of lead and lead alloys. Both
Type I and Type II superconductor behavior is found. One uses as
abscissa in these plots not M, but $-4\pi M$. One obtains for perfect diamagnetic material a slope of $45°$, since $\chi = M/H = -1/4\pi$ in the

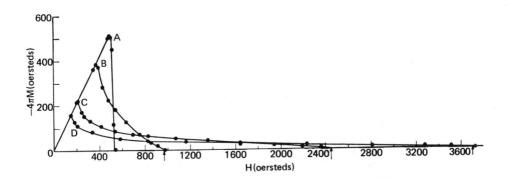

FIG. 4-94. Superconducting magnetization curves of annealed
polycrystalline lead and lead-indium alloys at 4.2°K. Lead (A);
lead -2.08 wt. percent indium (B); lead -8.23 wt. percent indium
(C); lead -20.4 wt. percent indium (D). After Ref. 157.

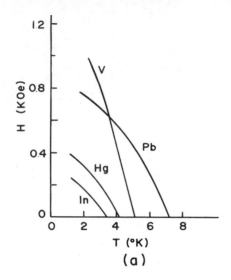

(a)

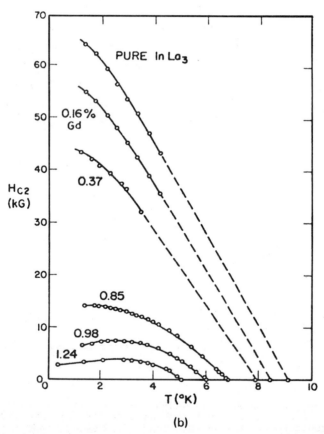

(b)

FIG. 4-95. Threshold curves of the critical field H(T) for several superconductors (a). A specimen is superconducting only below this field. After Ref. 158. Upper critical field versus temperature for InLa₃ and InLa₃–Gd$_c$ (b). After Ref. 216.

cgs system. The temperature-field dependence of the normal-super-
conducting phase transitions can be approximately given by:

$$(T/T_c)^2 + (H/H_c)^2 = 1 \qquad\qquad (4-98)$$

This is shown in Figs. 4-95a and b which give the phase diagrams of
superconducting materials.   Figure 4-95b shows H_{c2} as a function of T_c for several superconductors in bulk form.

A Type I superconductor will expel magnetic flux, since it is a
perfect diamagnetic material. Therefore, flux lines going through
a hole on a superconductor are trapped. Experiments show that flux
lines in such holes (including the penetration depth into the superconductor surrounding it) are quantized. These "fluxoids" can only
be multiples of a constant which is close to $hc/2e$. The factor of
one half in $hc/2e$ is an indication that electron pairs are responsible for superconductivity, since the original models suggested
fluxoids in multiples of hc/e. Since the magnetic field of a fluxoid will show up in a Bitter pattern, which is the pattern of a
powder of fine magnetic particles on a magnetic substrate, it is
possible to make these fluxoids visible. It has been shown that
fluxoids form a regular arrangement in thin films, just like a
planar crystal lattice. Not all electrons around a fluxoid in a
superconducting phase will move without resistance. One should
visualize the electron movement in such a way that there exists a
closed path around a fluxoid on which some electrons move without
resistance.

The transition from the normal to the superconducting state is
essentially due to changes in the electronic structure. This assumption seems to agree with all experimental evidence presently available. The jump in the specific heat at T_c (see Fig. 4-96), typical
for first order transitions, has been explained with changes in the
electronic specific heat. The specific heat in the superconducting
state, $c_{sc.}(T)$, decreases much more rapidly with temperature than
the electronic specific heat of the normal state. The $c/T = f(T^2)$

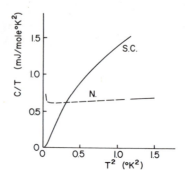

FIG. 4-96. The heat capacity of gallium in the normal (N) and superconducting (S.C.) state. The normal state (which is restored by a 200 G field) has electronic, lattice and (at low temperatures) nuclear quadropole contributions. After Ref. 293.

plot (4-96) gives results both for the superconductor in zero applied field and measurements where the applied magnetic field reduces the critical temperature drastically. For this last case, $c/T = f(T^2)$ gives the typical straight line plot except for $T^2 < 0.1°K^2$, as found for materials where the lattice specific heat is proportional to T^3, and the electronic specific heat is $\gamma_o T$. It is possible to calculate the entropy change from the normal to the superconducting state from the change in the specific heat. This would be the difference of the $\int (c/T)dT$ value of these curves. Since the entropy change of an order-disorder transformation of a system with N particles is of the order of $k_B \ln 2^N$, which is much larger than the entropy change found for the superconducting transition of a sample with N electrons, one has to conclude that only a small fraction of the conduction electrons participate in the superconductivity. A plot of the logarithm of the electronic specific heat in the superconducting state vs. the reciprocal of the absolute temperature follows a straight line (4-97). This is the typical Arrhenius plot where the slope of the line is proportional to the activation energy of the process. One would obtain such a plot if electrons are excited across an energy gap.

The probably infinitely high electrical conductivity of a superconductor has no parallel in an infinitely high thermal conductivity.

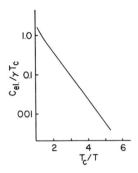

FIG. 4-97. The electronic part C_{es} of the heat capacity in the superconducting state is plotted on a log scale vs. T_c/T: the exponential dependence on $1/T$ is evident. Here $\gamma = 0.60$ mJ mole^{-1} deg^{-2}. After Ref. 293.

Experiments show that the thermal conductivity in the superconducting state can be either smaller or larger than the thermal conductivity of the normal state. No discontinuity in κ is found at T_c. However, $(d\kappa/dT)_{Tc}$ can show a discontinuity. Superconducting wires can sometimes be used as heat switches. Below T_c, κ may rapidly decrease. Since magnetic fields destroy superconductivity, magnetic fields affect κ markedly for $T < T_c(H = 0)$.

B.   BCS Theory and London Equation

The microscopic theories on superconductivity have been developed now to a state where one can understand superconductivity just as well as the electrical conductivity in the normal state, or as ferromagnetism [31]. A good deal of the properties of superconductors can be described with the theory by Bardeen, Cooper and Schrieffer (the BCS theory), which relies on an electron pair interaction with the lattice. The importance of the lattice for superconductivity was realized from the isotope effect, which shows for mercury that the critical transition temperature is proportional to the square root of $1/M$, where M is the mass of the mercury isotope under investigation: $T_c \propto M^{-1/2}$. This electron-lattice interaction can be

loosely described with a model in which one electron attracts surrounding ions . Such a system has a slightly more positive charge
than an electron surrounded by ions on undisturbed lattice positions.
The disturbed lattice section plus electron will repel another electron less than an undisturbed lattice section. In other words, the
two electrons seem to "attract" each other and form an electron
pair. The BCS theory shows that the interaction energy in such
pairs can modify noticeably the energy distribution of electrons at
low temperature. The BCS theory shows that the first excitation
state of electrons in this system is separated from the lowest state
by a finite energy gap, and experiments verify the predicted temperature dependence of this gap (see Fig. 4-98). The energy gap at
$0°K$ ε_0 is about $3.5 \cdot T_c k_B$.

   The BCS theory also gives a correlation between the density of
states at the Fermi level, $N(E_F)$, the Debye temperature T_D, and T_c:

$$T_c = T_D \exp[-1/V \cdot N(E_F)] \tag{4-99}$$

where V is the electron-lattice interaction energy. It represents
an electron attraction term. $V \cdot N(E_F)$ is much smaller than 1. This
theory leads also to the London equations, which correlate the current in the superconductor with the vector potential of the magnetic field and show that applied fields can only penetrate a thin
surface layer of a Type I superconductor. The penetration depth λ
(see Fig. 4-99) is given by

$$\lambda = (m \ c^2/4\pi \ n \ e^2)^{\frac{1}{2}} \tag{4-100}$$

which is typically of the order of a few hundred Angstroms. A
second parameter with the dimension of a length, which characterizes
a superconductor, is the coherence length, ξ. It indicates the
minimum distance over which the energy gap of the BCS theory can
change markedly. The energy gap of the electrons at the Fermi surface between lowest and first excited state is affected by magnetic

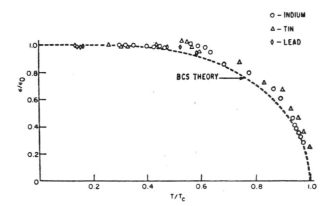

FIG. 4-98. Reduced values of the observed energy gap $\varepsilon/\varepsilon_0$ as a function of the reduced temperature T/T_c. The solid curve is drawn for the BCS theory. After Ref. 44.

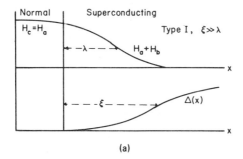

(a)

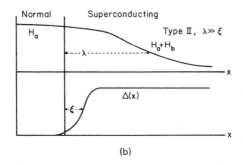

(b)

FIG.4-99. Schematic diagram of the magnetic field $H_a + H_b$ and energy gap parameter $\Delta(x)$ with position x near the interface of superconducting and normal regions for Type I (a) and Type II (b) superconductors. $\Delta(x)$ indicates the amount of electronic ordering, H_a is the applied field.

fields and can vary throughout the crystal. The BCS theory predicts for the coherence length ξ of a pure superconductor

$$\xi = 2 \hbar \nu/\pi E_g \sim 10^{-4} \text{ cm} \qquad\qquad (4\text{-}101)$$

Lattice defects and alloying reduce ξ. If $\xi \ll \lambda$, then the
material is a Type I superconductor, and $\lambda \ll \xi$ gives a Type II
superconductor.

Two phases in equilibrium are separated by an interface with a
surface energy. London pointed out that such a surface energy term
is necessary to make a Type I superconductor possible. Without
such an interface energy, the sample would break up into normal and
superconducting phases even for small magnetic fields with the applied field concentrated in the normal phase region. It would reach
high enough flux line densities to make the field larger than the
critical field. If the thickness of a thin film is λ, then the
maximum magnetic energy per unit area would be $\lambda H_c^2/8\pi = (1/2)\lambda H_c M_c$.
This is compensated by a surface energy term α. One can characterize α by a length β' (the length β' is equal to the thickness
of a layer where the magnetic energy at H_c is equal to the surface
energy). $\alpha = \beta' H_c^2/8\pi$. β' has to be larger than the penetration
depth.

C. Critical Temperatures of Elements, Compounds and Alloys

There is a close correlation between the critical temperature
of an element and its position in the periodic system, as Fig.
4-100 shows (see also Table 4-9). Noble metals show no superconductivity above 10 mºK, the detection limit. Of rare earth metals,
only lanthanum, which has a completely empty f-shell, is superconductive. Superconducting elements are found in two groups. One
contains elements with closed d-shells, and has conduction (s- plus
p-) electron concentrations between 2 and 6 per atom. The other
group contains transition elements. Ferromagnetic materials show

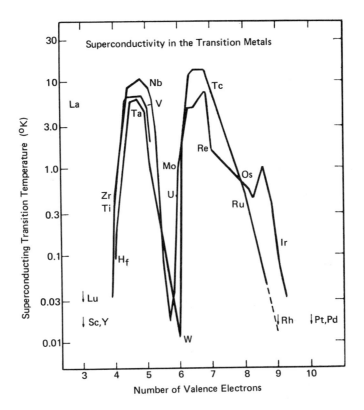

FIG. 4-100. Variation of T_c in transition metals. After Ref. 162.

no superconductivity. Equation 4-99 shows that there should be a correlation between $\gamma \propto N(E)$ and T_c. The low T_c values near el. /at.= 6 could be attributed to low $N(E)$ values of these alloys. However, for el./at.$\simeq$ 3, T_c becomes very small, in spite of the fact that γ_o is still rather large. Naturally, it may be possible to attribute low T_c values to low V values in Eq. 4-99. Figure 4-101 gives a plot of V of elements, as calculated from Eq. 4-99 and known experimentally determined T_c, T_D, and γ_o values.

Figure 4-102 gives a plot of the superconducting coupling strength, $g = V \cdot N(E)$, as a function of the electronic specific heat coefficient γ_o. g should be a straight line if V is constant. The data points show a smooth increase of g with increasing γ_o.

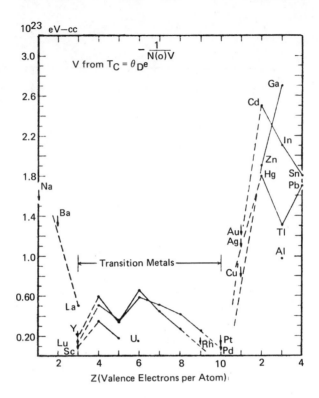

FIG. 4-101. BCS net attractive interaction among the elements obtained from experimental T_c, T_D and γ data. After Ref. 162.

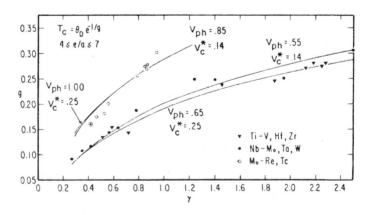

FIG. 4-102. Superconducting coupling strength g (for 4 < Z < 7) obtained from experimental T_c and $T_D = \theta_D$ data. g is plotted against the electronic specific heat coefficient γ. The BCS weak-coupling theory would give a straight line through the origin (notice shift in scale). We see that for Z = el./at. < 6 constant $V_{ph.}$ and V_c^* are in good agreement with experiments. The fit is not as good for Z > 6, even given a different choice (larger $V_{ph.}$) of interaction strength. After Ref. 162, with permission.

The full lines represent calculations, where V was separated into
$V = V_{ph.} + V_c^*$ $V_{ph.}$ represents a lattice interaction term, and V_c^*
is a constant pseudopotential term. Enhancement effects were taken
into account. The diagram shows that γ_o is not proportional to g.
Therefore, V cannot be a constant. Reasonable agreement between
the BCS theory and data is obtained only for some samples with
el./at. < 6, but it is not so good for larger el./at. values. For
$8 \leq$ el./at. $\leq$ 10, even the correct trend between γ_o and T_c is
not found. In solid solutions of Ru, Os, Ir and Rh, T_c decreases
if one adds Ir to Os, whereas γ_o increases. However, even if one
does not find the correlation between T_c and γ_o as expected from
the theory, the experiments show a systematic correlation between
T_c and el./at. in Ir-alloys, provided solute and solvent have a
similar electronic configuration. Figure 4-103 shows that for such
alloys T_c is a unique function of the el./at. ratio and can be
given by log $T_c \propto$ (el./at.) for $8.5 \leq$ el./at. $\leq$ 9.25. Here the
rigid band model gives a good correlation between experimental results obtained in different alloy systems. Naturally, one would not
expect a unique T_c = f(el./at.) relation if impurity and host differ markedly, as in Ir-alloys with W, Mo, V, Nb and Ta impurities
(4-104). The systematic trend of T_c, the electronic specific heat
coefficient γ and the susceptibility with the electron density
for some transition element alloys is given in Figs. 3-18 and 4-104.
The correlation of their curves requires a much more detailed model
on the various contributions to V [162].

TABLE 4-9

Superconducting Transition Temperature $T_c(^\circ K)$

Zn	Ti	V	Cd	Zr	Nb	Mo	Tc	Ru
0.85	0.39	5.03	0.52	0.55	9.1	0.92	11.2	0.49

Cd	β-La	Hf	Ta	W	Re	Os	Ir
0.52	6.0	0.16	4.48	0.01	1.7	0.66	0.14

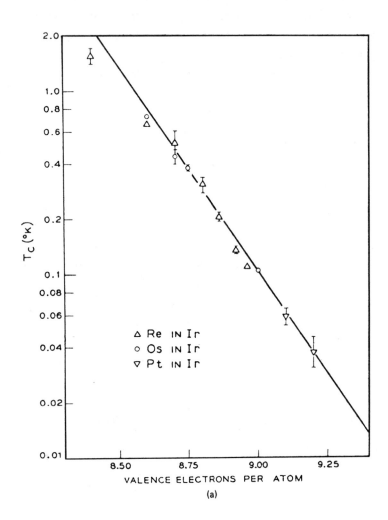

FIG. 4-103a. Variation of the superconducting transition temperature with the valency electron per atom ratio in fcc Re-Ir,
Os-Ir and Pt-Ir alloys. After Ref. 294, with permission.

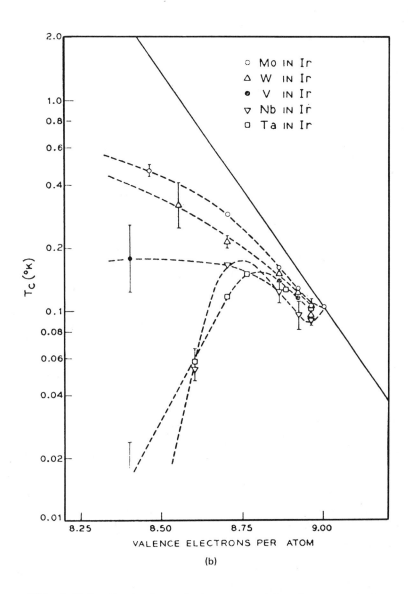

FIG. 4-103b. Variation of the superconducting transition temperature with the valency electron per atom ratio in fcc Mo-Ir, W-Ir, V-Ir, and Ta-Ir alloys. After Ref. 162, with permission.

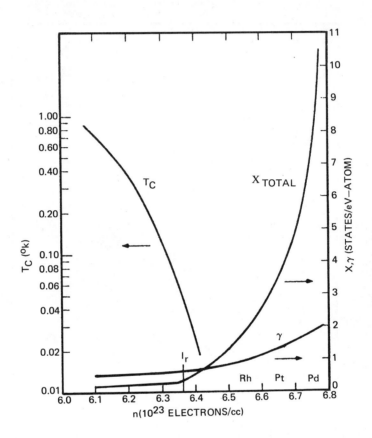

FIG. 4-104. General trend of the superconducting transition
temperature T_c, the magnetic susceptibility χ and the electronic
specific heat coefficient γ as a function of the valency electron
density in fcc transition element alloys. After Ref. 162.

Chapter 5

OPTICAL PROPERTIES AND ELECTRON EMISSION

I. INTRODUCTION

The investigation of the interaction of light with solids and gases helped to lay the foundation of modern quantum physics. During the end of the 19th century and at the beginning of the 20th century, a wealth of experimental information on free atoms and molecules was obtained from optical spectra, and quantum mechanical models were developed to explain the system of spectra with electron energy levels and the transition probabilities between energy states.

As Hertz discovered in 1886, light absorption by solids can lead to an emission of electrons. The maximum energy transfer from light to such electrons is independent of the light intensity. This cannot be explained with classical electromagnetic theories. Einstein proposed that the energy of light is quantized in units of $h\nu$, and that electrons can absorb light energy only in quantum values of:

$$E = h\nu \ . \tag{5-1a}$$

It is convenient to say that the light consists of photons with this energy $h\nu$. The electron emitted from a metal after it absorbs a photon has as maximum energy a lower value than $h\nu$ since a certain amount of energy, $W = e\phi$, is needed to remove the electron from the metal:

$$E_{max.} = h\nu - e\phi \ . \tag{5-1b}$$

The work function $W = e\phi$ depends markedly on the surface properties of the sample. One would expect that the number of electrons
which will be emitted from the metal will depend on the transition
probability for a jump from its energy state, and on the density of
states for electrons in the initial state and the final state. Therefore, electron emission experiments should give information on the
density of states curve not only at the Fermi energy level, but also
on energy levels below and above the Fermi energy. This is much more
than electron transport phenomena measurements can give; those measurements give information only on electrons with energies close to
the Fermi energy level. Details of $N(E)$ as a function of E can
be obtained only indirectly with assumptions of special band models.

Several techniques aside from electron emission experiments
have been used to determine energy states in metals over an extended
energy range. One can, for instance, remove an inner core electron by
X-rays and determine the energy absorption as a function of wave
length. Absorption edges reveal energy states of core electrons.
Then an outer electron will drop into the lower core state level and
the energy change will lead to the emission of a photon. The energy
distribution of photons will give information on the $N(E)$ values
of the initial and final states of the electron jumping into an inner
shell. One would expect that the inner shell has a rather sharp energy level. The energy distribution curve of the emitted photons
(which yield in this case soft X-ray radiation) reflects the density
of states of the band from which the electrons originate. In copper
this is the conduction and d-band.

Unfortunately, the emission of an electron from a lower lying
energy band modifies the energy states of this inner band. It cannot
be assumed to be a sharp line. Further, the transition probability
of electrons from one energy band to another may depend in a complex
way on initial and final energy states for these electron transitions.
This makes it difficult presently to determine $N(E)$ quantitatively
from soft X-ray experiments.

Optical measurements are also used to study the density of states curve of metals and alloys. Light illuminating a surface will be partly absorbed, partly reflected. The fraction of light absorbed, A, is a function of the wavelength. Since some of the absorbed light excites electrons to higher energy states, the light absorptivity depends on the ability of the electron to jump to higher energy levels. Therefore, the light absorptivity, or the reflectivity of light, depends on the electron transition probability, which again depends on the density of states in the initial and final state.

Optical reflectivity measurements may not seem to be able to yield as much information on the electron energy states as photoemission studies. However, reflectivity measurements have the advantage of a higher reproducibility than electron emission studies. This may be due to the fact that light under normal incidence is absorbed in surface layers several hundred atomic layers thick and the amount of absorbed light is not critically dependent on sample preparation, whereas electron emission experiments are markedly affected in some energy ranges by even minor surface contamination.

II. ABSORPTION AND REFLECTION

A. Electromagnetic Waves in Solids

The interaction of electromagnetic waves with solid materials can be described macroscopically with solutions of Maxwell's equations. These equations have in the cgs. system the form (this system is usually used in the analysis of optical properties):

$$\nabla \times \vec{H} = (\varepsilon/c)(\partial \vec{F}/\partial t) + (4\pi\sigma/c)\vec{F}, \qquad \nabla \vec{F} = 0, \qquad (5\text{-}2a)$$

$$\nabla \times \vec{F} = -(\mu/c)(\partial \vec{H}/\partial t), \qquad\qquad \nabla \vec{H} = 0, \qquad (5\text{-}2b)$$

where $\vec{H}$ is the magnetic field, $\vec{F}$ the electric field, ε the dielectric constant, c the velocity of light in free space, σ the
electrical conductivity, and μ the permeability. $(\varepsilon - 1)\vec{F} = \vec{P}$
is the polarization or the electrical dipole moment per unit volume
of the sample. Equivalent equations to Eqs. 5-2a and b in the SI
system are obtained if one sets c equal to one.

μ can be set equal to one for para- and diamagnetic systems.
Then Eq. 5-2a + b leads to:

$$\nabla^2\vec{F} = (\varepsilon/c^2)(\partial^2\vec{F}/\partial t^2) + (4\pi\sigma/c^2)(\partial\vec{F}/\partial t) \qquad (5\text{-}2c)$$

A solution for this equation is an electromagnetic wave with:

$$F = F_o \exp\{i\vec{K}\cdot\vec{r} - i\omega t\}. \qquad (5\text{-}3)$$

Inserting Eq. 5-3 into Eq. 5-2c gives for the propagation constant K:

$$K^2 = \varepsilon\omega^2/c^2 + i4\pi\sigma\omega/c^2$$
$$K = (\omega/c)(\varepsilon + i4\pi\sigma/\omega)^{1/2}. \qquad (5\text{-}4a)$$

One obtains for an electromagnetic wave in free space, where $\sigma = 0$
and $\varepsilon = 1$:

$$K = \omega/c. \qquad (5\text{-}4b)$$

For an insulator with $\sigma = 0$ one obtains:

$$K = (\omega/c)\cdot\varepsilon^{1/2}. \qquad (5\text{-}4c)$$

One can obtain Eq. 5-4a formally from Eq. 5-4c by replacing ε,
which is a real number, by the complex dielectric constant $\hat{\varepsilon}$:

$$\hat{\varepsilon} = \varepsilon + i(4\pi\sigma/\omega)$$
$$= \varepsilon_1 + i\varepsilon_2 \qquad (\varepsilon_1 \text{ and } \varepsilon_2 \text{ are real numbers}). \qquad (5\text{-}5)$$

Instead of characterizing the wave propagation in a solid by the wave propagation constant K, one can use the refractive index N, which can be defined as:

$$N = (\varepsilon + i4\pi\sigma/\omega)^{1/2} \qquad (5\text{-}6a)$$

which gives, together with Eq. 5-4a:

$$K = (\omega/c) \cdot N. \qquad (5\text{-}6b)$$

N is a complex number. It separates into:

$$N = n + ik, \qquad (5\text{-}7)$$

where n and k are real numbers. n is the real refractive index, and k is the attenuation index. In an insulator, n is the ratio (wavelength in vacuum)/(wavelength in insulator). k characterizes the exponential damping of the wave. This can be seen be rewriting Eq. 5-3 for the case of light propagation in the x_3-direction:

$$F = F_o \exp(i\omega n x_3/c - i\omega t) \cdot \exp(-k\omega x_3/c). \qquad (5\text{-}8a)$$

One sets $k\omega/c = 1/\lambda^*$, where λ^* is the penetration depth. λ^* in metals is of the order of 10^{-6} cm. The light intensity is proportional to $I \propto |F|^2$. Therefore, I also decays exponentially with penetration depth in a metal. One obtains:

$$I \propto |F|^2 \propto \exp(-2k\omega x_3/c) = \exp(-K^*x_3). \qquad (5\text{-}8b)$$

$K^* = 2k\omega/c = 2/\lambda^*$ is the absorption coefficient.
Equations 5-4b to 5-7 give:

$$n^2 + 2ink - k^2 = \varepsilon + i(4\pi\sigma/\omega) = \varepsilon_1 + i\varepsilon_2. \qquad (5\text{-}9)$$

Separating real and imaginary components gives:

$$\varepsilon_1 = \varepsilon = n^2 - k^2$$
$$\varepsilon_2 = 2nk = 4\pi\sigma/\omega. \tag{5-10}$$

The current induced in the material is $j = (-i\varepsilon\omega/c + 4\pi\sigma/c)F$. This is the right hand side of Eq. 5-2a. The power absorption of Joule heat is the real part of:

$$\vec{j} \cdot \vec{F} = -(i\omega/c)N^2 F^2. \tag{5-11}$$

Therefore, the absorption coefficient η, which is the fraction of energy absorbed in a unit volume of the material, is given by:

$$\eta = \text{Re}(\vec{j} \cdot \vec{F})/|F|^2 = 2nk\omega/c = 4\pi nk/\lambda = \varepsilon_2\omega/c. \tag{5-12}$$

The physical interpretation of these parameters for specific electronic models is given in the following sections. There pairs n and k or ε_1 and ε_2 are associated with the excitation and energy absorption of a dipole or with the polarization of electrons or ions due to electromagnetic radiation.

There exist several techniques to determine the dielectric constants. For metals, one now frequently measures the reflectance R under normal or nearly normal incidence. R is the ratio of the reflected to total incident light intensity. R is related to N by:

$$R = |(1 - N)/(1 + N)|^2$$
$$= \{(n - 1)^2 + k^2\}/\{(n + 1)^2 + k^2\}. \tag{5-13}$$

The real and imaginary parts of $N^2 = (n^2 - k^2) + i2nk$ are not independent. This makes it possible to calculate both $(n^2 - k^2)$ and $2nk$ with the Kramers-Kronig relation, provided the reflectivity R is measured over a large frequency interval. Reference 31 gives more details.

B. Absorption Mechanisms in Solids

The absorption of electromagnetic radiation in solids is due to
several mechanisms. For high energy photons in the X-ray range, the
energy absorption mechanism is partly due to the excitation of electrons from lower lying orbits to higher states, One would therefore expect that the absorption coefficient changes suddenly at critical photon energy values. At low photon energies, a classical model
can give a reasonable description of the absorption process. One
would assume that the electromagnetic wave interacts with ions or
electrons. These particles may oscillate around equilibrium positions. In first approximation, one may assume that the force is proportional to the displacement of the ion or electron.

This seems to be a reasonable model for an ion in a lattice, or
an electron tied to an individual atom. This concept cannot be used
to describe directly properties of an electron gas where the electrons
have no potential energy. The electromagnetic wave will affect all
conduction electrons at the same time. Therefore, their combined
charge will oscillate with the same frequency as the frequency
of the electromagnetic wave. Their average charge will be displaced
from the average positive charge of the heavy ions, which can only
move much more slowly than the electrons. Typically, one assumes
that the positive charges are smeared out to a uniform "jellium".
If one assumes that an electron has an equilibrium position inside
this jellium, one can show that the force on the electron is proportional to the displacement. One should have a "harmonic oscillator"
in the classical model. Since the electrons are called a "plasma",
their oscillation is defined as "plasma oscillation".

One can calculate the frequency of plasma oscillations with the
following model. First one displaces the electrons of the electron
gas from their equilibrium position. This leads to a force on these
electrons. If electrons of volume density N are moved a distance
r with respect to the fixed positive charge, one obtains for the

polarization P:

$$P = Ner. \tag{5-14}$$

This leads to an electrical field F:

$$F = -4\pi P. \tag{5-15}$$

The force on each electron is then:

$$m(\partial^2 r/\partial t^2) = eF = 4\pi Ne^2 r. \tag{5-16}$$

This is the equation of a harmonic oscillator. Its frequency is:

$$\nu_{pl.} = (Ne^2/\pi m)^{1/2}. \tag{5-17}$$

$h\nu_{pl.}$ in metals is of the order of 10 to 100 eV.

B.   Interaction between an Electromagnetic Wave and an
Electric Dipole

Electrons or ions in an electromagnetic field are discussed with
the model of a harmonic oscillator with damping. Therefore, energy
will be absorbed. This energy may be reemitted as secondary radiation. The oscillator will follow the primary radiation with a phase
shift δ. The secondary radiation of the oscillator follows the oscillation of the oscillator by 90°. The phase shift δ will tend to
zero if ω/ω_o approaches zero. ω_o is the Eigenfrequency of the oscillator. δ will be close to 90° for $\omega = \omega_o$. For $\omega \gg \omega_o$, the
phase shift δ approaches 180°. The energy absorption of the oscillator reaches its highest value if δ is close to 90°, since the
energy absorption reaches its highest value if force and displacement
are 90° out of phase. k, ε_2 and η should reach a maximum

near ω_o. The energy absorption curve is bell-shaped for a damped oscillator.

The phase shift between primary and secondary radiation should lead also to a shift in the phase velocity of the resulting radiation. For low ω/ω_o values, δ will be small. Therefore, the resulting phase velocity of the primary beam will be lower than the light velocity outside the sample (keeping in mind the 90° phase delay between oscillator and secondary radiation). $c(vacuum)/c(materie) = n$ is therefore slightly larger than one for low ω/ω_o values. It increases with ω/ω_o. For $\omega/\omega_o = 1$, the phase shift between primary and secondary radiation is close to 180° and n is therefore close to one. A further increase of ω/ω_o leads to decreasing n values. However, since $\delta + 90^o \rightarrow 270^o$ for $\omega/\omega_o \rightarrow \infty$, n approaches again the value of one. Figures 5-1 to 3 show schematically the frequency dependence of n, ε_1 and ε_2.

Oscillating charges lead to a fluctuating polarization P, with $P = (\varepsilon - 1)F$. The polarization of one atom is $\alpha = \ell q/F$ with $\pm q$ the positive and negative charge of the dipole, and ℓ the distance between charges. The polarization should be equal to αNF, where N is the number of dipoles per unit volume. One has to keep in mind that the electric field as seen by the atom is equal to the applied field only for infinitely dilute systems. In solids, F acting on one atom is the field in a small hole inside the sample. The field in vacuum $F_{vac.}$ and the field in the hole $F_{ho.}$ differ by the factor $1/(\varepsilon + 2)$. One therefore obtains with this "Lorentz correction" $(\varepsilon - 1)/(\varepsilon + 2) = P/F_{appl.}$ instead of $(\varepsilon - 1) = P/F$. This gives:

$$(\varepsilon - 1)/(\varepsilon + 2) = P/F_{appl.} = \alpha(\omega) = e\ell/F$$
$$= (e^2/m)/(\omega_o^2 - \omega^2) \qquad (5\text{-}18)$$

if the oscillator is not damped. With damping, the maximum displacement of an electron is ℓ_o with:

$$\ell_o \propto \{(\nu^2 - \nu_o^2)^2 + (\Lambda^2/\pi)\nu_o^2\nu^2\}^{-1/2} \qquad (5\text{-}19)$$

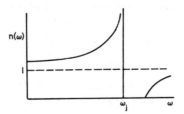

FIG. 5-1. Real part n of the refractive index for a system of electrons tied to ions. The singularity at ω_j will be smoothed out in real systems due to relaxation processes.

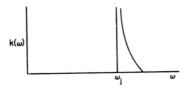

FIG. 5-2. Imaginary part k of refractive index. As in Fig. 5-1, relaxation processes will smooth out the singularity at ω_j.

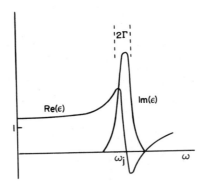

FIG. 5-3. $\varepsilon_1 = R(\varepsilon)$ and $\varepsilon_2 = Im(\varepsilon)$ of a system of electrons tied to ions. This is the case for intraband transitions. Again, relaxation processes due to scattering and radiation lead to broadening in the "dispersion curve".

where Λ is the logarithmic decrement. $1/\Lambda$ is the number of oscillations required to decrease the amplitude of the wave to 37%.

One can describe with similar equations the interaction of electrons with an electromagnetic wave in a quantum mechanical model.
Let us discuss first the case of an insulator, where an electron in
the ground state with wave function ϕ_o is excited to higher states
with the wave function ϕ_j. One obtains for the polarization $\alpha(\omega)$
(see Ref. 31 for more details):

$$\alpha(\omega) = \Sigma \, e^2 |x_{oi}|^2 2\omega_j / \hbar(\omega_j{}^2 - \omega^2) \qquad (5\text{-}20)$$

with $x_{oj} = \int \phi_j^* r \phi_o \, dx_1 dx_2 dx_3$ as the average displacement of the electron. This can be written as $\alpha(\omega) = (e^2/m) \Sigma f_j / (\omega_j{}^2 - \omega^2)$, where
$f_j = (2m/\hbar^2) \hbar\omega_j |x_{oj}|^2$ is the oscillator strength. The sum of Σf_j
is equal to one. This gives, if one takes the Lorentz correction
into account:

$$(\varepsilon - 1)/(\varepsilon + 2) = 4\pi N\alpha(\omega) = \Sigma f_j \omega_p{}^2 / (\omega_j{}^2 - \omega^2) \qquad (5\text{-}21)$$

where $\omega_{pl.}^2 = (2\pi\nu_{pl.})^2 = 4\pi Ne^2/m$.

D. Interaction between an Electromagnetic Wave and a Metal

One assumes for a metal that the current contribution due to the
polarization of ions can be neglected. This gives for ε of a metal:

$$\varepsilon = 4i\pi\sigma/\omega \qquad (5\text{-}22)$$

or:

$$n + ik = (2\pi\sigma/\omega)^{1/2}(1 + i). \qquad (5\text{-}23)$$

The reflecting power is therefore:

$$R = 1 - (2\omega/\pi\sigma)^{1/2}. \tag{5-24}$$

which is known as the Hagen–Rubens relation. It is applicable only if σ is independent of frequency. In other words, it is required that electrons make frequent collisions and that $\omega\tau < 1$ so that a relaxation process can take place. A calculation of the interaction of an electromagnetic wave with a free electron gas over a large frequency range (this goes beyond the case discussed above) gives the following result:

$$n^2 - k^2 = 1 - \omega_{pl.}^2 \tau^2 / (1 + \omega^2 \tau^2)$$
$$2nk = \omega_{pl.}^2 \tau / \omega (1 + \omega^2 \tau^2). \tag{5-25}$$

This is the Drude theory. For low frequencies, with $\omega\tau \ll 1$, the

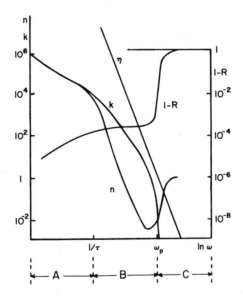

FIG. 5-4. Schematic behavior of optical properties of metals, showing real (n) and imaginary parts (k) of the dielectric constant, reflection coefficient (R) and absorption coefficient (η). $\omega_p = \omega_{pl.}$ is the plasma frequency. After Ref. 31.

metal is strongly reflecting. The absorption coefficient is proportional to the electrical conductivity and is more or less independent
of the frequency ω [31]. For $1/\tau \ll \omega \ll \omega_{pl.}$, the absorption coefficient η drops rapidly with increasing angular frequency. The
reflectivity is still rather high and is given by:

$$R \simeq 1 - 2/\omega_{pl.}\tau. \tag{5-26}$$

For $\omega \gg \omega_{pl.}$, the reflectivity approaches zero, and the metal
becomes nearly transparent, with the absorption coefficient equal to:

$$2\omega n(\omega)k(\omega)/c \simeq \omega_{pl.}^2/\omega^2\tau c. \tag{5-27}$$

Figure 5-4 gives a schematic survey of the frequency dependence of
n, k, η, and $A = 1 - R$ for metals.

E. Intraband and Interband Transitions

The interaction of light and electrons in real metals is a complex process. Fortunately, simple concepts show the essential features of possible electron excitations. Three general types of these
excitations are of importance: intraband transitions, interband
transitions, and plasma oscillations. In "intraband transitions",
electrons in one band stay in this band. In parabolic bands with
$N(E) \propto E^{1/2}$, photon assisted transitions are possible only if some
type of multi-particle interaction occures, typically with phonons.
It is not possible to satisfy momentum and energy conservation laws
if one electron is excited by only a photon to a higher energy state
in the same band. This explains the high reflectivity of metals at
low wavelengths. The dielectric constant for a system in which only
intraband transitions occur is given by:

$$\epsilon(\omega) = 1 - \omega_{pl.}^2/\omega(\omega + i/\tau), \tag{5-28}$$

with $\omega\tau > 1$. τ is the relaxation time associated with the scattering mechanism. $\varepsilon(\omega)$ for intraband transitions is similar to
$\varepsilon(\omega)$ obtained for the interaction of light with the classical free
electrons in a solid (Drude formula).

In "interband transitions", a photon lifts an electron from one
energy band to another energy band. These transitions are "vertical"
(5-5). The change of k with this transition is small. Lorentz
was the first to propose the simple model of an electrical insulator
for interband transitions in which an electron is tied with a spring
to a fixed point in a solid. The multielectron model of modern solid
state physics gives similar results for the dielectric constant due
to interband transitions, However, it is always useful to keep the
Drude and Lorentz models in mind for the interpretation of the properties of metals, since the models reveal the essential features of
electron-photon interactions.

Typical behavior of the reflectivity R and the real and imagi-
nary constants ε_1 and ε_2 of metals are given in Fig. 5-6. Free
electrons and intraband transitions give simple correlations between
R or ε_2 and the photon energy $E_{phot.}$. Both parameters decrease
with increasing $E_{phot.}$ = $h\nu$ values (5-6a). This behavior should be
found for a metal with a single energy band. If this metal has a
conduction and a d-band, as shown schematically in Fig. 5-7, one
would expect that the energy absorption would rapidly increase at a
critical photon energy, sufficient to lift electrons from the lower
lying d-band to the Fermi energy level. This leads to an increase
in ε_2 as shown in Fig. 5-6b + c. Intuitively, one would expect
also a decrease in the reflectance R for this case, since a larger
amount of light is absorbed if intraband transitions occur. This
behavior is shown in Fig. 5-6b. However, R can show a sharp peak
for interband transitions. This is shown in Fig. 5-6c. The interpretation of ε_2 data is much more straightforward for this case
of interband transitions. The sections d to f in Fig. 5-6 show
the behavior of R, ε_1 and ε_2 in the region of plasma excitations
and for excitations of deeper bands. These phenomena are usually
important for high photon energies.

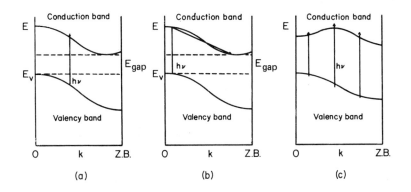

FIG. 5-5. Direct transitions (a), phonon-assisted transitions (b), and vertical interband transitions (c) in a semiconductor. After Ref. 31, with permission.

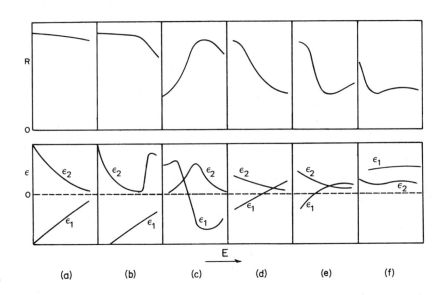

FIG. 5-6. Typical behavior of the optical constants ε_1 and ε_2 and of R with various excitation processes in metals and insulators: free electron range (a); low-energy interband transitions in metals (b). Semiconductor and metals: interband transitions (c); plasma region (d); plasma region modified by interband transitions (e); excitations of deeper bands above plasma frequency (f). After Ref. 295.

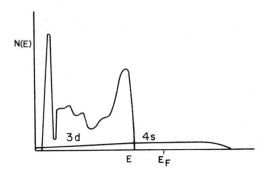

FIG. 5-7. Schematic diagram of the energy band of copper.

F. Metallic Elements

Some typical examples of pure elements will now be discussed. We will start with aluminum which comes very close in its behavior to a metal with free electrons. The selection of aluminum as a first example differs noticeably from our usual schedule, in which we discuss the case of copper first. This change in procedure is due to the fact that for most electron transport measurements only electrons at the Fermi energy level are important. Therefore, copper, which has a nearly spherical Fermi surface, can be used as an example of a simple system. However, in optical and electron emission studies, electrons in a large energy interval participate in the photon-electron interaction process. Aluminum, which is a good metal with high electrical conductivity and cubic symmetry, is therefore a better choice than copper for a simple metal in optical studies. Further, it has only six s-electrons and seven p-electrons.

Optical properties of aluminum are given in Fig. 5-8. Aside from small peaks in ε_1 and ε_2 close to 2 eV, which are due to minor interband transitions, the dielectric constants follow closely the prediction of the Drude free electron model. ε_1 seems to pass through zero at 15.8 eV, where ε_2 is very small. This is the typical behavior of plasma resonance. Figure 5-8b shows

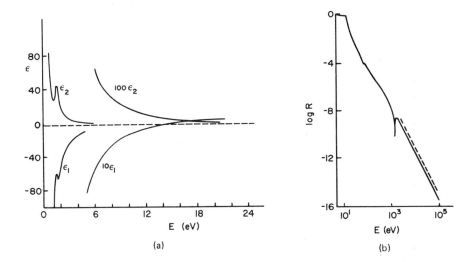

FIG. 5-8. Frequency dependence of the real (ε_1) and imaginary (ε_2) parts of the dielectric constants of aluminum (a). The frequency dependence of the reflectance R of aluminum (b). The theoretical curve with $R = \omega_{p1}^4/16\omega^4$ is given by the dashed line. After Ref.295.

results of calculations for R for high photon energies which agree well with experiments.

The essential features of the copper energy band were shown in Fig. 1-60. It shows that copper has a very broad conduction band with s- and p-electrons. The top of the d-band is more than 1 eV below the Fermi energy level. The density of states curve is very steep near the top of the d-band since the E(k) curves of the d-electrons are very flat near the top of the d-band. The calculations show further that unfilled bands exist above the Fermi energy level.

$2nk/\lambda = \varepsilon_2/\lambda$ of copper, as obtained by Pells and Shiga [296], is given in Fig. 5- 9. The maximum in ε_2/λ slightly above 2 eV indicates a sharp onset of interband transitions, and other interband transitions are found near 4 eV. The change in the absorption spectrum with increasing temperature shows that the peak between 4 and 5 eV at room temperature is really a superposition of several transitions. Neglecting the details of the transition mechanisms, one can say that the main absorption edge at 2 eV is due to

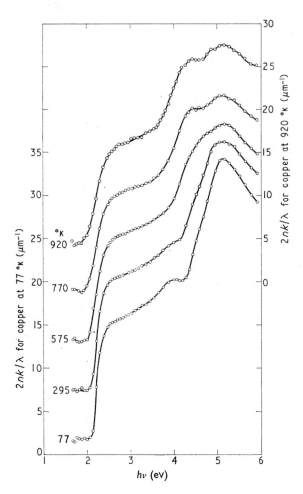

FIG. 5-9. The absorption spectrum for copper between 77 and 920°K. The values of $2nk/\lambda$ axis refer to copper at 77°K, the curves for other temperatures are successively displaced upwards by 5(μm^{-1}). After Ref. 296, with permission.

transitions from the nearly flat, highest d-bands to the Fermi energy level, whereas the absorption peak is due to transitions between full d-states or s-states to several empty bands above the Fermi energy level. A more detailed band model, obtained from electron emission experiments, will be discussed later.

The optical properties of gold are similar to those of copper
[295]. This is already indicated from the similar color of both
elements in their pure state. It shows that the main absorption
edge, responsible for the onset of the interband transitions, is
found at nearly the same position for both elements. Pells and
Shiga found that the peak in the absorption curve in gold between 4
and 5 eV is associated with a superposition of several transitions,
because this peak splits at elevated temperatures.

Silver has an appearence different from both copper and gold.
Since the reddish yellow color of these two elements is due to the
onset of interband transitions to the Fermi energy level for d-electrons which requires an energy of 2 eV (this corresponds to a wavelength of about 6000 Å), one would suspect from the bright, colorless appearance of silver that this transition from the top of the
d-band to the Fermi energy level is shifted out of the range of the
visible spectra into the ultraviolet. Figure 5-10 gives the reflec-
tivity, ε_1 and ε_2, of silver. It shows that the reflectivity of
this metal is close to one for $0 < h\nu < 3$ eV. A sharp dip in R
occurs at 3.8 eV. This corresponds to a wavelength of 2900 Å in
the ultraviolet. This explains (together with the possibility of
easily obtaining nearly atomically clean surfaces) the bright appearance of silver.

The imaginary dielectric constant of silver also shows a sharp
edge at 3.8 eV, typical for the onset of interband transitions. By
assuming that the intraband transition of electron follows the theoretical prediction from the Drude free electron model, it is possible
to separate ε_2 into the free and bound electron structure. ε_2, as
deduced from this calculation, can be described in the following
way (5-10c). The initially decreasing ε_2 values follow the free
electron model up to $E_{phot.}$ = 3.7 eV. This part of the ε_2 curve
can be extrapolated to higher photon energies by a smoothly decreasing
ε_2 function. The increase of ε_2 above this extrapolated curve for
photon energies near 4 eV is due to the excitation of bound elec-
trons.

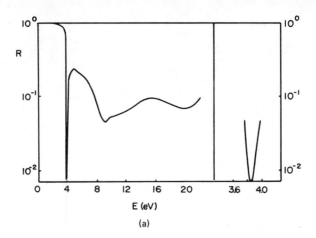

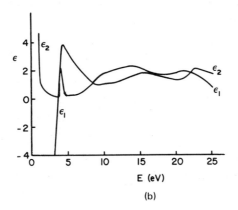

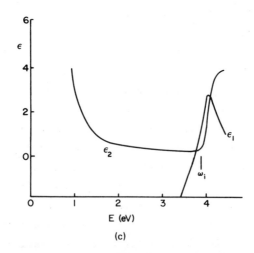

FIG. 5-10. Frequency dependence of the reflectance for Ag (a). Frequency dependence of ε_1 and ε_2 of Ag (b). Free (intraband) and bound (interband) contributions to ε in Ag(c). The threshold for interband transitions is given by ω_i. After Ref. 295.

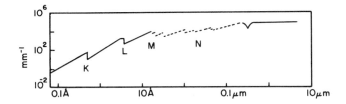

FIG. 5-11. Absorption coefficient K*/2 of silver. After Ref. 76.

The absorption coefficient of silver is given in Fig. 5-11 over an extended photon energy range from 10^{-1} eV to 10^5 eV. It shows the characteristic absorption edges for the excitation of electrons from core states.

Optical properties of nickel, an example of a transition element, are given in Fig. 5-12. The E(k) diagram of nickel is sometimes obtained from the E(k) diagram of copper by assuming that that the nickel subband with the larger number of electrons in the ferromagnetic state is shifted upwards by about 0.8 eV. This shift gives the right number of holes in the d-band to explain the saturation magnetization of nickel. In this model, transitions of electrons from states below the Fermi energy level to states above this level should be similar to those of copper. However, the sharp absorption edge of copper at 2 eV should not show up in nickel (this edge in copper is due to electron transitions from the top of the filled d-band to the Fermi level). Therefore, d-band to the Fermi level transitions are possible for very small photon energies. The $2nk/\lambda$ curve (5-12a) shows not the sharp edge at 2 eV as the same curve for copper (5-9).

The main absorption edge of nickel at 4 eV (5-12a) associated with transitions from d-band states to higher states should not be affected by the position of the Fermi energy level. This peak is slightly temperature dependent, with $\Delta E/\Delta T = 4.5 \cdot 10^{-4}$ eV/$^\circ$K. Noticeable changes in the optical absorption of nickel with temperature should be expected, since ferromagnetic ordering decreases with increasing temperature in this metal. The band splitting of 0.4 to 0.8 eV due to magnetic exchange interaction at magnetic saturation is zero at the Curie temperature of 632°K. One could therefore expect

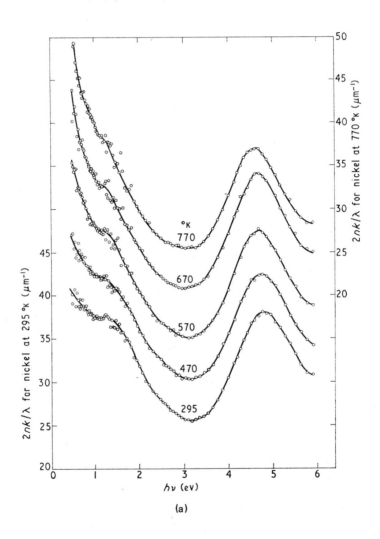

(a)

FIG. 5-12a. The optical absorption $(2nk/\lambda)$ of electropolished nickel at several temperatures. The ordinate scale is applicable only to the room temperature data. Successive higher temperatures are displaced upwards by $2nk/\lambda = 5$ μm^{-1}. After Ref. 297, with permission.

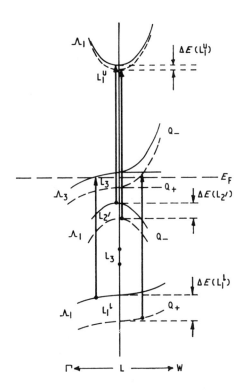

FIG. 5-12b. The band structure near the L point. Full curves
are spin down bands, broken curves are spin up bands. The defini-
tion of $\Delta E(L_1^u)$, $\Delta E(L_2)$, and $\Delta E(L_1^{1})$ are given in this figure.
The vertical arrows indicate transitions. After Ref. 297, with
permission.

changes in the position of edges and peaks of the order of 5 to
$10 \cdot 10^{-4}$ eV/$^\circ$K. The study by Shiga and Pells [297] suggested that
some of the electron transitions associated with this peak originate
below E_F and end at states above E_F. They are not affected by
changes in the position of the Fermi energy level, and should therefore be the same for copper and nickel. According to Shiga and Pells,
this peak in nickel is associated with the transition from L_2 to
L_1,. Figure 5-12b shows the details of the energy band of nickel
near L with electron transitions suggested from these studies.

The effect of magnetic ordering on optical properties has also
been studied on chromium, which changes from the antiferromagnetic
to the paramagnetic state at 311°K. Figure 5-13 gives R^3 as a
function of wavelength λ for several temperatures. This third power
plot of R was used because it allowed easy plotting of the experimental data. It shows a reflectance minimum close to λ = 10 µm at

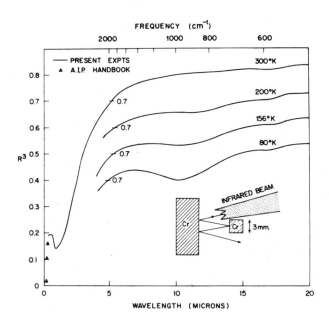

FIG. 5-13. Cube of the reflectivity of chromium. The three
lower curves are vertically displaced. The insert shows the experimental arrangement. After Ref. 298, with permission.

low temperatures. This corresponds to a photon energy of 0.08 eV, which is equal to 5.1 $k_B T_N$ (T_N is the critical or Néel temperature for the onset of antiferromagnetic ordering). As mentioned before, the energy gap due to antiferromagnetic ordering is, according to Overhauser, 3.5 $k_B T_N$ [125]. One may therefore conclude that the minimum in R at $\lambda = 10$ μm is due to antiferromagnetic ordering, especially since it desappears close to 300°K, where antiferromagnetic ordering is nearly destroyed.

$2nk/\lambda$ as a function of λ of several transition elements as obtained by Lenham is shown in Fig. 5-14. It follows no simple pattern even in the low photon energy or long wavelength range. The separation into a Drude (free electron) and Lorentz term is not easy.

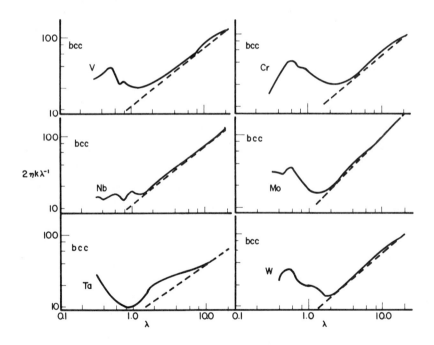

FIG. 5-14. Absorption $2nk/\lambda$ of transition elements. λ is given in units of μm. After Ref. 299, with permission.

Lenhams's evaluation shows that $2nk/\lambda$ will not follow the λ^2 dependence predicted from the Drude model even for the longest wavelength used in this investigation.

G. Semiconducting Elements and Their Alloys

Optical measurements on semiconductors have given a detailed picture of energy differences between separate segments of the energy band. Naturally, since the electrical conductivity of these elements is much lower than that found in metals, one would not expect high reflectivity values approaching one for photon energies approaching zero. The low energy effects should be dominated by interband transitions. Intraband transitions can be neglected. The interband transitions may be of the indirect type, in which electrons from the valency band can be lifted to an empty state in the conduction band not directly above the original state in the E(k) diagram ($k \neq 0$); or of the direct transition type, as shown schematically in Fig. 5-5. The indirect transition requires an interaction between electron and lattice, since the indirect transition has to be associated with either the creation or annihilation of one or more phonons.

The analysis of the optical properties of semiconductors, just as in metals, is not a straightforward problem [66]. It is possible to analyze the optical properties only if one knows a good deal about the energy band structure. Then the optical data will help in the determination of details of the energy band structure. It will give energy differences between points in the E(k) diagram. It will make it possible to select the correct band model from different possible models. Essentially, one uses a self-consistent approach. One tries out possible combinations of E(k) diagrams and electron transitions and compares them with the experiments. In metallic systems, systematic shifts of reflectance edges help in the correlation of experiments and theoretical models. A similar approach can be used in semiconductor or semimetal alloy systems.

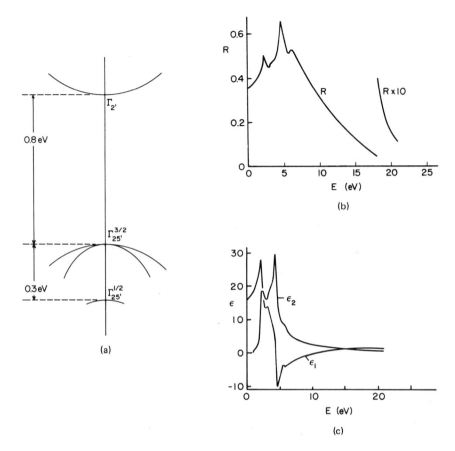

FIG. 5-15. Schematic structure of the valency band states and conduction band states at the Γ-point in germanium (a). Reflection spectrum (b) and ε_1 and ε_2 (c). After Ref. 66.

As shown in the discussion of the energy band structures of germanium and silicon, the energy gaps which determine the electronic transport process in these elements are formed by three orbitally degenerate states in the middle of the Brillouin zone at Γ. This is shown for germanium in Fig. 5-15a. Lowest energy transitions are expected from the $\Gamma_{25'}^{3/2}$ point, which corresponds to the top of the valency band, to the $\Gamma_{2'}$ point of the conduction band. The optical transition associated with these states is very sharp, with $h\nu = 0.8$

eV at 290°K. Figure 5-15c shows the dielectric constants of germanium. ε_2 shows two peaks, one at 2 eV, the other at 3 eV. Phillipp and Ehrenreich [300] attributed the first peak to a transition from $L_{3'}$ to L_1, and the second peak to the X_4 to X_5 transition. This selection gives the right value for the energy changes.

Good agreement between a few measured and predicted energy values is not always good enough to make firm predictions for the validity of proposed energy bands. One usually needs additional evidence to correlate E(k) diagrams with measurements. Investigations by Tauc and Antocik [302] showed that the first peak is a doublet. It is known that the $L_{3'}$ state is split by spin-orbit coupling, and the theory predicts that splitting should be 0.2 eV, in good agreement with experimental data. A refined calculation of ε_2 confirmed that the proposed transition associated with the first peak was associated with a transition along the [1 1 1] direction in reciprocal space, and that it should be found between Λ and Λ_1. Λ is associated with a saddle point. At saddle points, the joint density of states (density of states containing both the initial and final states) is usually very high, higher than found for extrema in the joint density of states with only maxima and minima.

The band structure of silicon is very similar to that of germanium. The first peak in R and ε_i at 3.4 eV corresponds to a different electron transition than the first peak in germanium. This can be easily seen from germanium-silicon alloy data. Both elements form a complete series of solid solutions. The reflectivity of some of the alloys is shown in Fig. 5-16. One can see that germanium and the alloy with low silicon concentration have the double peak, which is due to the spin-orbit splitting of the top of the valency band. The pure silicon sample shows no splitting in the first peak. One concludes therefore that the peak in silicon should be associated with $\Gamma_{25'} \rightarrow \Gamma_{15}$ transitions. Figure 1-77 shows the proposed positions of the symmetry points for the germanium-silicon alloys as a function of composition. One sees that some of the lines intercept. Therefore, transitions to Γ or L for pure germanium may be

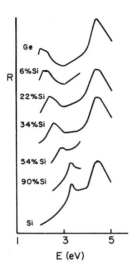

FIG. 5-16. Reflectivity of Ge-Si alloys.   After Ref. 66.

replaced by transitions to Δ for pure silicon.   For photon energies
above the range of interband transitions, the reflectance   R   and
ε_2 decrease with increasing photon energies (5-15).

H.   Semiconducting Compounds

The optical properties of III-V and II-VI semiconductor compounds are very similar to those of the group IV elements.   The reflectance curves show peaks due to interband transitions in the same
energy range as found for germanium, and at higher photon energies
the plasma oscillations.   Figure 5-17 gives the reflection data of
GaAs, GaP, InAs, and InSb between 12 and 28 eV.   It shows the effects
of electron transitions from d-bands.   The sharp raise in R   should
be due to exciton states.

Herman [51] proposed that one can describe the properties of the
isoelectric compounds by the potential of the group IV element, plus
a perturbation potential $\lambda V_{pert.}$. $V_{pert.}$ would be the same for the
whole series, and λ represents the strength of the perturbation.

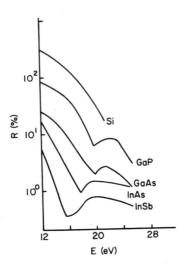

FIG. 5-17   Reflection spectra at 300°K showing the transitions
from the d-band of the cation. The ordinate should be multiplied
by 2 for InSb, 1 for InAs, 1/2 for GaAs, 1/4 for GaP, 1/10 for Si.
After Ref. 327.

λ gives the degree of ionicity. λ = 1 for III-V compounds, and
λ = 2 for the   II-VI compounds. The perturbation energy vanishes
in first approximation for all non-degenerate states in a horizontal
series, and also for some degenerate states whenever the perturbation
does not split the degeneracy. Herman proposed that energy changes
should be proportional to λ^2. Figure 5-18 shows that this is correct in the series Sn, InSb, and CdTe. A plot of energy gaps as a
function of λ^2 gives a nearly straight line.

Different results are obtained if the perturbation potential
lifts the degeneracy of the energy states. Then the perturbation is
non-zero in first order. Splitting for such a system is proportional
to λ. Low temperature measurements by Greenaway[303] showed that
the splitting of the X_1 conduction band state, which is degenerate
in the element and splits into X_1 and X_3 in the compounds, is
0 eV for germanium, 0.43 eV for GaAs, and 0.9 eV for ZnSe. This
is essentially a splitting linear in λ.

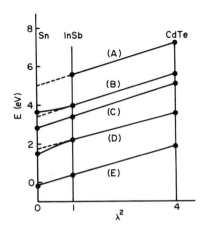

FIG. 5-18. Interband energy differences in the horizontal series grey Sn, InSb. CdTe plotted agains λ^2. $<L_3> \rightarrow L_3$ (A); $X_4 \rightarrow X_1$ (B); $\Gamma_{15} \rightarrow \Gamma_{25}^{3/2}$ (C); $<\Lambda_3> \rightarrow \Lambda_{1i}$ (D); $<\Gamma_{25}> - \Gamma_2$, (E). After Ref. 66.

Another useful tool used to identify transitions is the observation of spin-orbit splitting [300]. This is predominantly determined by the charge distribution of the atomic cores, which are not changed very much if free atoms form a solid. The splitting in the free atom can therefore be used to estimate the splitting found in the solid. The predicted splitting of the $L_{3'}$, L_3 or Λ_3 valency bands is given by:

$$\Delta_{III-V} = 0.35 \; \Delta_3 + 0.65 \; \Delta_5$$
$$\Delta_{II-VI} = 0.20 \; \Delta_2 + 0.80 \; \Delta_6$$

(5-29)

Δ_i is the spin orbit splitting observed on the atom "i". The first equation means that the electron spends 35% of its time on atom III, and 65% of its time on atom V. For the second equation, this would correspond to 20% and 80% respectively.

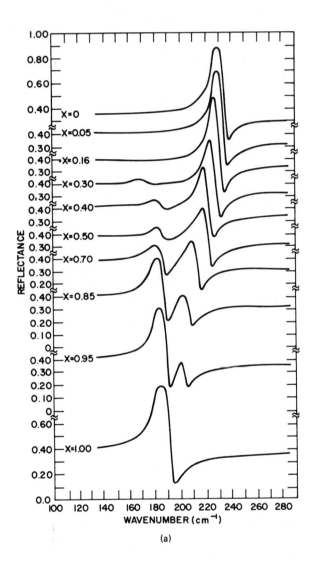

FIG. 5-19a. Reflectance as a function of wave number for the alloy system $Ga_{1-x}In_xSb$. After Ref. 304, with permission.

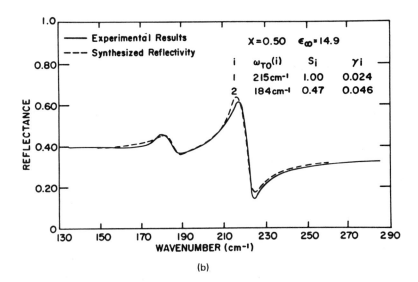

FIG. 5-19b. Theoretical fit to experimental reflectivity data for x = 0.50. The oscillator parameters and high-frequency dielectric constant ε_∞ are shown in the figure. After Ref. 304.

I. Semiconducting Mixed Crystals and Ionic Crystals

The reflectivity of spectra in the infrared has been investigated by M. H. Brodsky et al. [304] on the mixed crystal $Ga_{1-x}In_xSb$. Experiments in this area have been traditionally explained with one-mode or two-mode behavior. One-mode behavior has been defined as the occurrence of one strong transverse optical phonon mode over the entire composition range. The frequency and the oscillator strength change continuously over the entire composition range. This type of phenomena has been found, for example in alkali-halides like $Na_{1-c}K_cCl$, in which the concentration of the two types of cation changed, and in the system $KCl_{1-c}Br_c$, where the concentrations of two types of anions changed. Other examples have been found for mixed crystals of the form $Me(1)_{1-c}^{II} Me(2)_c^{II} (Non-Me)^{VII}$.

Two-mode behavior is characterized by two dominant transverse optical phonon modes. Typically, each of the pure compounds would

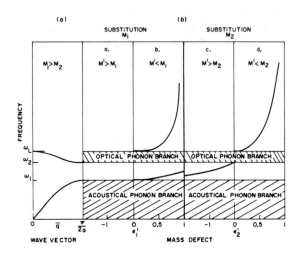

FIG. 5-20. Optical (upper curve) and acoustical (lower curve) phonon branches of the dispersion relation for a diatomic linear lattice (a). Frequencies of the local and gap-mode frequencies in a diatomic linear lattice as a function of the mass-defect parameter, characterizing the isotropic substitution M' (b). $M_1 > M_2$. See text for details. After Ref. 305.

have a one-mode transverse optical phonon mode. These are found at different frequencies. The mixed crystal would show both modes, and the strength of each mode would increase with the concentration of the compound responsible for it. However, this model does not explain the properties of the $Ga_{1-x}In_xSb$ alloy system. There the reflectance spectrum for the alloys shows two peaks and the position of these peaks is composition dependent (5-19). The shape of the peak follows closely the prediction of the damped Lorentzian oscillator. The results can be described very well with a single oscillator excitation for alloys with x = 0.05. Figure 5-19b shows good agreement between calculated and measured reflectance measurements. Experiments follow a two-oscillator model closely for alloys with x = 0.3 to x = 0.95.

The model description for this alloy behavior closely follows the discussion of the dispersion curves of binary alloys, except that it has to be expanded into the three element system found in the

gallium-indium-antimony compound. Simple binary systems with small
amounts of impurities of lighter mass show, as indicated in Fig. 1-
36, an extra peak above the Debye frequency of the host, or a (usually
rather weak) maximum below the Debye frequency if the impurity is
heavier than the host (if the spring constants are the same). The
dispersion curve of a diatomic system with two different masses shows
a well separated optical and acoustical branch. One would expect
that the ternary compound would have a dispersion curve which shows
the features of both types of curves. Results of calculations for
such a system for the simplifying case of a linear chain are shown in
Fig. 5-20. The results for a simple diatomic chain are given in Fig.
5-20(a). If the heavier atom on this chain is replaced by an even
heavier impurity, no new separate energy states become available
(5-20(b)-a).   However, if the heavy matrix element is replaced by a
lighter impurity, new states above the acoustical branch will be found
(5-20(b)-b).   If the lighter host atom is replaced by a heavier
impurity atom, new states below the optical branch of the frequency
are formed (5-20(b)-c), but a lighter impurity atom leads to states
above the optical branch (20b-d). States between the acoustical and
optical branchs are called gap-states, and those above the optical
branch are called "local states".

   Optical properties of ionic semiconductors and ionic crystals
are characterized by a very high reflectivity in the far infrared.
This cannot be attributed to a free electron gas, as is found
in metals. The high reflectivity is due to lattice vibrations. As
discussed in Chapter 1, the lattice frequency spectrum of a crystal
AB, in which the two types of atoms have different masses, contains
two sections, the optical and the acoustical branchs. The frequencies
in the optical branch are due to the relative motion of adjacent atoms
of different masses. If these also have different charges, they form
an oscillating dipole. In this lattice frequency range, strong interactions with electromagnetic radiation of the same frequency (in
this case of infrared light) takes place. This leads to a high
reflectivity. If light is reflected several times on surfaces of such

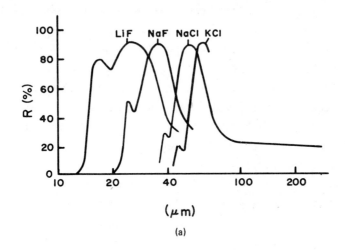

(a)

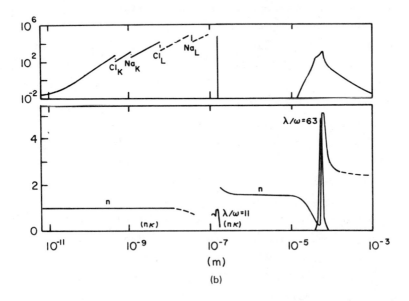

(m)

(b)

FIGS. 5-21a and b. Reflectivity of alkali-halides (a). Absorption coefficient nk (top diagram) and n (lower diagram) of NaCl (b). After Ref. 76.

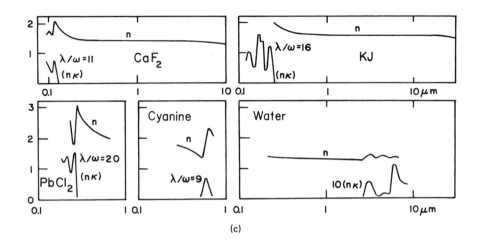

FIG. 5-21c. n, and $n\kappa$ as a function of wavelength for CaF_2, KJ, $PbCl_2$, Cyanine and Water.

crystals, essentially only light in the frequency range of the optical branch of the lattice frequency spectrum remains. This multireflection method was used around 1900 to obtain a narrow band of infrared light which was called "Reststrahlen" for "remaining radiation".

It is easy to give an estimate of the value of the "Reststrahlen" frequency of ionic crystals. In the case of NaCl, the distance between adjacent planes of Na ions and Cl ions is   d = 2.81 A.   The sound velocity is $3.3 \cdot 10^5$ cm. This gives for the oscillatory frequency ν = v/2d a value of $5.9 \cdot 10^{12}$ sec^{-1}, whereas the frequency of the "Reststrahlen" is $5.8 \cdot 10^{-12}$ sec^{-1}(Fig. 5-21a). The absorption coefficient of NaCl over the phonon energy range from 10^{-3} to 10^5 eV is given in Fig. 5-21b. It shows very low light absorption from the "Reststrahlen" regime ($\lambda > 10\mu m$) to the near ultra violet ($\lambda <$ 0.3μm). Such "windows" in ionic or covalent crystals make it possible to detect very small amounts of impurities when their absorption bands lie in this window. The absorption constant is again large in the x-ray region, with the characteristic absorption edges due to the

excitation from low lying electron levels in ion cores. Infrared absorption curves similar to those for NaCl are found in other ionic
crystals. Figure 5-21c shows experimental results. A quantitative
discussion of the reflectivity due to lattice vibrations is obtained
if one assumes that the atoms react like damped harmonic oscillators.

J. Impurity States in Insulators

Small variations in the chemical composition of metals will have
only a minor effect on optical properties. Small amounts of impurities can change the appearance of a perfect aluminum oxide single
crystal, which in its pure state is colorless. 0.5% of Cr^{3+} impurities make it dark red, it is then a ruby, whereas Ti^{3+} impurities color
it blue. Then it is a sapphire. Not only impurities, but also defects
produced by irradiation produce color centers in, for example, diamond.

These are only a few examples of the large variety of lattice
defects which influence optical properties of solids. Essentially one
would expect that electrons of impurities or those associated with crystal defects can be excited to higher states. This leads to an energy
absorption at well defined photon energies. For instance, the elec-
tron-hole pair, the exciton (See Chapter 1) has a potential energy
due to the Colomb attraction, just as the electron and ion of the
hydrogen atom. The exciton pair in a semiconductor shows similarly
separate energy levels. Complete ionization in hydrogen would correspond to complete separation of electron and hole. The electron would
then be inside the conduction band, and the hole in the valency band.
Figure 5-22a shows similar energy levels in a semiconductor; Fig. 5-
22b gives the absorption coefficient of gallium arsenide near 1.5
eV, which shows the absorption edge near the energy gap at 1.522 eV.
Then the formation of an exciton leads to a peak in the absorption
coefficient. This is an example of weakly bound excitons.

Typical examples of tightly bound exciton systems are F-
centers (see Chapter 1). They may be prepared by exposing NaCl
to a sodium vapor at $500^{\circ}C$, which leads to an excess sodium atom
contentration.

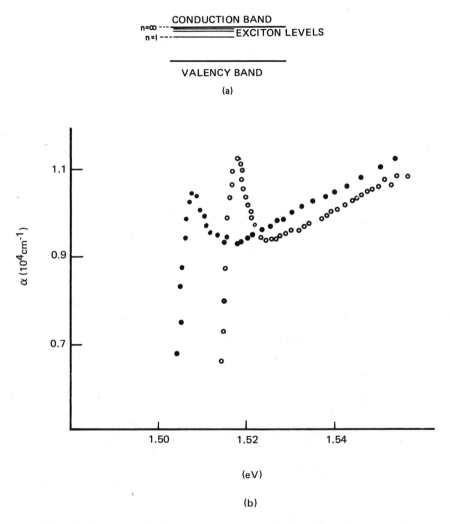

FIG. 5-22. Energy levels of an exciton (schematic, a). Exciton absorption in GaAs (b). Full circles: 90°K, open circles: 21°K. After Ref. 11, with permission.

The light absorption in alkali-halides with excess metallic atoms has been studied in detail. Figure 5-23a shows the light absorption in NaCl with excess sodium atoms. The energy diagram associated
with this F-center is given in Fig. 5-23b. The light absorption corresponds to a Frank-London transition. (Transition from the ground
state A to the excited state B in Fig. 5-23b.) The system is not in
equilibrium immediately after the absorption of a photon. Rearrang-
ing of the ions leads to a reduction of the energy of the system
(position C in Fig. 5-23b) to the equilibrium position. A phonon is
created during this process. Then the electron drops in another
Frank-London transition from C to D, creating an F'-center. The
transition from C to D is associated with fluorescent light emission.
A second relaxation process, again associated with the creation of a
phonon, moves the electron back into the ground state A in Fig. 5-23b.

Light absorption studies are especially simple in the NaCl system, since every wavelength shows equal quantum efficiency. In other
words, the number of electrons excited is proportional to the photon
number and independent of the frequency. In NaCl, with excess sodium
atoms (each is an F-center), one can study the electron distribution
with electrical current measurements. If the sample is exposed to
light of the wavelength of 4,700 $\overset{\circ}{A}$, the current is proportional to

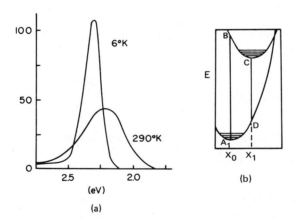

(a)

(b)

FIG. 5-23. F-center absorption (a). Configuration coordinate
diagram (b). After Ref. 11.

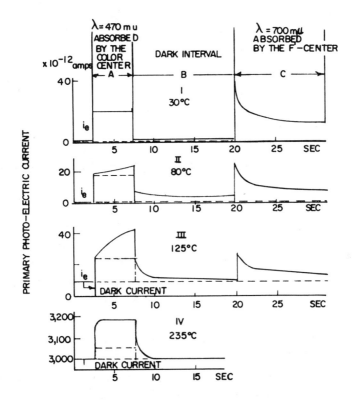

FIG. 5-24. Photoelectric current in a sodium-chloride cyrstal as a function of time. The electrodes were 4.2 mm. apart and the field was 1.07 x 10⁵ volt/cm. The crystal was illuminated first with blue and subsequently with red light, the red light being absorbed by F' centers. After Ref. 76.

the light intensity. This implies that the mean range of the excited electrons is independent of the number of electrons ejected from the initially neutral sodium atoms. This means that the ejected electrons are caught in a metastable position which existed in the sample before the illumination. The current tends to zero at room temperature as soon as the light is turned off (see Fig. 5-24). A dark interval current is found at higher temperatures. The current in the next period under illumination at 7000Å again shows a high initial value which drops rapidly with time. One interprets this behavior with a

model in which one assumes that the electrons are caught in semi-
stable configuration. These are F'-centers. On suspects that these
F' centers are F-centers which trap one electron. Such an F-center
is naturally not electrically neutral, and the potential energy of
the extra electron is much lower than that of the first electron,
which leads to the charge neutrality of the metallic impurity. Therefore, the ionization energy of the outer electron of the F'-center
is much smaller. This model also explains why the distance Λ , which
electrons emitted from F-centers can travel before being trapped,
is inversely proportional to the F-center concentration. Λ should
be inversely proportional to the concentration of traps, and the F-
centers act as traps. Quantum efficiency measurements [306] show
that the adsorption of one quantum between $-80^{\circ}C$ and $-100^{\circ}C$ leads to
the destruction of two F-centers. This would be expected from the
above model, since one F-center loses one electron, another one
absorbs an electron. Both F-centers change their status. The first
F-center transforms to a vacancy, the second to an F'-center.

The photoelectric current study on sodium chloride is temperature dependent. Figure 5-24 shows that the dark-current increases
with temperature. This is explained with the assumption that
thermal excitation frees electrons of F-centers, which are responsible
for the "dark current". Therefore, the current under illumination
with monochromatic light of wavelength $\lambda = 7000 \overset{\circ}{A}$ is much smaller
than in the first experiment, since the number of F-centers after
20 sec. in their experiment is smaller at these higher temperatures.

The absorption spectra of F and F' bands in KCl (5-23a) closely
follows the predicted form for an exponentially damped harmonic oscillator. Such a model cannot be used, however, for all F-centers, indicating that other kinds of damping other than exponential damping
character have to be taken into account.

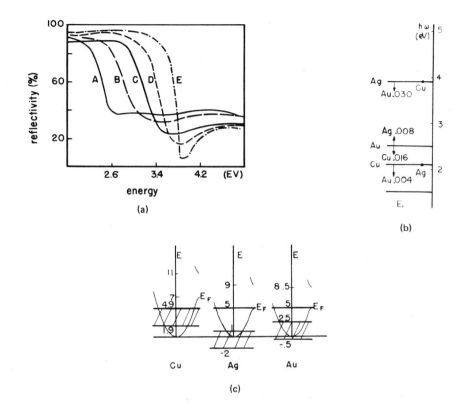

FIG. 5-25. Reflectivity of silver-gold alloys (a). (A) 100% Au; (B) 50% Au; (C) 20% Au; (D) 10% Au; (E) 100% Ag. After Ref. 307. Summary of the directions and rates of shift of the E_1 edge in the noble alloys (b). After Ref. 308. Schematic diagram of the band structure of copper, silver and gold (c). After Ref. 308.

K. Metallic Alloys

Figure 5-25a shows the reflectivity of Ag-Au alloys. The re-flectance edge shifts with increasing silver concentration to higher energy values. This should imply that the distance between Fermi energy level and top of the d-band increases gradually. Optical pro-perties of Cu-Ag, Cu-Au and Ag-Au noble metal alloys were measured by Beaglehole and Erlach in the photon energy range from 1.5 to 10 eV

[308]. Only copper-gold and silver-gold form a complete series of
solid solutions, whereas only a few percent silver will dissolve
in copper or vice versa. The authors tried to find if optical
data could explain these solubility ranges which are very difficult
to analyze with valency effects or from the atomic radii of the
three elements. The authors determined the onset of interband
transitions from the position of the first maximum in the ε_2-curve
and found for copper a value of 2.1 eV, for silver the value of
3.8 eV, and for gold the value of 2.5 eV, in good agreement with
other studies. This peak was again associated with transitions
from the top of the d-band to the Fermi energy level. Figure 5-25b
shows how these values are affected by alloying. The energy gap
$\Delta E = [$(top of d-band) to $E_F]$ does not change if silver is added to
copper and vice versa. The authors suggest therefore that the d-
electron bands of these two elements in the alloy will not interact
with (or modify) each other. They will not form a common band.
Therefore, the solubility in the copper-silver system is very small.
ΔE decreases if small amounts of copper are added to gold and
vica versa. The shift of ΔE, when gold is added to copper, can
be partly accounted for by the shift of the Fermi energy level due
to changes in the lattice parameter. The remainder is attributed
by Beaglehole and Erlach to changes in the d-band structure. However, it is also possible that changes of the Fermi surface due to
alloying, as deduced from electronic specific heat measurements on
copper-gold alloys, may be responsible for the change in ΔE. The
specific heat measurements show that the area of the Fermi surface
decreases if either a small amount of copper is added to gold, or
vica versa. This was attributed to "spheroidization effects" due
to a decrease in the energy gap across the first Brillouin zone
boundary with increasing alloying.

Reflectance measurements by Biondi and Rayne [309] on copper-
zinc alloys in the fcc phase are shown in Fig. 5-26. The reflectance
edge of copper at 6000 $\overset{\circ}{A}$ ($\sim$ 2 eV) moves to a lower wavelength (higher
photon energies) with the addition of zinc to copper. This is

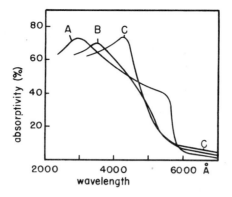

FIG. 5-26. Absorptivity of Cu and CuZn alloys. Pure Cu: A; 10 at.% Zn: B; 30 at.% Zn: C. After Ref. 309, with permission.

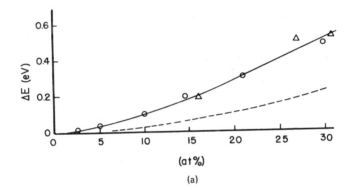

(a)

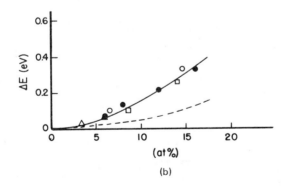

(b)

FIG. 5-27. Fermi energy shifts in CuZn alloys (a). Experimental points from Biondi and Rayne [309] given as open circles, from Pells and Montgomery [311] as triangles. Calculated energy shift with the rigid band model by Friedel: dashed line. Calculation by Yen: full line. After Ref. 310. Fermi energy shifts of copper base alloys with group IIIA elements (b). Open circles: CuAl; squares: CuGa; triangles: CuIn; full circles: CuGa. Cu data from Ref. 311. Dashed line: rigid band calculation. Full line: calculation by Yen. After Ref. 310.

expected if one associates this edge with d-band to E_F transitions.
Figure 5-27a shows this shift as determined from these experiments:
data obtained by Pells and Montgomery; Biondi and Rayne; as
calculated with the Friedel band model; and as calculated with the
assumption of screening electron clouds around Zn-impurities. The
latter is based on Friedel's calculation and a higher order correction
calculation by Yen [310]. The rigid band model predicts a more
rapid shift than found experimentally, whereas the screening assumption as predicted by Friedel underestimates the shift. However, the
more detailed calculation by Yen showed that good agreement with
experiments can be obtained if one takes higher order correction
terms into account. Calculations and experimental data on alloys
with B III and B IV impurities are given in Fig. 5-27b.

The rigid band model fails to explain reflectance measurements
of Cu-Ni alloys if one assumes that each substitutional nickel
atom subtracts one electron (or even a large fraction of one electron) from the conduction electron band. This electron reduction
reduces the E_F level, the "d-band"-"E_F" distance decreases, and
the reflection edge at 2.2 eV moves to lower photon energy values.
The reflection edge at 2.2 eV in CuNi alloys, however, does not
move to lower values with increasing nickel concentration [312].

One may wonder if the optical properties of the Cu-Ni system
could be described just as a superposition of the reflectance (or
dielectric constant) curves of pure copper and pure nickel. This,
however, is not possible because nickel at room temperature and
above the Curie temperature has electron transitions at 1.4 eV and
close to 5 eV. Other concepts have to be used to explain the optical properties of noble metals with transition element impurities.
The more recent models make extensive use of the concept of virtual
energy states as proposed by Friedel and Anderson (see Chapter 1).
These authors proposed that electrons of transition element impurities form localized electron states with a resonance energy, which
can differ from the energy level of the impurity atom in its pure
state.

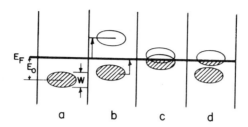

FIG. 5-28. Schematic energy level diagrams for resonant bound
states associated with a transition metal impurity in a noble metal.
Probable arrangement for Pd in Ag (a); Mn in Ag (b); Mn in Cu (c)
and (d).   After Ref. 313.

Transition element impurities in noble metals have been
studied by Myers, Walden, and Karlson [313].   These authors wanted
to find out if optical measurements could be used to determine the
characteristic parameters of transition element impurities in noble
metals.   The parameters are the distance between the Fermi energy
level and the virtual energy state, E_o, and the width of this state
W.   Figure 5-28 gives probable correlations of these parameters for
palladium in silver, manganese in silver, and manganese in copper,
as expected by these authors.   The open circles in this figure
would indicate empty virtual energy states.   Energy states of manganese may split because manganese carries a localized magnetic
moment.   These diagrams with the general position of impurity states
are obtained from electrical and magnetic measurements.   However,
the exact values of the two parameters E_o and W cannot be obtained from such measurements.

Figure 5-29 shows the interband absorption of some silver-
palladium and gold-palladium alloys in which ε_2^b/λ is plotted as a
function of photon energy.   It is possible to separate the contribution of the free electrons to ε_2 from the total energy loss function if one assumes that the Drude model is adequate to describe
this free electron contribution, which is given by:

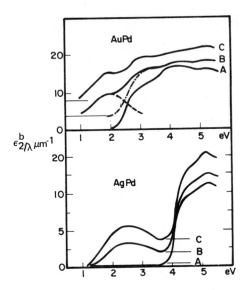

FIG. 5-29. The interband optical absorption ε_2^b/λ for some Au-Pd and Ag-Pd alloys. Top diagram, pure Au: A; 10% Pd: B; 25% Pd: C. Bottom diagram, pure Ag: A; 10% Pd: B; 23% Pd: C. After Ref. 313.

$$\varepsilon_2/\lambda = \lambda^2/[\ell\lambda_p^2(1 + \lambda^2/\ell^2)] \qquad\qquad (5\text{-}30)$$

ℓ is a parameter proportional to the mean free path. The interband contribution to ε_2 is $\varepsilon_2^b = \varepsilon_2$ (experiment) $- \varepsilon$ (calculated from Drude model of free electrons). ε_2^b shows some major peaks. The maximum of this peak in the silver alloys is clearly visible between 4 and 6 eV. Minor peaks extend from about 1 eV to 3 eV. They are attributed to a virtual energy state band with a width of about 3 eV.

The separation of ε_2 of gold into components of band-transition is difficult since the top of the d-band lies only 2 eV below the Fermi energy level. The separation of these two terms given in Fig. 5-29 for $Au_{90}Pd_{10}$ shows that the peak of the virtual energy states band lies about 2 eV below the Fermi energy level and its band width is 2 eV.

It should be noted that pure palladium shows no discernible
features at the peaks associated with palladium impurities in the
noble metal base alloys. This confirms that palladium impurities
have a different electronic structure than pure metallic palladium
(this was also deduced from susceptibility measurements in Pd-noble
metal alloys), and that the resonance between impurity states and
the noble metal matrix leads to new virtual energy states. Figure
5-29 shows that E_o and W are essentially independent of composition to at least 25 at.% Pd. It shows further that the energy
structure of the solvent is essentially unchanged even if 25% impurity atoms are added. The data indicate no change in the absorption edge of the solvent within a limit of about 0.1 to 0.2 eV
(this sensitivity is not sufficient to apply corrections as discussed in the interpretation of CuZn alloys).

The interpretation of optical properties of alloys with ferromagnetic impurities in a non-ferromagnetic matrix follows similar
arguments. Studies on the imaginary dielectric constant of copper-
manganese [313] show no evidence of virtual energy states, in spite
of the fact that manganese should carry localized moments in this
alloy (5-30). Measurements on Ag-Mn alloys (5-30c) reveal shallow
maxima between 1 and 4 eV, which should be associated with virtual
energy states. The main absorption edge in these alloys at 3.8 eV
is not shifted noticeably by the addition of even 40 at.% Mn, indicating clearly that the rigid band model cannot be used for the
interpretation of optical data of these alloys.

The same conclusion is reached in studies of copper-nickel
alloys. A superposition of this structure on the pure copper curves
could explain why the main reflectance edge at 2.2 eV is hardly
changed, but it would not explain why the reflectance minimum of
pure copper at 4 - 5 eV splits into two minima. Nickel atoms in a
copper matrix may create either virtual energy states about 4 eV
below the Fermi energy level, or the resonance interaction between
impurity and matrix will split new energy levels in the same way as
an increase in temperature will separate levels in pure copper.

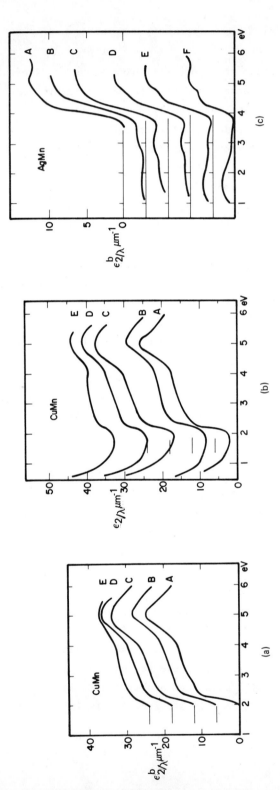

FIG. 5-30. The interband optical absorption ε_2^b/λ for Cu-Mn alloys (a). Pure Cu: A; 5%Mn: B; 15% Mn: C; 20% Mn: D; 30% Mn: E. The total optical absorption ε_2/λ for Cu-Mn alloys (b). Pure Cu: A; 5% Mn: B; 15 % Mn: C; 20% Mn: D; 30% Mn: E. The interband optical absorption $\varepsilon_2 b/\lambda$ for Ag-Mn alloys (c). Pure Ag: A; 5% Mn: B; 10% Mn: C; 15% Mn: D; 30% Mn: E; 40% Mn: F. After Ref. 313, with permission.

Reflectance measurements on some ternary noble base alloys
show that optical measurements can give some insight into the interactions of impurities in a noble metal matrix. For instance, investigations of CuZnNi [314] alloys show that the position of the main
adsorption edge near 2 eV depended only on the Zn concentration,
whereas the position of this edge for CuZnPd alloys indicated that
palladium seems to subtract some of the extra electrons of zinc
responsible for the shift of the reflectance edge [315]. One may
assume in this alloy that zinc and palladium interacted, so that
the deficit of one electron in palladium with respect to copper is
partially compensated by an extra electron of the zinc impurity.

Little data on optical properties of transition element alloys
are available. This is due to the fact that even for low photon
energies interband transitions can be found. A large number of
electron transitions are found, which makes the analysis of data
very difficult. However, since the energy band structure of
chromium alloys has been discussed in previous sections, a few comments on optical properties of chromium alloys may be in order.
Chromium shows an onset of interband transitions at 0.424 eV, as
Lenham's investigation showed (5-14). This is the same photon energy
range where the reflectance curve shows a steep slope which,
therefore, should be associated with interband transition. Schroder
and Crandell proposed that this absorption edge could be attributed
to transitions from the steep slope of the density of states curve
which lies 0.2 to 0.8 eV below the Fermi level of chromium. Measurements of the reflectance of Cr-V, Cr-Mn and Cr-Fe alloys [316] indicate that the Fermi level of chromium decreases with increasing
vanadium concentration, as expected from the rigid band model.
Since R of Cr Fe and Cr Mn alloys is independent of impurity concentration, Fe and Mn impurities should create virtual energy states.
These results agree well with the interpretation of electron transport properties of these alloys. Calculations show that electron
wavefunctions of d-electrons in bonding bands in vanadium and
chromium are similar to wavefunctions of conduction electrons. The

s and bonding d-electrons form essentially a single band in the
alloys so that a rigid band model may be appropriate. Since iron
has localized d-electrons, this is not to be expected for iron in
chromium.

III. PHOTOEMISSION

A. Electron Excitation in Solids

Light penetrating the first few hundred atomic layers of a
metal will excite electrons. The energy which an electron can absorb is equal to hν. For sufficiently large hν values, electrons
gain enough energy to escape from the metal. It should be possible
to determine from the energy distribution of escaped electrons the
energy distribution or density of stages of electrons inside the
sample. However, there exist some effects which complicate the determination of the density of states curve from such electron emission experiments. For instance, before an electron can escape from
the solid it must travel a distance inside the metal which is of
the order of the penetration depth of the light, namely a few 10 to
100 $\overset{\circ}{A}$. Since the mean path of an electron is typically 25 to 100 $\overset{\circ}{A}$,
most electrons will undergo one or more scattering processes before
leaving the metal. This leads to a "randomization" of the initial
electron velocity.

Electron emission experiments depend on the surface state. A
monolayer of impurity atoms can change the work function (energy
to remove electron with Fermi energy from metal) markedly. A
contamination atmosphere (pressure x time) of 10^{-6} Torr·sec can
lead to such a monolayer if the sticking coefficient is one, since
10^{-6} Torr corresponds to an exposure in which each surface atom is
hit on the average just once per second by an atom of the gas phase.
Recent experiments were conducted at pressures well below 10^{-9} Torr.

Problems with impurity monolayers can probably be neglected in these
new studies since the total contaminating atmosphere was far below
10^{-6} Torr·sec.

It is reasonable to assume that the electron transfer is not
only a function of the density of states of the initial and the
final state (where the final state is a vacuum state inside the
material), but it also depends on the transition probability. This
transition probability is not known. For simplicity, it is frequently assumed to be constant. Another difficulty in the analysis
of electron emission data is due to the fact that one does not know
if the electron transition is direct, with the conservation of the
electron momentum k, or indirect, with a change in k.

The number and energy of emitted electrons as a function of the
wavelength of monochromatic light, which is measured in electron
emission experiments as a function of light intensity, can be evaluated in different ways. Krolikowski and Spicer [317], for instance,
used the following approach for the analysis of their data. They
calculated the probability of exciting an electron below the Fermi
level E_F to an energy E in the conduction band above E_F and
the probability that this electron will then travel to the surface
without suffering an electron-electron collision before escaping to
the vacuum. The probability of exciting the electron is then

$$P(E, h\nu) = M \, N_c(E) \, N_v(E - h\nu) \qquad\qquad (5\text{-}31)$$

where M is the matrix element for the transition probability,
$N_c(E)$ is the density of states curve of the conduction band (this
is the energy band above the Fermi level), and $N_v(E)$ is the density
of states of the valency band (the energy band below E_F). Both
$N_v(E)$ and $N_c(E)$ are called optical density of states curves (ODS),
since they can differ from the true density of states curve. Not
all steps in the calculation may be accurate. In particular the use
of a constant transition probility matrix element is most likely
not very good.

The essential steps in the calculation of N(E) are as follows.
Initially, one calculates the escape probability. The probability Q
that the electron does not scatter is given by $Q = \exp\{-x_1/L(E)\cos\theta\}$
where θ is the angle between electron velocity and the normal to
the surface, x_1 is the distance of the electron to the surface, and
L(E) is the electron-electron scattering length. One may wonder,
keeping in mind that both $N_c(E)$ and $N_v(E - h\nu)$ are showing up as a
product in Eq. 5-22, how one can determine both these parameters
and the threshold function $T_F(E)$ from Q and P(E,hν) in a self-
consistent way. The electron-electron scattering length L(E) is not
known even in a case like copper.

The self-consistent process used by Krolikowski and Spicer to
determine the optical density of states consists of the following
phases. First, the location of peaks in the valency band are deduced from the peaks in the curve of the number of electrons with
energy E per energy interval which gives the "photoelectric energy
distribution curve" (EDC curve). Then one assumes that the shape
of the valency band has a shape similar to the EDC curve a few
electron volts above the threshold energy. The shape of the conduction band is then determined by arbitrarily adjusting the location of the bottom of the free electron band and superimposing conduction band structure upon this free electron envelope. The relative peak heights of the conduction band structure were chosen in
such a way as to obtain the best overall agreement with experimental
data.

This procedure gave the first approximation of the optical
density of states curve. This curve, which gave both $N_c(E)$ and
$N_v(E)$, was then used to recalculate the various data points. Only
slight modifications in the N(E) curve were usually needed to improve agreement between the calculated curves and the measurements.
In a first approximation one may, therefore, use the EDC curve as
optical density of states curve.

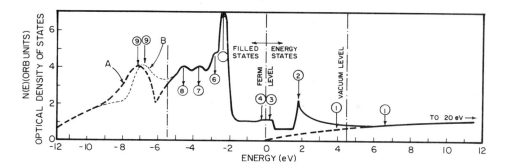

FIG. 5-31. ODS states for Cu, extending from $\sim$ - 11 to $\sim$ + 20 eV above the Fermi level. The curve below - 5 eV is dashed because this is determined with less accuracy than that at higher energy. Curves A and B in this region indicate two possible ODS curves. See [318] for details. Reproduced with permission from Ref. 317.

B. Elements

Figure 5-31 gives the optical density of states curve (ODS curve) of copper as obtained by Krolikowski and Spicer [317]. It shows the rather flat N(E) section near the Fermi energy level, extending from -2 eV to + 1.5 eV, if we neglect the subdivision into three horizontal lines of slightly different heights. High N(E) values found between - 2 eV and -9 eV are due to d-electron states. The sharp increase of N(E) at - 2 eV would be responsible for the optical absorption edge of ε_2 or for the reflection edge of R at 2 eV for copper which, as suggested by Mott and Jones [29], should be associated with transitions from the top of the d-band to the Fermi energy level. However, the prediction that the maximum of ε_2 or the minimum in R at 4 eV is associated with transitions from the Fermi level to higher states as originally proposed cannot be confirmed from these ODS curves. There is no peak in $N_v(E)$ at E_F + 4 eV. The distance between the sharp edge of the N(E) curve at - 2 eV and the first sharp peak above E_F is about 4 eV, just where the peak in the imaginary dielectric constant and its change in shape with temperature reveals transitions from the top of the d-band to states above E_F .

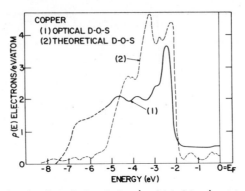

FIG. 5-32. Optical D-O-S of Cu (solid line) obtained from photoemission studies by Krolikowski. The D-O-S for Cu calculated by
Snow is given by the dashed line. After Ref. 318.

The ODS curve is not very accurately determined for energy
levels below E_F - 5.5 eV. It reveals high density of states values
over an energy range of 10 eV. These photoemission measurements
make it possible to determine the difference between the Fermi
energy level and the vacuum level, so that one can obtain the value
of the work function as shown in Fig. 5-31.

Figure 5-32 gives a comparison of calculated N(E) curves [319]
with experimental data by Krolikowski and Spicer [317]. The position of the upper maximum of the d-band, and the first steep slope
in N(E) below E_F near the Fermi energy level is found by these authors in about the same position. High N(E) values are always found
for the d-band; however, quantitative agreement of detailed features
in the density of states curves is not obtained. The low lying
parts of the electron emission curves can be affected markedly by
sample surface state [320]. Some of the features of the EDC curves
should be associated with impurity states of absorbed atoms, as both
experiments and calculation indicated. This may explain why measurements by Nielson [321] on copper show a narrower d-band than obtained
by Krolikowski and Spicer.

The density of states curve of transition elements, as obtained
from electron emission experiments, generally conforms with older
models. Figure 5-33 shows results obtained on nickel, a very

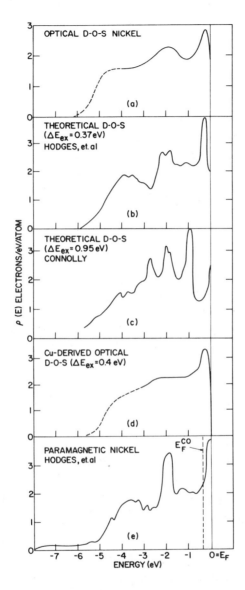

FIG. 5-33. Optical D-O-S of Ni (a), D-O-S calculated by Hodges et al. [322] (b), D-O-S calculated by Connolly [323] (c), Optical D-O-S derived from Cu using rigid band approximation (d), D-O-S calculated for nonmagnetic Ni by Hodges et al. [322] (e). The Fermi level for nonmagnetic Co using the rigid band model is also indicated. After Ref. 318, with permission.

frequently studied transition element. This element contains itinerant magnetic electrons. It has only one electron less than copper. Therefore, one frequently uses the energy band structure of
copper as a basis for determining the energy band structure of
nickel, as discussed before. In this approach, one splits the band
into subbands with spin-up and down, and shifts these two sections
with respect to each other by $\Delta E_{ex.}$, the exchange energy. The results of theoretical calculations for the case of nickel with a
splitting between bands with spin-up and spin-down of $\Delta E_{ex.}$ =
0.37 eV to 0.95 eV are shown in Fig. 5-33. Whereas the smaller
shift gives a sharp maximum slightly below the Fermi energy level,
the energy split of 0.95 eV gives a pronounced minimum in N(E) about
0.5 eV below the Fermi energy level. Such a minimum cannot be reconciled with experimental results. Instead of calculating the density of states curve from the theoretical curve of copper, with a
band splitting $\Delta E_{ex.}$, one can also try to obtain the density of
states curve of nickel from the measured optical density of states
curve of copper by splitting this experimental band, and shifting
the two subbands by $\Delta E_{ex.}$. Results of such an approach are given
in Fig. 5-33 which agree well with the experimental curve of nickel.
The last figure in this diagram gives the density of states for
nickel in the paramagnetic state as calculated by Hodges et al.
[322]. The calculation leads to a marked increase in the density
of states at the Fermi level, compared with the calculation by the
same group of authors for the ferromagnetic case. The maximum in
N(E) about 2 eV below the Fermi level is also much more pronounced
in the paramagnetic state.

Figure 5-34 shows the photoelectric energy distribution curve
of cobalt. It shows a broad maximum over 5 to 6 eV with a minor
peak close to E_F and a shallow maximum about 2 eV below E_F. The
dashed line in this curve gives results of calculations, which
started with the energy band of nickel, and (using the rigid band
concept) then shifted the two subbands by $\Delta E_{ex.}$ = 1.05 eV in
respect to each other. This calculation reveals three sharp peaks

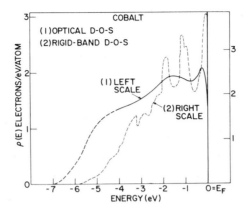

FIG. 5-34. The optical D-O-S of Co is shown by the solid line. The dashed line shows the D-O-S estimated for ferromagnetic Co using Connolly's bands for fcc Ni with $\Delta E_{ex.}$ = 1.05 eV. After Ref. 318.

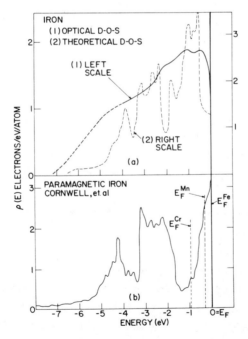

FIG. 5-35. Optical D-O-S of ferromagnetic Fe (solid line) and calculated D-O-S of ferromagnetic Fe based on Wakoh and Yamashita's energy bands ($\Delta E \sim 1.8$ eV) (a), calculated D-O-S of nonmagnetic Fe [324](b). Fermi levels for nonmagnetic Cr (6 electrons) and Mn (7 electrons) using the rigid-band model are also indicated. After Ref. 318.

separated by approximately one eV, which could not be detected in
the electron emission experiments. These peaks are also well above
experimental peaks found in older experiments at 4 to 5 eV below
the Fermi energy level, attributed by Eastman to impurity states due
to absorbed atoms. The calculation by Connolly [323] again assumed
a rigid band. The relative shift of sub-bands with opposite spin
is obtained in such a way that it gives 1.56 holes per atom.
This gives the right magnetization. This calculation gives a
slightly larger exchange splitting than a previous estimate by
Hodges and Ehrenreich, who obtained a value of $\Delta E_{ex.} \sim 0.86$ eV.

Results of experiments on iron (see Fig. 5-35) obtained by
Eastman shows the absence of a hump at E_F - 5 eV, disagreeing with
older data which were obtained under less stringent vacuum conditions. Calculated N(E) curves in Fig. 5-35 have been given which
were based on calculations of energy bands by Yamashita and
Wakoh. The exchange splitting in this calculation of the energy
bands was found to be energy dependent, and a $\Delta E_{exch.} \approx 1.8$ eV at
the Fermi level gave the right magnetization. One obtains general
agreement between calculated $N(E)$ and measured curves. Experimentally dominant peaks at -0.5 and -1.1 eV are in excellent agreement with theoretical peaks at -0.5 eV and -1.0 eV, respectively.
The predicted minimum at 2 eV, however, is not found experimentally.

Results of measurements and experiments on chromium are given
in Fig. 5-36. The electron emission DOS curve shows a shoulder at
-0.4 eV and peaks at -1.2 eV and -2.3 eV, with a high optical density of state curve extending to 4 eV below the Fermi energy level.
The peaks at -1.2 eV and -2.3 eV agree well with calculated peaks
except that the experiments yield much shallower curves than experiments. The band width in both types of investigation is about
the same, namely about 4 eV. The calculation gives, however, low
N(E) values from E_F to E_F -0.8 eV, in disagreement with optical
measurements on pure chromium, where Lenham's measurements show
already an increase in intraband transitions from 0.24 eV on.
This would be in better agreement with the electron emission data

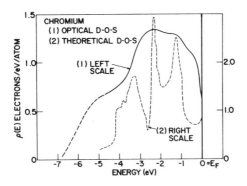

FIG. 5-36. The optical and the calculated D-O-S curve of
chromium. Reproduced with permission from Ref. 318.

than with calculations. The electron emission data agree also with
low temperature specific heat measurements if one uses the rigid
band model to calculate N(E) from the electronic specific heat coefficient of chromium-vanadium alloys.

The discussion of optical and electron emission data shows that
density of states curves obtained by different techniques for pure
elements give agreement in the general features of the N(E) curves.
However, they differ in the details of the structures. It is frequently necessary to use a series of different types of measurements
to obtain reliable data on the density of states curves. In comparing these curves, one should keep in mind that the resolution of
the N(E) curves is of the order of 0.1 to 0.3 eV for electron emission studies, whereas N(E) obtained from electron specific heat
measurements is determined in the energy interval of $k_B T < 10^{-3}$ eV.
Naturally, possible enhancement effects, uncertainties due to the
contribution from magnetic ordering, and other effects may easily
lead to errors in the evaluation of specific heat data. The resolution of the N(E) curves obtained from these calculations depends
mostly on the computer program. The width of individual steps indicates the resolution.

These differences in resolution explain partly why electron
emission, x-ray data, and calculations disagree, but they cannot

explain why even the position of major peaks may differ by more than
1 eV, as has shown up in some investigations.   A major difference
between calculations and measurements is due to the fact that calculations of N(E) yield bulk properties, whereas measurements may be
markedly affected by the conditions of the surface.   The N(E) curves
obtained from electron emission experiments give the same result
for bulk samples and evaporated films, because electrons emitted from
the bulk sample come from the surface layer.   These experiments show
only that the N(E) curves of a thin film are similar to the N(E)
curve in the surface region of a bulk sample.   Calculations by Pendry
and Forstmann   [320] indicate that surfaces may be responsible for
additional states in the N(E) curve well below the Fermi energy
level.

C. Alloys

The electron emission curves of disordered copper-gold and
silver-gold alloys are shown in Fig. 5-37.   In both systems, the
electron emission curves change smoothly with alloying.   The rather
narrow d-band for pure silver broadens if gold is added.   The addition of gold to copper leads to a continuous narrowing of the d-band.
The edge in the electron emission curve, associated with the top of
the d-band, is sharp for all alloys and elements.   The slope of this
edge in the alloys is similar to the slope for pure Au and Cu, even
for the Ag_3Au_1 alloy, in spite of the fact that pure Ag shows a
steep slope.   There is no evidence of virtual energy states in the
section just below the Fermi energy level.   It is tempting to associate individual peaks in the d-band section to specific d-wave functions.   However, the positions of these peaks shift rather irregu-
larly with alloy concentration.   More detailed theoretical studies
are needed before the compositional dependence of N(E) of alloys,
and possible ordering effects, can be worked out.
An extensive investigation of the electron emission characteristics has been conducted on copper-nickel alloys, since these

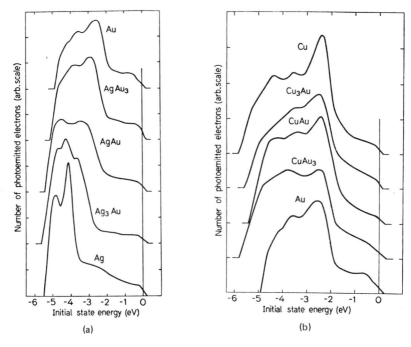

FIG. 5-37. Electron energy distribution curves for the Ag-Au
system (a). Electron energy distribution curves for the Cu-Au system (b). After Ref. 321.

elements form a complete series of solid solutions, and since both
elements have been studied in detail. For copper-nickel alloys,
specific heat and susceptibility measurements were originally explained with a rigid band model. Optical reflectance measurements,
however, indicated initially that nickel in this alloy system is responsible for virtual energy states. Earlier measurements of the
dielectric constant did not have enough resolution to settle the
question of whether the position of the main absorption edge (or
the reflectance edge) was composition dependent or not.

The position of E_F with respect to the top of the d-band in a
$Cu_{75}Ni_{25}$ alloy for the rigid band model is given in Fig. 5-38. The
rigid band model predicts a decrease of the distance "E_F" to "top of
the d-band" by 1 eV, whereas this distance can stay constant for the
virtual bound state model. There one should keep in mind that the

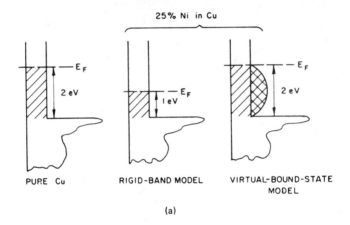

(a)

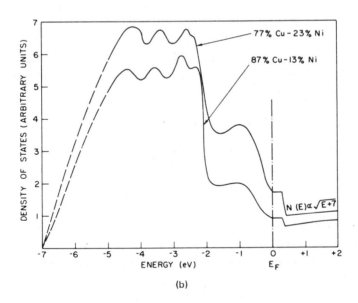

(b)

FIG. 5-38. Illustration of the filled density of states in Cu
and Cu-Ni alloys, showing the expected behavior for the rigid-band
and virtual-bound-state model (a). ODS for $Cu_{77}Ni_{23}$ and $Cu_{87}Ni_{13}$(b).
After Ref. 325, with permission.

"E_F" - "top of the d-band" distance should be constant for the copper matrix. The states associated with nickel atoms may be found
above, below, or at the Fermi level. The electronic specific heat
of the CuNi alloys increases only slightly with increasing nickel
concentration, and the d-band of nickel reaches the pure Fermi level.
Therefore, the schematic diagram of the alloy shows a slight bulge
in the density of virtual bound states near E_F. The optical density
of states curves as obtained for $Cu_{77}Ni_{23}$ and $Cu_{87}Ni_{13}$ are given
in Fig. 5-38. It shows that the position of the top of the d-band
at E_F - 2 eV is independent of small amounts of nickel impurities,
and even the detailed shape of the d-band for both alloys is the
same. The major change, compared with pure copper, is found in the
region from E_F - 2 eV to E_F where the addition of nickel adds new
states, which should be the virtual energy states as predicted by
Friedel and Anderson. The distance between the top of the d-band
and E_F, obtained from the assumption that the d-states width of
nickel changes linearly with composition, is given by Seib and
Spicer as 0.95 ± 0.05 eV, in good agreement with thermopower and
specific heat data. Investigations on nickel-rich nickel copper
alloys again show that the rigid band model cannot describe the
electron emission experiments properly. The data indicate that
alloys with less than 20 at.% Cu have N(E) curves very similar to
nickel. It is possible to associate electron states in alloys with
up to 39 at.% Cu impurities with similar states in pure copper, if
the energy is less than E_F - 2 eV.

Photoemission experiments on silver-rich silver-palladium
alloys lead to results similar to those for CuNi alloys, in spite of
the fact that palladium is paramagnetic, and nickel is ferromagnetic
in its pure state [326]. The d-states in silver should be responsible
for the high number of photon emitted electrons 4 eV below the Fermi
level. Additions of palladium increase the height of the peak but
do not shift its position. However, additional states are added
in the flat section of the number of photoemitted electrons from
E_F - 3.7 eV to E_F, indicating that the virtual energy states are

connected with palladium impurities in silver close to the Fermi
level. Figures 5-30a to c show that the addition of manganese to
both silver and copper leads to the same result.

APPENDIX

Selected Constants

Avogadro's Number	0.6023×10^{24} mole^{-1}
Gas Constant R	1.987 cal/mole $^\circ$K
Planck's Constant h	6.626×10^{-34} Joule·sec
Boltzmann Constant k_B	1.3810×10^{-23} Joule/$^\circ$K
	0.86×10^{-4} electron Volt/$^\circ$K
Permittivity of Free Space ε_o	8.854×10^{-12} Coul/Volt · m
Inductance of Free Space μ_o	$4\pi \times 10^{-7}$ Ohm sec/m
Velocity of Light c	2.998×10^8 m/sec
Charge of Electron	1.602×10^{-19} Coul.
Electron Mass	9.11×10^{-28} gm
Mass of Hydrogen Atom	1.6617×10^{-24} gm
Bohr Magnetone μ_B	9.273×10^{-24} Amp · m^2
Bohr Radius	0.529×10^{-10} m

1 Joule = 1 Volt · Coul. = 10^7 erg = 0.2340 cal. = 0.625×10^{19} eV

1 eV = 23.06 kcal/mole

1 cal = 4.184 Joule

1 Gauss = 10^{-4} Weber/m^2 = 10^{-4} Volt sec/m^2

1 Oersted = $10^3/4\pi$ Amp/m

1 Newton = 10^5 dyn

1 Ryd = 13.60 eV

REFERENCES

1. Moffatt, W.G., Pearsall, G.W., and Wulff, J., "The Structure and Properties of Materials," Vol. 1, Wulff, Ed., J. Wiley, New York, 1964.

2. Ziman, J.M., "Electrons and Phonons," Clarendon Press, Oxford, England, 1963.

3. Kittel, C., "Introduction to Solid State Physics," 3rd Edition, J. Wiley, New York, 1968.

4. Masing, G., "Handbuch der Metalphysik," Akademische Verlags-buchhandlung, Leipzig, 1935.

5. Callaway, J., "Solid State Physics," Vol. 7, Seitz and Turnbull, Eds., Academic Press, New York, 1966, pp. 99.

6. King, H.W., "Physical Metallurgy," Cahn, Ed., North Holland, New York, 1970, 33.

7. Massalski, T.B. and King, H.W., _Progr. Mat. Science_, _10_, 1 (1961).

8. Massalski, T.B., "Physical Metallurgy," Cahn, Ed., North Holland Pub. Co., Amsterdam, 1970, pp. 159.

9. Masing, G., "Lehrbuch der Allg. Metallkunde," Springer, Berlin, 1950.

10. Wernick, J.H., "Physical Metallurgy," Cahn, Ed., North Holland Pub. Co., Amsterdam, 1970, pp. 229.

11. Pick, H., "Optical Properties of Solids," Abeles, Ed., North Holland Pub. Co., Amsterdam, 1972, pp. 653.

12. Isobe, M., _Rept. Tohoku Univ._, _A3_, 78 (1951).

13. Neighbours, J.R. and Alers, G.A., _Phys. Rev._, _111_, 707 (1958).

14. Gerlich, D., _Phys. Rev._, _135_, A1331 (1964).

15. Debye, P., _Ann. d. Physik_, _39_, 789 (1912).

16.   Einstein, A., Ann. d. Physik, 22, 180, 800 (1907).

17.   Walker, C.B., Phys. Rev., 103, 547 (1956).

18.   DeLaunay, J., "Solid State Physics," Vol. 2, Seitz and
      Turnbull, Eds., Academic Press, New York, 1956, pp. 219.

19.   Born, M. and von Karman, T., Physik Z., 13, 297 (1912).

20.   Woods, A.D.B., Brookhouse, B.N., Cowley, R.A. and Cochran, W.,
      Phys. Rev., 131, 1025 (1963).

21.   Fuchs, K., Proc. Roy. Soc. (London), A153, 622 (1935); A151,
      585 (1936).

22.   Fuchs, K., Proc. Roy. Soc. (London), A157, 444 (1936).

23.   Kaufmann, A.R., Pan, S.T. and Clark, J.A., Rev. Mod. Phys., 17,
      87 (1945).

24.   Fine, P.C., Phys. Rev., 56, 355 (1939).

25.   Nicklow, R.M., Gilat, G., Smith, H.G., Raubenheimer, L.J. and
      Wilkinson, M.K., Phys. Rev., 164, 922 (1967).

26.   Madelung, O., "Grundlagen der Halbleiterphysik," Springer,
      Berlin, 1970.

27.   Leath, P.L. and Goodman, B., Phys. Rev., 181, 1062 (1969).

28.   Powell, P.M., Martel, P. and Woods, A.D.B., Phys. Rev., 171,
      727 (1968).

29.   Mott, N.F. and Jones, H., "The Theory of the Properties of
      Metals and Alloys," Dover Pub., New York, 1958.

30.   Mott, N.F., Adv. in Physics, 13, 325 (1964).

31.   Ziman, J.M., "Principles of the Theory of Solids," Cambridge
      University Press, 1972.

32.   Kittel, C., "Introduction to Solid State Physics," 4th Edition,
      J. Wiley, New York, 1971.

33.   Ziman, J.M., "Solid State Physics," Vol. 26, Ehrenreich, Seitz
      and Turnbull, Eds., Academic Press, New York, 1971, pp. 1.

34.   Dimmock, J.O., "Solid State Physics," Vol. 26, Ehrenreich,
      Seitz and Turnbull, Eds., Academic Press, New York, 1971, pp. 104.

35.   Kronqvist, E., Arkiv Fysik, 5, 453 (1952).

36.   Overhauser, A.W., Phys. Rev., 156, 844 (1967).

37.   Snow, E.C., Phys. Rev., 171, 785 (1968).

38.   Gray, D. and Brown, E., Phys. Rev., 160, 567 (1967).

39.   Arlinghaus, F.J., Phys. Rev., 186, 609 (1969).

40.   Kimball, J.C., Stark, R.W. and Mueller, F.M., Phys. Rev., 162,
      600 (1967).

41. Ketterson, J.B. and Stark, R.W., _Phys. Rev._, 156, 748 (1967).

42. Falicov, L.M., _Phil. Trans. (London)_, A 255, 55 (1962).

43. Harrison, W.A., _Phys. Rev._, 118, 1182 (1960).

44. Fetter, A.L. and Hohenberg, P.C., "Superconductivity," Parks, Ed., M. Dekker, New York, 1969, pp. 817.

45. Hoffstein, V. and Boudreaux, D.S., _Phys. Rev._, B2, 3013 (1970).

46. Mueller, F.M., Freeman, A.J., Dimmock, J.O. and Furdyna, A.M., _Phys. Rev._, B1, 4617 (1970).

47. Mattheiss, L.F., _Phys. Rev._, 139, A1893 (1965).

48. Lomer, W.M., _Proc. Phys. Soc. (London)_, 84, 327 (1964).

49. Kane, E.O., _J. Phys. Chem. Sol._, 1, 249 (1957).

50. Smith, R.A., "Semiconductors," Cambridge University Press, 1959.

51. Herman, F., _J. Electronics_, 1, 103 (1955).

52. Hilsum, C. and Rose-Innes, A.C., "Semiconducting III-V Compounds," Pergamon Press, New York, 1961.

53. Collins, T.C., Stuckel, D.J. and Euwema, R.N., _Phys. Rev._, B1, 724 (1970).

54. Ih, C.S. and Langenberg, D.N., "The Physics of Semimetals and Narrow Gap Semiconductors," Carter and Bate, Eds., Pergamon Press, Oxford, 1971.

55. Friedel, J., _Can. J. Phys._, 34, 1190 (1956).

56. Anderson, P.W., _Phys. Rev._, 124, 41 (1961).

57. Daniel, E. and Friedel, J., "Proc. IX Intl. Conf. on Low Temperature Physics," Columbus, Ohio, 1964, LT 9, pp. 933, Plenum Press, New York, 1965.

58. Cohen, M.H. and Heine, V., _Adv. in Phys._, 7, 395 (1958).

59. Friedel, J., _Adv. in Phys._, 3, 447 (1954).

60. Beeby, J.L., _Phys. Rev._, 135, A130 (1964).

61. Velicky, B., Kirkpatrick, S. and Ehrenreich, H., _Phys. Rev._, 175, 747 (1968).

62. Levin, K. and Ehrenreich, H., _Phys. Rev._, B3, 4172 (1971).

63. Amar, H., Johnson, K.H. and Sommers, C.B., _Phys. Rev._, 153, 655 (1967).

64. Segall, B., _Phys. Rev._, 125, 109 (1962).

65. Kirkpatrick, S., Velicky, B. and Ehrenreich, H., _Phys. Rev._, B1, 3250 (1970).

66. Tauc, J., "Proceedings of the Intl. School of Physics, Enrico Fermi Course XXXIV, 1965, The Optical Properties of Solids," Academic Press, New York, 1966, pp. 63.

67. Stuckel, D.J., _Phys. Rev._, _B3_, 3347 (1971).

68. Buot, F.A., "The Physics of Semimetals and Semiconductors," Carter and Bate, Eds., Pergamon Press, Oxford, 1971, pp. 102.

69. McGroddy, J.C., McAlister, A.J. and Stern, E.A., _Phys. Rev._, _139_, A1844 (1965).

70. Tracey, J.M. and Stern, E.A., _Phys. Rev._, _B8_, 582 (1973).

71. Anderson, J.R. and Hines, D.C., _Phys. Rev._, _B2_, 4752 (1970).

72. Mott, N.F., _Proc. Cambridge Phil. Soc._, _32_, 281 (1936).

73. Brailsford, A.D., _Phys. Rev._, _149_, 456 (1966).

74. Shepherd, J.P.G. and Gordon, W.L., _Phys. Rev._, _169_, 541 (1968).

75. Higgins, R.J. and Markus, J.A., _Phys. Rev._, _141_, 553 (1966).

76. Pohl, R.W., "Einführung in die Optik," Springer, Berlin, 1948.

77. Honda, K. and Kaya, S., _Sci. Rep. Tohoku Univ._, _15_, 721 (1926).

78. Kaya, S., _Sci. Rep. Tohoku Univ._, _17_, 639 (1928).

79. Bozorth, R.M., "Ferromagnetism," Van Nostrand, New York, 1951.

80. Van Vleck, J.H., "Theory of the Electrical and Magnetic Susceptibilities," Clarendon Press, Oxford, 1932.

81. Henry, W.E., _Phys. Rev._, _88_, 559 (1952).

82. Morrish, A.H., "The Physical Principles of Magnetism," J. Wiley, New York, 1966.

83. Gerstenberg, D., _Ann. d. Physik_, _2_, 236 (1958).

84. Stoner, E.C., _Proc. Roy. Soc. (London)_, _A-165_, 327 (1938).

85. Stoner, E.C., _Proc. Roy. Soc. (London)_, _A-169_, 339 (1939).

86. Stoner, E.C., _Phil. Mag._, _25_, 899 (1938).

87. Wohlfarth, E.P., _Rev. Mod. Phys._, _25_, 211 (1953).

88. Peierls, R.E., _Z. f. Physik_, _80_, 763 (1933).

89. Peierls, R.E., _Z. f. Physik_, _81_, 168 (1933).

90. Wannier, G.H. and Upadhyaya, U.N., _Phys. Rev._, _136_, A803 (1964).

91. Vogt, E., "Magnetism and Metallurgy," Academic Press, New York, 1969.

92. Garber, M., Henry, W.G. and Hoeve, H.G., _Can. J. Phys._, _38_, 1595 (1960).

93. Markus, J.A., _Phys. Rev._, _76_, 413, 621 (1949).

94. Shoenberg, D., _Proc. Roy. Soc. (London)_, _A170_, 39, 341 (1939).

95. Kriessman, C.J. and Callen, H.B., _Phys. Rev._, _94_, 837 (1954).

96. Shimizu, M., Katsuki, A. and Takahashi, T., _J. Phys. Soc. Japan_, _18_, 240 (1963).

97. Hoare, F.E. and Yates, B., <u>Proc. Roy. Soc. (London)</u>, <u>240A</u>, 42 (1957).

98. Shimizu, M., Takahashi, T. and Katsuki, A., <u>J. Phys. Soc. Japan</u>, <u>17</u>, 1740 (1962).

99. Vogt, E., <u>Z. angew. Phys.</u>, <u>29</u>, 241 (1968).

100. Hahn, A. and Treutmann, W., <u>Z. angew. Phys.</u>, <u>26</u>, 129 (1969).

101. Freeman, A.J., Furdyna, A.M. and Dimmock, J.O., <u>J. Appl. Phys.</u>, <u>37</u>, 1256 (1966).

102. Wannier, G.H., <u>Rev. Mod. Phys.</u>, <u>17</u>, 50 (1945).

103. Katsuki, A., <u>J. Faculty of Science, Shinshu Univ.</u>, <u>2</u>, 19 (1967).

104. Newmann, M.M. and Stevens, K.W.H., <u>Proc. Phys. Soc. (London)</u>, <u>74</u>, 290 (1959).

105. Clogston, A.M., Matthias, B.T., Peter, M., Williams, H.J., Corenzwit, E. and Sherwood, R.C., <u>Phys. Rev.</u>, <u>125</u>, 541 (1962).

106. Martin, D.H., "Magnetism in Solids," The MIT Press, Cambridge, Mass., 1967.

107. Madelung, O., "Physics of III-V Compounds," J. Wiley, New York, 1964.

108. Wehrli, L., <u>Phys. kondens. Materie</u>, <u>8</u>, 87 (1968).

109. Stevens, D.K. and Crawford, J.H., Jr., <u>Phys. Rev.</u>, 92, 1065 (1953).

110. Tylor, F., <u>Phil. Mag.</u>, 11, 596 (1931).

111. Sucksmith, W. and Pearce, R.R., <u>Proc. Roy. Soc. (London)</u>, <u>A167</u>, 189 (1938).

112. Bertaut, E.F. and Pauthenet, R., <u>Proc. Inst. Elec. Eng.</u>, <u>Pt B 104</u>, 261 (1957).

113. Lidiard, A.B., <u>Rept. Progr. Phys.</u>, <u>25</u>, 441 (1962).

114. Kneller, E., "Ferromagnetismus," Springer, Berlin, 1962.

115. Chikazumi, S., "The Physics of Magnetism," J. Wiley, New York, 1964.

116. Onsager, L., <u>Phys. Rev.</u>, <u>65</u>, 117 (1944).

117. Domb, C., <u>Adv. in Phys.</u>, <u>19</u>, 339 (1970).

118. Domb, C., "Magnetism," Vol. IIA, Rado and Suhl, Eds., Academic Press, New York, 1965.

119. Fisher, M.E., <u>Rept. Progr. Phys.</u>, <u>XXX</u>, Pt. 2, 615 (1967).

120. Vonsovskii, S.V., <u>J. Phys. (USSR)</u>, <u>10</u>, 468 (1946).

121. Zener, C., <u>Phys. Rev.</u>, <u>81</u>, 440 (1951).

122. Zener, C., <u>Phys. Rev.</u>, <u>83</u>, 299 (1951).

123. Asano, S. and Yamashita, J., J. Phys. Soc. Japan, 23, 714
     (1967).

124. Bastow, T.J. and Street, R., Phys. Rev., 141, 510 (1966).

125. Overhauser, A.W., Phys. Rev., 128, 1437 (1962).

126. Koehler, W.C., Moon, R.M., Trego, A.L. and Mackintosh, A.R.,
     Phys. Rev., 151, 405 (1966).

127. Oguchi, T. and Obokato, T., J. Phys. Soc. Japan, 27, 1111
     (1969).

128. Robbins, C.G., Claus, H. and Beck, P.A., Phys. Rev. Lett., 22,
     1307 (1969).

129. Mishra, S., Beck, P.A. and Foner, S., J. Phys. Chem. Sol., 32,
     1979 (1971).

130. Néel, L., Ann. Phys. (Paris), 1, 163 (1949).

131. Pauthenet, R. and Bochirol, L., J. Phys. Rad., 12, 249 (1951).

132. Guillaud, C., J. Phys. Rad., 12, 239 (1951).

133. Néel, L., Ann. Phys. (Paris), 3, 137 (1948)

134. Phillips, N.E., CRC Critical Rev. in Solid State Sci., 2, 467
     (1971).

135. Chevenartd, P., Revue de Metallugie, 25, 14 (1928).

136. Masumoto, H., Sci. Rep. Tohoku Univ., 23, 265 (1934).

137. Néel, L., C.R. Acad. Sci. (Paris), 237, 1613 (1953).

138. Néel, L., J. Phys. Rad., 15, 525 (1954).

139. Taniguchi, S. and Yamamoto, Y., Sci. Rep. Tohoku Univ., A6,
     330 (1954).

140. Graham, C.D., Jr., "Magnetic Properties of Metals and Alloys,"
     ASM, Cleveland, 1959, pp. 288.

141. Nernst, W. and Lindemann, F., Preuss. Akad. d. Wiss. (Berlin),
     Sitzungsber., 22, 494 (1911).

142. Debye, P., Ann. d. Phys. (Leibz.), 39, 739 (1912).

143. Martin, D.L., Phys. Rev., 141, 576 (1966).

144. Kagan, Yu.M. and Iosilevski, Ya.A., Zh. Eksp. i. Teor. Fiz.,
     45, 819 (1963).

145. Maradudin, A.A., "Solid State Physics," Vol. 18, Seitz and
     Turnbull, Eds., Academic Press, New York, 1966, pp. 273.

146. Simmons, R.O. and Balluffi, R.W., Phys. Rev., 117, 52 (1960).

147. Brooks, C.R. and Bingham, R.E., J. Phys. Chem. Sol., 29, 1553
     (1968).

148. Christy, R.W. and Lawson, A.W., J. Chem. Phys., 19, 517 (1951).

149. Clarebrough, L.M., Hargreaves, M.E. and West, G.W., Proc. Roy. Soc. (London), A232, 256 (1953).

150. Stoner, E.C., Phil. Mag., 21, 145 (1936).

151. Hindley, N.K. and Rhodes, P., Proc. Phys. Soc., 81, 717 (1963).

152. Giannuzzi, A., Tomaschke, K. and Schröder, K., Phil. Mag., 21, 479 (1970).

153. Davis, T.H. and Rayne, J.A., Phys. Rev., B6, 2931 (1972).

154. Haga, E., Proc. Phys. Soc. (London), 91, 169 (1967).

155. Stern, E.A., Phys. Rev., 144, 545 (1966).

156. Isaacs, L.L. and Massalski, T.B., Phys. Rev., 138, A134 (1965).

157. Livingston, J.D., Phys. Rev., 129, 1943 (1963).

158. Sevin, R., "Encyclopedia of Physics," Vol. XV, pp. 210, Springer, Berlin, 1956.

159. Ziman, J.M., Adv. in Phys., 10, 1 (1961).

160. Montgomery, H., Pells, G.P. and Wray, E.M., Proc. Roy. Soc., A301, 261 (1967).

161. Heiniger, F., Bucher, E. and Muller, J., Phys. kondens. Materie, 5, 243 (1966).

162. Gladstone, G., Jensen, M.A. and Schrieffer, J.R., "Superconductivity," Parks, Ed., Marcel Dekker, New York, 1969, pp. 665.

163. Goodenough, J.B., Phys. Rev., 120, 67 (1960).

164. Cheng, C.H., Wei, C.T. and Beck, P.A., Phys. Rev., 120, 426 (1960).

165. Gupta, K.P., Cheng, C.H. and Beck, P.A., "Metallic Solid Solutions,", Friedel and Guinier, Eds., W. A. Benjamin, New York, 1963.

166. Schröder, K., Phys. Rev., 125, 1209 (1962).

167. Schröder, K. and Shabel, B., J. Phys. Chem. Sol., 27, 253 (1966).

168. Christian, J.W., "The Theory of Transformations in Metals and Alloys,", Pergamon Press, Oxford, 1965.

169. Crawford, J.H., Jr. and Stevens, D.K., Phys. Rev., 94, 1415 (1954).

170. Bragg, W.L. and Williams, E.J., Proc. Roy. Soc., A145, 699 (1934).

171. Bragg, W.L. and Williams, E.J., Proc. Roy. Soc., A151, 540 (1935).

172. Mato, T. and Takagi, Y., "Solid State Physics," Vol. 1, Seitz and Turnbull, Eds., Academic Press, New York, 1955, pp. 194.

173. Bethe, H., _Proc. Roy. Soc._, _A150_, 552 (1935).

174. Sykes, M.F., Hunter, D.L., McKenzie, D.S. and Heap, B.R., _J. Phys. A: Gen. Phys._, _5_, 667 (1972).

175. Norvell, J.C. and Als-Nieson, J., _Phys. Rev._, _B2_, 277 (1970).

176. Chipman, D.R. and Walker, C.B., _Phys. Rev. Lett._, _26_, 233 (1971).

177. Ashman, J. and Handler, P., _Phys. Rev. Lett._, _23_, 642 (1969).

178. Keating, D.T. and Warren, B.E., _J. Appl. Phys._, _22_, 286 (1951).

179. Connelly, D.L., Loomis, J.S. and Maphoter, D.E., _Phys. Rev._, _B3_, 924 (1971).

180. Gschneidner, K.A., "Rare Earth Alloys," Van Nostrand, Princeton, 1961.

181. Takahashi, T. and Shimizu, M., _J. Phys. Soc. Japan_, _23_, 945 (1967).

182. Pawel, R.E. and Stansbury, E.E., _J. Phys. Chem. Sol._, _26_, 607 (1965).

183. Orehotsky, J. and Schröder, K., _Phys. kondens. Materie_, _17_, 37 (1973).

184. MacInnes, W.M., Dissertation, Syracuse University, Syracuse, New York, 1970.

185. MacInnes, W.M. and Schröder, K., "Dynamical Aspects of Critical Phenomena," Budnick and Kawatra, Eds., Gordon and Breach, New York, 1972, pp. 305.

186. van der Hoeven, B.J.C., Jr., Teaney, D.T. and Moruzzi, V.L., _Phys. Rev. Lett._, _20_, 719 (1968).

187. McCoy, B.M. and Wu, T.T., _Phys. Rev. Lett._, _21_, 549 (1968).

188. Weber, R. and Street, R., _J. Phys. F_, _2_, 873 (1972).

189. Pepperhoff, W. and Ettwigg, H.H., _Z. angew. Phys._, _24_, 88 (1968).

190. Van Kranendonk, J. and Van Vleck, J.H., _Rev. Mod. Phys._, _30_, 1 (1958).

191. Baum, N.P. and Schröder, K., _Phys. Rev._, _B3_, 3847 (1971).

192. Schröder, K., _J. Appl. Phys._, _32_, 880 (1961).

193. Livingston, J.D. and Bean, C.P., _J. Appl. Phys._, _32_, 1964 (1961).

194. Scurlock, R.G. and Wray, E.M., _Phys. Lett._, _6_, 28 (1963).

195. Robbins, C.G., Claus, H. and Beck, P.A., _J. Appl. Phys._, _40_, 2269 (1969).

196. Meissner, W., _Handb. d. Experimentalphysik_, _11_, 338 (1935).

197. Lindemann, F., _Phys. Z._, _11_, 609 (1910).

198. Blatt, F.J., "Solid State Physics," Vol. 4, Seitz and Turnbull, Eds., Academic Press, New York, 1957, pp. 199.

199. "Metals Handbook," Vol. 1, Properties and Selection of Metals, ASM, Metals Park, 1961.

200. Kawatra, M.P. and Budnick, J.I., "Dynamical Aspects of Critical Phenomena," Budnick and Kawatra, Eds., Gordon and Breach, New York, 1972, pp. 257.

201. Kohler, M., _Ann. d. Physik (Leibz)_, _32_, 211 (1938).

202. Kohler, M., _Z. Physik_, _124_, 722 (1948).

203. Justi, E. and Scheffers, H., _Metallwirtschaft_, _17_, 1359 (1938).

204. Becker, R. and Doring, W., "Ferromagnetismus," Springer, Berlin, 1939.

205. Gibbons, D.F., "Physical Metallurgy," Cahn, Ed., North Holland, Amsterdam, 1970, pp. 77.

206. Kierspe, W., Kohlhaas, R. and Gonska, H., _Z. angew. Phys._, _24_, 28 (1967).

207. Craig, P.P., Goldburg, W.I., Kitchens, T.A. and Budnick, J.I. _Phys. Rev. Lett._, _19_, 1334 (1967).

208. Coles, B.R., _Adv. in Phys._, _7_, 40 (1958).

209. Linde, J.O., Thesis, U. Lund., Sweden, 1939.

210. Pearson, G.L. and Bardeen, _J._, _Phys. Rev._, _75_, 865 (1949).

211. Prince, M.B., _Phys. Rev._, 92, 681 (1953).

212. Prince, M.B., _Phys. Rev._, _93_, 1204 (1954).

213. Ludwig, G.W. and Watters, R.L., _Phys. Rev._, _101_, 1699 (1956).

214. Ryder, E.J., _Phys. Rev._, _90_, 766 (1953).

215. Folberth, O.G., Madelung, O. and Weiss, H., Z. _Naturf._, _9a_, 954 (1954).

216. Crow, J.E., Guertin, R.P. and Parks, R.D., _Phys. Rev. Lett._, _19_, 77 (1967).

217. Ehrenreich, H., _J. Appl. Phys._, _32_, 2155 (1961).

218. Bass, J., _Adv. in Phys._, _21_, 431 (1972).

219. Johansson, C.H. and Linde, J.O., _Ann. d. Phys._, _25_, 1 (1936).

220. Linde, J.O., _Ann. d. Phys._, _10_, 52 (1931).

221. Linde, J.O., _Ann. d. Phys._, _14_, 352 (1932).

222. Linde, J.O., _Ann. d. Phys._, _15_, 219 (1932).

223. Pearson, W.B., _Phil. Mag._, _46_, 911 (1955).

224. Gerritsen, A.N. and Linde, J.O., _Physica_, _18_, 877 (1952).

225. Kondo, J., _Progr. of Theor. Phys._, _32_, 37 (1964).

226. Coles, B.R. and Taylor, J.C., Proc. Roy. Soc., A267, 139 (1962).

227. Busch, G. and Yuan, S., Helv. Phys. Acta, 32, 465 (1959).

228. Yessik, M., Dissertation, Syracuse University, Syracuse, New York, 1966.

229. Coles, B.R., Proc. Phys. Soc., B65, 221 (1952).

230. Sousa, J.B., Chaves, M.R., Pinto, R.S. and Pinheiro, M.F., J. Phys. F, 2, 183 (1972).

231. Arajs, S. and Dunmyre, G.R., J. Appl. Phys., 37, 1017 (1966).

232. Arajs, S., Dunmyre, G.R. and Dechter, S.J., Phys. Rev., 154, 448 (1967).

233. Komura, S., Hamagushi, Y. and Kunitomi, N., J. Phys. Soc. Japan, 23, 174 (1967).

234. Thompson, N., Proc. Roy. Soc. (London), A155, 111 (1936).

235. Jain, A.L., Phys. Rev., 114, 1521 (1959).

236. Busch, G. and Vogt, E., Helv. Phys. Acta, 33, 451 (1960).

237. Berman, R., Z. f. Phys. Chem. (Neue Folge), 16, 10 (1958).

238. Slack, G.A., Phys. Rev., 105, 829 (1957).

239. Eucken, A. and Kuhn, G., Z. f. Phys. Chem., 134, 193 (1928).

240. Devyatkova, E.D. and Stilbans, L.S., J. Tech. Phys. USSR, 22, 968 (1952).

241. Ioffé, A.F., Can. J. Phys., 34, 1342 (1956).

242. Schröder, K., and McCain, C.E., Phys. Rev., 135, A149 (1964)

243. Berman, R., Simon, F.E., Klemens, P.G. and Fry, T.M., Nature, 166, 864 (1950).

244. Berman, R., Proc. Phys. Soc., A65, 1029 (1952).

245. Sondheimer, E.H., Proc. Roy. Soc., A203, 75 (1950).

246. Ziman, J.M., Proc. Roy. Soc., A226, 436 (1954).

247. White, G.K., Proc. Roy. Soc. (London), A66, 559 (1953).

248. Schröder, K. and Shabel, B., Z. f. Metallkunde, 58, 727 (1967).

249. Smith, A.W., Phys. Rev., 30, 1 (1910).

250. Jan, J.P. and Gijsman, H.M., Physica, 18, 339 (1952).

251. Jan, J.P., Helv. Phys. Acta, 25, 678 (1952).

252. Brooks, H., "Advances in Electronics and Electron Physics," Vol. VII, Marton, Ed., Academic Press, New York, 1955, pp. 87.

253. Morin, F.J. and Maita, J.P., Phys. Rev., 94, 1525 (1954).

254. Madelung, O. and Weiss, H., Z. f. Naturforschung, 9a, 527 (1954).

255. Ehrenreich, H., J. Phys. Chem. Sol., 2, 131 (1957).

256. Ehrenreich, H., J. Phys. Chem. Sol., 9, 129 (1959).

257. Rode, D.L., Phys. Rev., B2, 1012 (1970).

258. Rode, D.L. and Knight, S., Phys. Rev., B3, 2534 (1971).

259. Roaf, D.J., Phil. Trans. Roy. Soc., A25, 135 (1963).

260. Hagmann, D. and Saeger, K.E., Z. f. Metallkunde, 54, 650 (1963).

261. Cooper, J.R.A. and Raimes, S., Phil. Mag., 4, 145, 1154 (1959).

262. Ziman, J.M., "The Fermi Surface," Proc. of Conference Held at Cooperstown, N.Y., 1960; Harrison and Webb, Eds., J. Wiley, New York, 1961.

263. Allgaier, R.S., J. Phys. Chem. Sol., 28, 1293 (1967).

264. Pugh, E.M., Phys. Rev., 97, 647 (1955).

265. Blatt, F.J., Flood, D.J., Rowe, V., Schroeder, P.A. and Cox, J.E., Phys. Rev. Lett., 18, 395 (1967).

266. Sondheimer, E.H., Proc. Roy. Soc. (London), A193, 384 (1948).

267. Sondheimer, E.H. and Wilson, A.H., Proc. Roy. Soc. (London), A190, 435 (1947).

268. MacInnes, W.M. and Schröder, K., Phys. Rev., B4, 4091 (1971).

269. Cusack, N. and Kendall, P., Proc. Phys. Soc., 72, 898 (1958).

270. Huebener, R.P., "Solid State Physics," Vol. 27, Ehrenreich, Seitz and Turnbull, Eds., Academic Press, New York, 1972, pp. 63.

271. Rudnitskii, A.A., "Thermoelectric Properties of the Noble Metals and Their Alloys," Publishing House, Academy of Sciences, USSR, 1956.

272. McDonald, D.K.C., "Thermoelectricity, An Introduction to the Principles," J. Wiley, New York, 1962.

273. Rowe, V. and Schroeder, P.A., J. Phys. Chem. Sol., 31, 1 (1970).

274. De Vroomen, A.R., Van Baarle, C. and Cuelenaere, A.J., Physica, 26, 19 (1960).

275. Schröder, K. and Giannuzzi, A., Phys. Stat. Sol., 34, K133 (1969).

276. Nagy, I. and Pal, L., Phys. Rev. Lett., 24, 894 (1970).

277. Schröder, K. and Tomaschke, K., Phys. kondens. Materie, 7, 318 (1969).

278. Schröder, K., and Otooni, M., J. of Phys. D, 4, 1612 (1971).

279. Middleton, A.E. and Scanton, W.W., Phys. Rev., 92, 219 (1953).

280. Airoldi, G., Asdente, M. and Rimini, E., Phil. Mag., 10, 43 (1964).

281. Blatt, F.J. and Lucke, W.H., Phil. Mag., 15, 649 (1967).

282. Domenicali, C.A. and Otter, F.A., Phys. Rev., 95, 1134 (1954).

283. Kimura, H. and Shimuzu, M., J. Phys. Soc. Japan, 19, 1632 (1964).

284. Sondheimer, E.H., Proc. Roy. Soc., A193, 484 (1948).

285. Dugdale, J.S. and Guénault, A.M., Phil. Mag., 13, 503 (1966).

286. Fletcher, R. and Greig, D., Phil. Mag., 17, 21 (1968).

287. Schröder, K. and Yessik, M., J. Phys. Chem. Sol., 28, 1713 (1967).

288. Schröder, K., Yessik, M.J. and Baum, N.P., J. Appl. Phys., 37, 1019 (1966).

289. Kolomoets, N.V. and Vedernikov, M.V., Soviet Physics-Solid State, (English transl.), 3, 1996 (1962).

290. Kamerlingh Onnes, H., Commun. Kamerlingh Onnes Laboratory, Univ. Leiden, Suppl., 346, 55 (1913).

291. File, J. and Mills, R.G., Phys. Rev. Lett., 10, 93 (1963).

292. Meissner, W. and Ochsenfeld, R., Naturwissenschaften, 21, 787 (1933).

293. Phillips, N.E., Phys. Rev., 134, A385 (1964).

294. Andres, K. and Jensen, M.A., Phys. Rev., 165, 533 (1968).

295. Ehrenreich, H., "Proceedings of the Intl. School of Physics. Enrico Fermi Course XXXIV, 1965, "The Optical Properties of Solids," Tauc, Ed., Academic Press, New York, 1966.

296. Pells, G.P. and Shiga, M., J. Phys. C, Ser. L, 2, 1835 (1969).

297. Shiga, M. and Pells, G.P., J. Phys. C, Ser. L, 2, 1847 (1969).

298. Barker, A.S., Jr., Halperin, B.I. and Rice, T.M., Phys. Rev. Lett., 20, 384 (1968).

299. Lenham, A.P., J. Opt. Soc. Am., 57, 473 (1967).

300. Philipp, H.R. and Ehrenreich, H., Phys. Rev., 129, 1550 (1963).

301. Crangle, J. and Hallam, G.C., Proc. Roy. Soc. (London), A272, 119 (1963).

302. Tauc, J. and Antocik, E., Phys. Rev. Lett., 5, 253 (1960).

303. Greenaway, D.L., Phys. Rev. Lett., 9, 97 (1962).

304. Brodsky, M.H., Lucovsky, G., Chew, M.F. and Plasket, T.S., Phys. Rev., B2, 3303 (1970).

305. Lucovsky, G., Brodsky, M.H. and Burstein, E., Phys. Rev., B2, 3295 (1970).

306. Mott, N.F. and Gurney, R.W., "Electronic Processes in Ionic Crystals," Dover, New York, 1964.

307. Wessel, P.R., Phys. Rev., 132, 2062 (1963).

308. Beaglehole, D. and Erlbach, E., Solid State Commun., 8, 255 (1970).

309. Biondi, M.A. and Rayne, J.A., Phys. Rev., 115, 1522 (1959).

310. Yen, E., Dissertation, Syracuse University, Syracuse, New York, 1972.

311. Pells, G.P. and Montgomery, H., J. Phys. C, 3S, 330 (1970).

312. Schröder, K. and Önengüt, D., Phys. Rev., 162, 628 (1967).

313. Myers, H.P., Walden, L. and Karlsson, A., Phil. Mag., 18, 725 (1968).

314. Mamola, K., McCain, C.E. and Schröder, K., Phys. Rev., B1, 1411 (1971).

315. Mamola, K. and Schröder, K., Phys. Stat. Sol., 40, 81 (1970).

316. Crandel, M.E. and Schröder, K., J. Appl. Phys., 42, 1710 (1971).

317. Krolikowski, W.F. and Spicer, W.E., Phys. Rev., 185, 882 (1969).

318. Eastman, D.E., J. Appl. Phys., 40, 1387 (1969).

319. Snow, E.C., Phys. Rev., 171, 785 (1968).

320. Forstmann, F. and Pendry, J.B., Z. Physik, 235, 69, 75 (1970).

321. Nielson, P.O., Phys. kondens. Materie, 11, 1 (1970).

322. Hodges, L., Ehrenreich, H. and Lang, N.D., Phys. Rev., 152, 505 (1966).

323. Connolly, J.W.D., Phys. Rev., 159, 415 (1967).

324. Cornwell, J.F., Hum, D.M. and Wong, K.G., Phys. Lett., 26A, 365 (1968).

325. Seib, D.H. and Spicer, W.E., Phys. Rev., B2, 1676, 1694 (1970).

326. Norris, C., J. Appl. Phys., 40, 1396 (1969).

327. Philipp, H.R. and Ehrenreich, H., Phys. Rev. Lett., 8, 92 (1962).

AUTHOR INDEX

Numbers in brackets are reference numbers and indicate that an author's work is referred.

A

Airoldi, G. 460 [280]
Allgaier, R.S., 434 [263]
Als-Nieson, J., 306-307 [175]
Amar, H., 117, 127 [63]
Anderson, J.R., 129 [71]
Anderson, P.W., 105, 173 [56]
Andres, K., 488 [294]
Antocik, E., 518 [302]
Arajs, S., 390 [231], 390-391, 393-394 [232]
Arlinghaus, F.J., 86, 89 [39]
Asano, S., 221 [123]
Asdente, M., 460 [280]
Ashman, J., 306-307 [177]

B

Balluffi, R.W., 266-267 [146]
Bardeen, J., 368, 423 [210], 353
Barker, A.S., Jr., 514 [298]
Bass, J., 375 [218]
Bastow, T.J., 222-223 [124]
Baum, N.P., 329 [191], 473 [288]
Beaglehole, D., 533 [308]
Bean, C.P., 330 [193]
Beck, P.A., 233-234 [128], 235-236 [129], 292-293, 330 [164], 292, 294 [165], 332 [195]
Becker, R., 361 [204]

B (continued)

Beeby, J.L., 111-112, 117 [60]
Berman, R., 226, 406 [237], 408 [243], 407 [244]
Bertaut, E.F., 201 [112]
Bethe, H., 304 [173]
Bingham, R.E., 268 [147]
Biondi, M.A., 533, 535 [309]
Blatt, F.J., 351, 414 [198], 438 [265], 461 [281]
Bochirol, L., 242 [131]
Born, M., 43, 259 [19], 1
Boudreaux, D.S., 93 [45]
Bozorth, R.M., 309 [79]
Bragg, W.L., 304 [170], 301 [171]
Brailsford, A.D., 130 [73]
Brodsky, M.H., 522-523 [304], 524 [305]
Brookhouse, B.N., 46 [20]
Brooks, C.R., 268 [147]
Brooks, H., 421 [252]
Brown, E., 85-86 [38]
Bucher, E., 288, 294 [161]
Budnick, J.I., 362 [200], 363 [207]
Buot, F.A., 190-191 [68]
Burstein, E., 524 [305]
Busch, G., 189 [227], 401 [236]

C

Callen, H.B., 165-168 [95]
Chaves, M.R., 388-389 [230]

SUBJECT INDEX